AF443446

# The Chemical Physics of Food

# The Chemical Physics of Food

*Edited by*
Professor Peter Belton

*Head of Chemistry*
*School of Chemical Sciences and Pharmacy*
*University of East Anglia, UK*

Editorial Offices:
Blackwell Publishing Ltd, 9600 Garsington Road, Oxford OX4 2DQ, UK
    Tel: +44 (0)1865 776868
Blackwell Publishing Professional, 2121 State Avenue, Ames, Iowa 50014-8300, USA
    Tel: +1 515 292 0140
Blackwell Publishing Asia, 550 Swanston Street, Carlton, Victoria 3053, Australia
    Tel: +61 (0)3 8359 1011



First published 2007 by Blackwell Publishing Ltd

ISBN: 978-14051-2127-9

Library of Congress Cataloging-in-Publication Data
The chemical physics of food / edited by Peter Belton.
        p.c.m.
    Includes bibliographical references and index.
    ISBN-13: 978-1-4051-2127-9 (hardback : alk. paper)
    ISBN-10: 1-4051-2127-0 (hardback : alk. paper)
    1. Food--Analysis. 2. Food--Preservation. 3. Food--Composition. I.
Belton, P. S.

    TX541.C435 2006
    664--dc22

A catalogue record for this title is available from the British Library

Set in 10/13pt Times by
by Sparks, Oxford – www.sparks.co.uk
Printed and bound in Singapore
by Markono Print Media, Pte, Ltd

The publisher's policy is to use permanent paper from mills that operate a sustainable forestry policy, and which has been manufactured from pulp processed using acid-free and elementary chlorine-free practices. Furthermore, the publisher ensures that the text paper and cover board used have met acceptable environmental accreditation standards.

For further information on Blackwell Publishing, visit our website:
www.blackwellpublishing.com

# Contents


*Preface*                                                                              *xi*
*Contributors*                                                                    *xiii*
*About the Editor*                                                              *xiii*

**1   Emulsions**                                                                    **1**
   *John N. Coupland*

1.1   Introduction                                                                      1
1.2   Emulsion structure                                                           2
   1.2.1   Size                                                            2
   1.2.2   Concentration                                          3
   1.2.3   Surface properties                                  4
      1.2.3.1   Modified surfaces                   5
      1.2.3.2   Types of interfacial material   6
   1.2.4   Interdroplet potentials                        8
1.3   Emulsion dynamics                                                        10
   1.3.1   Creaming                                                 11
   1.3.2   Flocculation                                          12
   1.3.3   Coalescence                                          13
1.4   Emulsion functionality                                                16
   1.4.1   Rheology                                                 16
   1.4.2   Chemical reactivity                              17
1.5   References                                                                      18

**2   Physicochemical Behaviour of Starch in Food Applications**                                                          **20**
   *Alain Buleon and Paul Colonna*

2.1   Introduction                                                                     20
2.2   Starch composition and chemical structure              21
   2.2.1   Granular structure                                   21
   2.2.2   Molecular composition                           24
      2.2.2.1   Amylose                                   25
      2.2.2.2   Amylopectin                            26
      2.2.2.3   Intermediate materials            30

            2.2.2.4   Minor components                                          31
2.3   Modifications of starch by hydrothermal treatments and shearing           32
      2.3.1   Gelatinization, pasting and melting                               33
              2.3.1.1   Structural changes                                      33
              2.3.1.2   Mechanisms of gelatinization-melting                    34
              2.3.1.3   Functional properties                                   40
      2.3.2   Gelation                                                          40
              2.3.2.1   Structural changes                                      40
              2.3.2.2   Mechanisms                                              42
              2.3.2.3   Functional properties                                   44
      2.3.3   Glass transition and plasticization by water                      45
      2.3.4   Physical ageing                                                   46
2.4   Interactions with other molecules                                        46
      2.4.1   Hydrocolloids and proteins                                        47
      2.4.2   Sugars                                                            49
      2.4.3   Amylose complexation with small molecules                         50
              2.4.3.1   Lipids                                                  50
              2.4.3.2   Alcohols, aroma and flavours                            53
2.5   Starch as a nutrient                                                      55
      2.5.1   Classification                                                    55
      2.5.2   Resistant starch                                                  56
2.6   Conclusions                                                              57
2.7   References                                                               59


**3   Water Transport and Dynamics in Food**                                   **68**
      *Brian Hills*

3.1   Introduction                                                             68
3.2   Statistical thermodynamics and the microscopic water distribution        70
3.3   Experimental probes of the microscopic water distribution                74
3.4   The water self-diffusion propagator                                      77
3.5   Experimental probes of the water self-diffusion propagator               78
3.6   Water transport in nonequilibrium microheterogeneous systems             80
3.7   The state of water in nanopores                                          82
3.8   Experimental probes of water–biopolymer interactions                     86
3.9   Molecular dynamics simulations of water–biopolymer interactions          94
3.10  The dependence of water dynamics on state variables                      95
      3.10.1  Low-water-content systems                                         95
      3.10.2  Nonfreezing water                                                99
      3.10.3  Diffusion studies of surface water                               100
      3.10.4  Water dynamics under high pressure                               101
3.11  Conclusion                                                              103
3.12  References                                                              104

**4  Glasses**                                                                                         **108**
*Roger Parker and Stephen G. Ring*

4.1   Introduction                                                                                    108
4.2   Glass transitions                                                                               109
        4.2.1   Low molecular weight liquids and glasses                                              109
        4.2.2   Biopolymer glasses and plasticization                                                111
        4.2.3   Colloidal glasses                                                                     113
4.3   Glassy state dynamics                                                                           114
4.4   Structural relaxation in low molecular weight organic liquids and biopolymers                  117
4.5   Mechanical stability – colloidal systems                                                        119
4.6   Chemical stability                                                                              119
        4.6.1   Chemical kinetics and the glassy state in single-phase systems                       120
        4.6.2   Chemical kinetics and the glassy state in multiphase systems                         125
4.7   Glassy carbohydrates as encapsulation matrices and solvents                                    125
        4.7.1   Flavour encapsulation in glassy carbohydrates                                         125
        4.7.2   Solvent properties of amorphous carbohydrates                                         126
4.8   Concluding remarks                                                                              129
4.9   References                                                                                      130


**5  Powders and Granular Materials**                                                                 **135**
*Gary C. Barker*

5.1   Introduction                                                                                    135
5.2   Packing                                                                                         139
5.3   Segregation                                                                                     142
5.4   Jamming                                                                                         145
5.5   Discussion                                                                                      148
5.6   References                                                                                      148


**6  Gels**                                                                                           **151**
*Victor J. Morris*

6.1   Introduction                                                                                    151
6.2   Polysaccharide gels                                                                             153
        6.2.1   What are polysaccharides?                                                             153
        6.2.2   How do polysaccharides form networks?                                                158
                   6.2.2.1   Point cross-links                                                        159
                   6.2.2.2   Block structures                                                         160
                   6.2.2.3   Higher-order helical aggregates                                          163
        6.2.3   What are fluid gels?                                                                  166
        6.2.4   Polysaccharide mixtures                                                               168
        6.2.5   Phase-separated networks                                                              169

|  |  |  |
|---|---|---|
| | 6.2.5.1 Starch | 169 |
| | 6.2.5.2 Semi-refined carrageenans | 172 |
| 6.2.6 | Swollen networks | 172 |
| 6.2.7 | Interpenetrating networks | 173 |
| 6.2.8 | Coupled networks | 173 |
| | 6.2.8.1 Pectin-alginate gels | 173 |
| | 6.2.8.2 Xanthan-glucomannan gels | 174 |
| | 6.2.8.3 Xanthan-galactomannan gels | 176 |
| | 6.2.8.4 Algal polysaccharide glucomannan or galactomannan mixed gels | 177 |
| 6.3 | Protein gels | 178 |
| 6.3.1 | What are proteins? | 178 |
| 6.3.2 | How do proteins form networks? | 179 |
| | 6.3.2.1 Globular proteins | 179 |
| | 6.3.2.2 Fibrous proteins | 181 |
| | 6.3.2.3 Casein gels | 183 |
| 6.3.3 | Protein mixtures | 184 |
| 6.3.4 | Interfacial protein networks | 184 |
| | 6.3.4.1 Interfacial gelatin networks | 185 |
| | 6.3.4.2 Globular protein networks | 185 |
| 6.3.5 | Interfacial protein networks in foods | 186 |
| 6.4 | Polysaccharide-protein gels | 189 |
| 6.5 | Conclusions | 191 |
| 6.6 | References | 191 |


## 7   Wheat-Flour Dough Rheology

*Robert S. Anderssen*


|  |  |  |
|---|---|---|
| 7.1 | Introduction | 199 |
| 7.1.1 | The two independent aspects of cereal science and technology: molecular biorheology and process biorheology | 200 |
| | 7.1.1.1 Genetics as the key to plant breeding: molecular biorheology | 200 |
| | 7.1.1.2 Process rheology as the key to efficiently maximizing end-product quality: process biorheology | 201 |
| 7.1.2 | The pervasive nature of wheat-flour dough rheology in cereal science and technology | 201 |
| 7.1.3 | The rheology perspective: the recovery of information from indirect measurements | 211 |
| 7.2 | Background, preliminaries and notation | 212 |
| 7.3 | The phenomenology of wheat-flour dough formation | 212 |
| 7.4 | Wheat-flour dough rheology modelling from an indirect measurement perspective: a plethora of models | 216 |
| 7.5 | The indirect measurement modalities that directly underpin the rheology of | |

wheat-flour dough formation                                                    219
  7.5.1  The walk-in-refrigerator experiments                        219
  7.5.2  Temperature measurements                                    220
  7.5.3  Mixograms                                                   222
      7.5.3.1  Qualitative and quantitative summaries of the global
      stress–strain dynamics in a mixogram                                223
      7.5.3.2  The hysteretic nature of the local structure in a mixogram    224
      7.5.3.3  A hysteretic summary of the global structure in a mixogram    227
  7.5.4  Uniaxial and biaxial extensions                             227
  7.5.5  The modalities that indirectly underpin the rheology         230
7.6  Modelling the viscoelasticity of wheat-flour dough formation        230
7.7  Some future challenges                                               233
Appendix 1: A brief literature summary                                         235
Appendix 2: Symbols and abbreviations                                          236
7.8  References                                                          237

*Index*                                                                        *241*

*The colour plate section appears after page 82*

# Preface

The idea of this book started out with my long-held conviction that the complexity of food materials, and the difficulty of describing them in a quantitative manner, has sometimes led food scientists to think that the rigour required of them was less than that required by the traditional physical sciences. However, a number of food scientists are trying to approach the treatment of materials with precisely this level of rigour. This is not an easy task. A working definition of food, which I have found useful, is that it is slightly decayed organic matter that somebody wants to eat. As such the material is intractable, highly variable and is not characterized by parameters that are easily measured. These challenges require a higher, not lower, level of rigour in thinking and experimental design in order to produce useful models of material properties.

The topics I have chosen for this book are areas where the authors exemplify the chemical physics approach. By this I mean a combination of the applications of chemical and physical methods, often of the most advanced kind, together with a clear quantitative consideration of the data. It is my hope that the approach taken here will come to be the norm in food science.

P.S. Belton
Norwich

# Contributors

**Robert S. Anderssen**, CSIRO Mathematical and Information Sciences, PO Box 664, Canberra, ACT 2601, Australia.

**Gary C. Barker**, Institute of Food Research, Norwich Research Park, Colney, Norwich NR4 7UA, UK

**Alain Buleon**, INRA, BP 71627, 44316 Nantes Cedex 3, France

**Paul Colonna**, INRA, BP 71627, 44316 Nantes Cedex 3, France

**John Coupland**, Pennsylvania State University, 103, Borland Lab, University Park, PA 16802, USA

**Brian Hills**, Institute of Food Research, Norwich Research Park, Colney, Norwich NR4 7UA, UK

**Victor J. Morris**, Institute of Food Research, Norwich Research Park, Colney, Norwich NR4 7UA, UK

**Roger Parker**, Institute of Food Research, Norwich Research Park, Colney, Norwich NR4 7UA, UK

**Stephen G. Ring**, Institute of Food Research, Norwich Research Park, Colney, Norwich NR4 7UA, UK

### *About the Editor*

Peter Belton is Head of Chemistry in the School of Chemical Sciences and Pharmacy at the University of East Anglia,UK

# Chapter 1
# Emulsions

*John N. Coupland*

## 1.1 Introduction

Oil and water are almost completely mutually insoluble yet commonly coexist in foods in the form of emulsions. The oil forms a separate dispersed phase in the aqueous material (or vice versa, although the topic of water-in-oil emulsions will not be considered here). There is a thermodynamic pressure for a complete phase separation but this is kinetically retarded largely by amphiphilic material adsorbed at the interface. Example food emulsions include beverage cloud emulsions, flavour emulsions, fluid milk, ice cream mix, mayonnaise, salad dressing and soups. Other foods including cake batter, hot-dog mix and frozen ice cream are not precisely emulsions but their behaviour can be understood in similar terms. Additionally, the language of dispersions as revealed in the study of emulsions can inform our understanding of foods as diverse as dough foams and fluid chocolate.

Despite the great diversity of foods containing emulsions, each with its own unique qualities, food emulsions have some common features that make them worth examining as a group. Firstly, the droplets present are of the order of a micrometre in diameter and so scatter light very efficiently, consequently most food emulsions appear opaque and white. Secondly, the presence of a dispersed phase increases the viscosity of the aqueous continuous phase, and interdroplet interactions can lead to dramatic non-Newtonian rheological properties and even gelation. Finally the presence of nonpolar domains within an aqueous continuum enables the partitioning of solutes and a change in their reactivity. All of these properties can be related quantitatively to the structure of the emulsions and will change as the emulsion structures changes.

To understand the functional properties of an emulsion, it is therefore necessary to start with a proper description of its structure and the mechanisms of its destabilization and then consider how these structures affect bulk properties. The study of food emulsions is a mature field and the subject of many books (see, e.g., refs 1–6). A comprehensive review of the depth of knowledge is beyond the scope of this work; instead I have tried to guide the reader through the core topics in a logical way and provide guidance to good sources for a more thorough treatment.

## 1.2 Emulsion structure

Emulsion structure can be described by a limited number of parameters. Changes in these parameters will reflect changes in the stability of the product and may affect its functional properties as discussed below.

### *1.2.1 Size*

The drive to minimize interfacial area makes emulsion droplets spherical under all but the most extreme conditions and thus they can be characterized by a single length dimension. Real emulsions contain a broad range of droplet sizes (i.e. polydisperse) typically ranging from about a tenth to some tens of micrometres. The range of particle sizes present can be represented as a particle size distribution in which the percentage of the dispersed phase volume (or area, length or number of particles) within a given size range is expressed as a histogram.[3] Very often the histogram is replaced with a scatter plot when the number of size bands is large. The distinction between volume, area, length and number distributions is important and each offers distinct insights into the properties of the emulsion. For example, Fig. 1.1 shows the properties of a typical food emulsion. When expressed on a volume basis the distribution appears bimodal, but as relatively few large droplets contain much of the oil, when the same distribution is expressed on an area or particularly a number basis the distribution appears unimodal. Very often it is preferable to express a distribution in terms of a mean, and some of the many useful means are listed in Table 1.1.

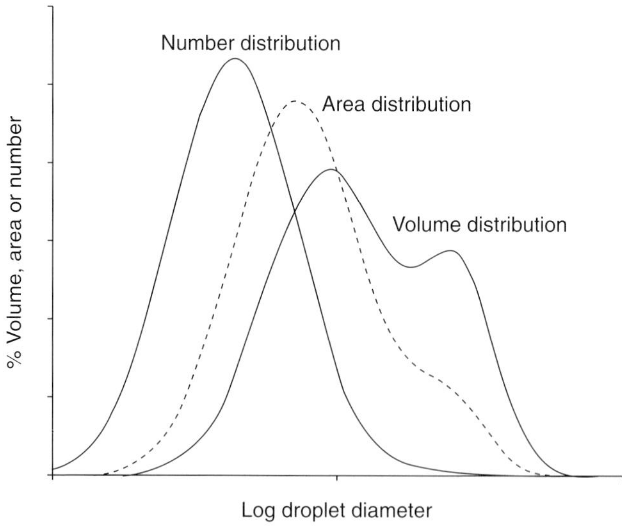


**Fig. 1.1**    Particle size distribution of a typical food emulsion. The same emulsion is represented as a volume-, area- and number-based distribution. Note: the x-axis is a logarithmic scale and is a representation of a histogram as a scatter plot; the y-axis shows the proportion of the droplets of a characteristic size.

**Table 1.1** Different means that can be used to describe emulsion distributions (adapted from Hunter[3]).

| Mean | Definition | Notes |
|---|---|---|
| Length | $d_{1,0} = \dfrac{\sum n_i d_i}{\sum n_i}$ | Typical ruler measurement |
| Area | $d_s = \sqrt{\dfrac{\sum n_i d_i}{\sum n_i}}$ | Important when the area of the emulsion is important, for example when droplets act as a catalyst. Often results for image analysis measurement |
| Volume | $d_v = \sqrt[3]{\dfrac{\sum n_i d_i^3}{\sum n_i}}$ | Volume (or weight) of particles. Measured by electrozone sensing |
| Surface-volume | $d_{32} = \dfrac{\sum n_i d_i^3}{\sum n_i d_i^2}$ | The Sauter mean is useful for surface-active material. The surface area per unit volume is given by $6\phi/d_{32}$ ($\phi$ = volume fraction) |
| Equivalent volume | $d_{43} = \dfrac{\sum n_i d_i^4}{\sum n_i d_i^3}$ | Typical output of some light-scattering instruments. Divide by length average to get a polydispersity index |

## *1.2.2 Concentration*

The concentration of droplets is typically expressed as a mass fraction of the dispersed phase, though in certain cases a volume fraction is more relevant and these can be readily inverted knowing the density of both phases. In foods the concentration of dispersed-phase oil can vary between a fraction of a percent (e.g. beverage flavour emulsions) and about 75% in mayonnaise. As the volume fraction increases, the particles increasingly interact with one another until they are close packed.[7] The maximum theoretical close packing of identical spheres is 0.7405, but in reality this type of highly organized structure does not occur and random close packing occurs at much lower volume fractions (~0.64). Polydisperse emulsions can pack to a higher volume fraction as the smaller droplets can fit within the gaps left between the large.

Droplet volume fractions beyond close packing are only attainable by deforming the spherical droplets and forming a highly concentrated emulsion, also known as a liquid foam. While in dilute systems the properties of the emulsions are governed by the interactions between the continuous and dispersed phases, as concentration is increased droplet–droplet interactions become increasingly important until in highly concentrated emulsions the droplet–droplet interactions dominate.[7] A striking example of this is mayonnaise, which is a viscoelastic solid formed from a concentrated dispersion of one Newtonian liquid in another. The elastic properties of the concentrated emulsion are due to the reversible deformation of the droplets in response to deformation.

### *1.2.3 Surface properties*

The short-range molecular interactions responsible for holding liquids together as separate phases also lead to interfacial tension. In an oil-in-water emulsion, water molecules are strongly attracted to other water molecules via a network of hydrogen bonds and so there is a net force acting on surface molecules pulling them back into the bulk. At a macroscopic scale this can be seen (and measured) as a force acting normal to a line of unit length drawn in the surface (i.e. the interfacial tension, $\gamma$). Alternatively surface tension can be seen as the proportionality constant linking interfacial area to the energy cost of generating it.[6,8]

Although at a molecular scale surfaces are inevitably somewhat diffuse, it is convenient to imagine them as an infinitesimally thin plane (Fig. 1.2b). The Gibbs definition of this plane is the surface drawn so that the deficit concentration on one side is equal to the excess concentration on the other (Fig. 1.2c).

The surface properties of emulsion droplets are further complicated by the fact that they are highly curved. Curved surfaces lead to more interfacial molecular contact than the corresponding planar interface and so are higher energy structures. At a molecular level this can be interpreted as the molecules at the interface being more exposed to the other phase whereas macroscopically this is seen as an increased pressure on the concave side of a curved surface – the Laplace pressure, $P_{\mathrm{L}}$, i.e:

$$P_{\mathrm{L}} = 2\gamma/r \tag{1.1}$$

where $r$ is the radius of curvature. High internal pressure makes small fluid droplets spherical and hard to deform and also leads to the phenomenon of capillary rise. The high pressure inside very fine droplets importantly increases the solubility of their contents (the logarithm of relative solubility increases with inverse radius – the Kelvin equation). If oil has some

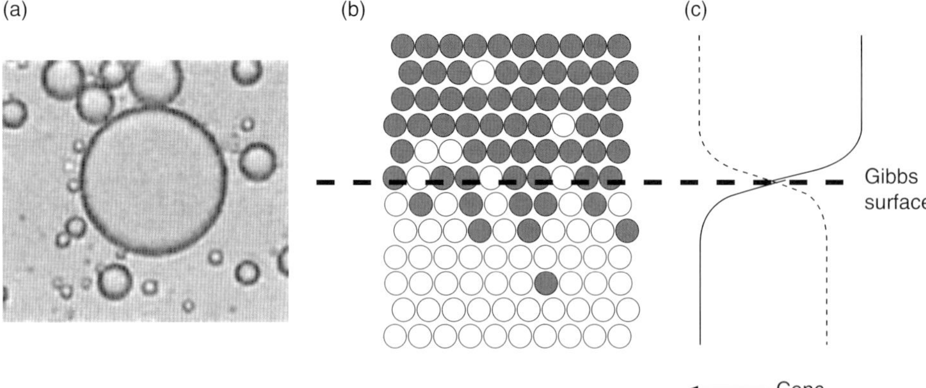


**Fig. 1.2**   Illustrations of the oil–water interface. (a) Optical micrograph of an oil droplet in water; at a bulk scale the interface is seen as a clean line (the thick line seen in most microscopy is an optical artefact). At a molecular scale (b) there is some diffusion in one phase into the other although the bulk solubility is negligible. This mixing can be drawn (c) as a decaying concentration of oil with distance moving into the aqueous phase and vice versa. The Gibbs surface is drawn so the deficit concentration on one side is equal to the excess on the other.

solubility in the continuous phase, the difference in surface curvature will favour the diffusion of oil from a small droplet to a larger (i.e. Ostwald ripening). The smaller droplets will
eventually disappear and there will be a net increase in average droplet size without any
direct contact between the droplets. In most food emulsions the solubility of triglycerides
in water is so low that the rate of this process is negligible; however, certain flavour oils
are moderately polar and may diffuse faster. Furthermore, surfactants may incorporate oil
molecules in their core and hence increase their effective aqueous phase solubility and the
rate of Ostwald ripening.[9]

### 1.2.3.1 Modified surfaces

Unmodified surfaces are only present for a very short period of time following emulsion
formation because amphiphilic material (see below) absorbs rapidly to reduce the interfacial free energy. The kinetics of accumulation of material is first limited by diffusion times
from the bulk and second by any energy barriers of adsorption. Important consequences
are the limited efficiencies of homogenization processes (i.e. droplets recoalesce before an
amphiphilic layer can form so limiting the minimum droplet size that can be achieved) and
the preferential accumulation of one surfactant over another (e.g. small molecule surfactants
diffuse faster than proteins).[10]

   For an individual amphiphilic molecule, the energetics of adsorption are governed by a
balance between loss of translational entropy and the largely enthalpic interactions that will
bind it to the surface. In food emulsions the hydrophobic forces acting on the nonpolar parts
of the molecule are largely responsible for the adsorption process but these can be modified
by attractive and repulsive interactions between adsorbed molecules. For example, sodium
dodecyl sulfate is a negatively charged surfactant. Its surface adsorption is driven by a need
to remove the dodecyl hydrocarbon chain from the aqueous environment but opposed by the
loss of molecular entropy and by the electrostatic repulsion between a molecule approaching
the surface and the negatively charged molecules already there. The relationship between
amount of surfactant present and amount adsorbed to the interface is given by the sorption
isotherm. Although the values will vary greatly between systems, some common salient
features can be seen in Fig. 1.3. First, the amount adsorbed increases at a decreasing rate
with bulk concentration over a limited range to a plateau above which the concentration
remains constant. The plateau value represents a surface saturated with surfactant, typically
at monolayer thickness. The monolayer value in most food emulsions is in the order of a
few milligrams per square metre. Adsorbed amphiphilic material reduces the interfacial free
energy by shielding the oil from the water. The amount the interfacial tension is reduced by
adsorbed surfactant is given as the surface pressure ($\pi$) where $\pi = \gamma - \gamma_0$ ($\gamma$ is the interfacial
tension of the modified surface and $\gamma_0$ the interfacial tension of the bare surface) (Fig. 1.3).
The relationship between the amount of bulk and adsorbed amphiphilic material and the
surface pressure is given by the Gibbs adsorption isotherm:

$$d\pi = -d\gamma = RT\Gamma\,d\ln a \tag{1.2}$$

where $\Gamma$ is the surface excess concentration of the amphiphilic material (i.e. amount adsorbed
per unit interfacial area), $a$ is its bulk activity (or concentration in some dilute systems), and

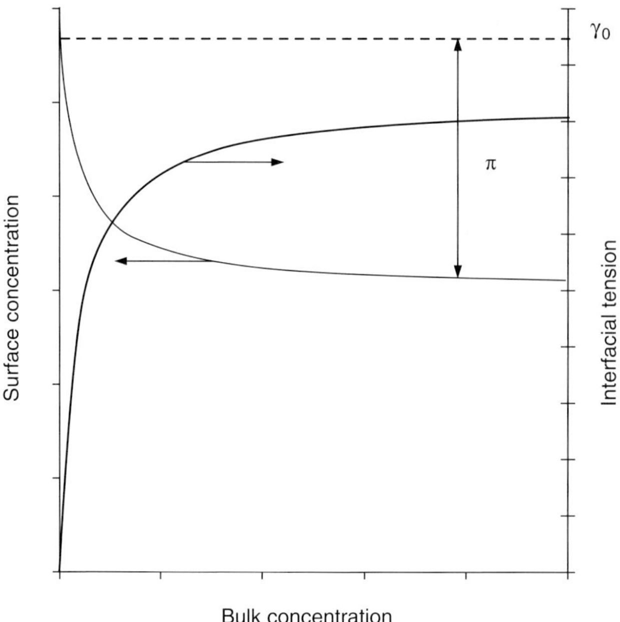


**Fig. 1.3**  Typical sorption isotherm. The amount of adsorbed material increases at a decreasing rate with bulk concentration until a plateau is reached corresponding to surface saturation. The surface tension decreases with surface coverage.

$T$ and $R$ are the absolute temperature and gas constant, respectively. Different materials have different surface tension lowering properties but one of the important distinctions in foods is that protein surfactants tend to lower the surface tension less than small molecules. A consequence of this is that small molecule surfactants can competitively displace polymers from the interface.[11] Recently similar exchanges between bound and unbound proteins have been observed.[12,13]

### 1.2.3.2 Types of interfacial material

Many food ingredients are surface active and can have a role in stabilizing food emulsions. Polymeric surfactants (MW > ~1000) are usually a linear chain whose monomers have varying water solubility. This group comprises chiefly proteins,[14] but some polysaccharides are also usefully surface active and have similar properties.[15] In proteins, the hydrophobic amino acids will tend to partition into the lipid phase while leaving the hydrophilic amino acids in the aqueous phase. Protein adsorption is largely irreversible because, although the energy cost of moving one hydrophobic amino acid from the oil phase is relatively low, the overall cost of simultaneously moving many residues is prohibitive. (However, as noted above, proteins can be displaced from a surface.)

The different affinities for the amino acid residues for the water and oil phases lead a disordered protein to form a series of loops, trains and tails at the interface (Fig. 1.4a). The thickness of a disordered protein film is often large (~10 nm) and in some cases the structure formed is in good accordance with the distribution of hydrophobic amino acids in the primary structure (e.g. $\alpha$ and $\beta$-casein[16]). In globular proteins, the strong secondary and tertiary structure opposes protein unfolding at the surface and the protein tends to adsorb more or less intact. The protein will then subsequently progressively unfold to improve its conformation. Although the gross changes in protein morphology are relatively small, enzymes particularly can be surface denatured. Films of globular proteins are thinner than disordered proteins (~2–3 nm) but often much more dense (Fig. 1.4b).

There are striking parallels between the three-dimensional reactivity of polymers in solution and polymers adsorbed at an interface.[17] For example, thermally treated solutions of $\beta$-lactoglobulin will gel through the formation of disulfide bonds (or through cross-links formed by transglutaminase) whereas the same protein adsorbed at a surface will form a cross-linked film stabilized by similar disulfide bonds[18] (or transglutaminase-catalysed bonds[19]). Another example is the observation that gum arabic and $\beta$-casein are thermodynamically incompatible in bulk and will also phase-separate when both are absorbed at an interface.[20]

The second important class of amphiphilic materials in foods are small-molecule surfactants (Fig. 1.4c). Whereas the hydrophobic portions of most surfactants are most often hydrocarbon-based, the hydrophilic group can be charged (anionic, cationic, zwitterionic) or merely polar. Their lower molecular weight means small-molecule surfactants diffuse more rapidly than proteins both in bulk and laterally at the surface of the droplet. Surfactant adsorption is also reversible because only one hydrophobic group need be detached from the surface to enable complete molecular desorption. Although each surfactant molecule is mobile, the average amounts at the surface and in bulk will remain constant. Many sur-

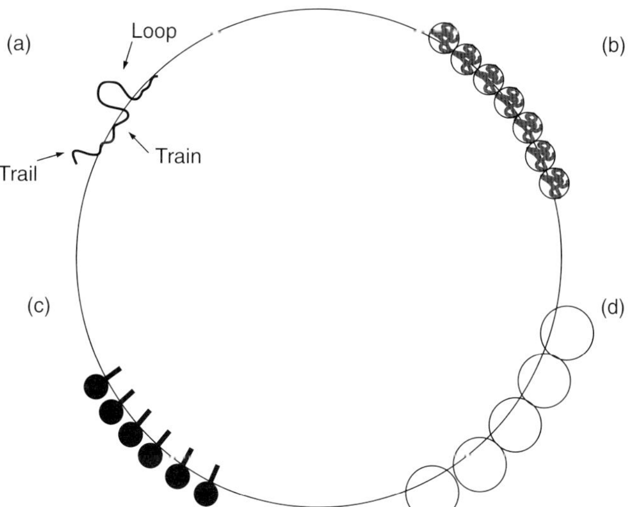


**Fig. 1.4** Diagrammatic representation of the conformation of (a) disordered polymer, (b) globular protein, (c) small-molecule surfactants and (d) fine particles at an interface. This diagram is not to scale; in reality the thickness of the surface layer would be three orders of magnitude smaller than the droplet (perhaps less for surface particles) and the curvature of the surface would not be detectable on the scale of the surfactant molecules.

factants will self-assemble in polar and nonpolar media to form a rich variety of dynamic yet thermodynamically stable structures.[5]

The final class of amphiphilic materials comprises some fine particles. Fine solid particles ($r \ll r_{droplet}$) (e.g. some flours, spices and crystals) are observed to adsorb to the interface and stabilize the emulsion (Fig. 1.4d). Although the theoretical basis of this functionality is less well developed, it is thought that the intermediate wetting properties of the solid forces it to adhere at a certain angle to the interface (i.e. Pickering stabilization).[21]

Figure 1.4 is a representation of the four types of amphiphilic material adsorbed to a single droplet. To facilitate the diagram the surface is shown highly curved relative to the scale of the surface layer. This representation shows that in a water-continuous dispersion the preponderance of the amphiphilic material is water soluble. The molecular asymmetry forces some surface curvature to the droplet and favours the formation of oil droplets rather than an oil continuous phase.

### *1.2.4 Interdroplet potentials*

The discussion of emulsion structure so far has treated the droplets as hard spheres, but in fact a number of intermolecular forces act between the droplet, the interface, the continuous phase and other droplets.[22] These are best understood in terms of attractive and repulsive forces acting around the droplets. The effects can be expressed as a force–distance function (Fig. 1.5a) showing the net forces acting on a droplet at a given distance from a second droplet or as an interdroplet pair potential (Fig. 1.5b), that is, the free energy change to bring one droplet to a given distance from a second droplet. Of course these are two representations of the same phenomena as energy is the integral of work over distance. As illustrated in Fig. 1.5, the net force field may be:

(I)    Repulsive at all ranges and the dispersion will be stable.

(II)   Attractive at all ranges and there will be rapid coalescence.

(III)  Repulsive at large separations but attractive at small ones. At infinite range the forces tend to zero and so the particles will tend to remain separated. If the height of the energy barrier is large enough (>several kT, where k is the Boltzmann constant and T absolute temperature; the product kT is a measure of the thermal energy of the system) the emulsion will be stable but as it decreases particles will 'jump' the barrier and begin to coalesce.

(IV)  Attractive at large separations but repulsive at short separations. Droplets will tend to accumulate in the energy minima; probably corresponding to flocculation rather than coalescence. Again the droplets may approach closer if the repulsive barrier is small enough.

The theoretical basis of these forces is reasonably well understood and can be measured experimentally with a colloidal force balance.[23] Some of the more important contributors to the stability of food emulsions are described below:

*Electrostatic forces*
Like-charge repulsion is largely a function of charge on adsorbed surfactant/protein. The amount of charge on a surface can be altered by pH or by specific ion binding. A charged

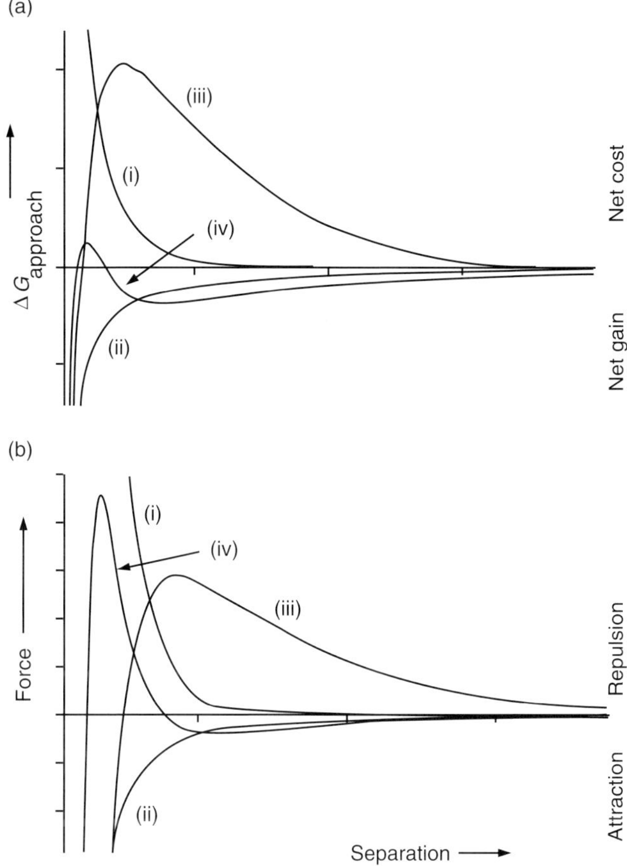



**Fig. 1.5**  Schematic representations of the (a) Gibbs free energy change required to bring a droplet from infinity to a given separation with another droplet, and (b) the forces acting on the approaching droplets as a function of distance during the approach. The four functions (I–IV) shown in each case represent forms of the interaction potential discussed in the text.

surface disturbs the ionic distribution in the continuous phase and sets up a double layer of an accumulation of oppositely charged counterions near the surface and similarly charged co-ions opposite. When the double layers of similar droplets overlap they repel one another. A simple formulation of the electrostatic potential between similar spheres ($w_{electrostatic}$) at separation $h$ is given by McClements:[4]

$$w_{electrostatic} = 2\pi\varepsilon_0\varepsilon_r\psi_0^2\ln(1 + e^{-\kappa h}) \tag{1.3}$$

where $\varepsilon_0$ and $\varepsilon_r$ are the dielectric constants of a vacuum and the relative dielectric constant on the continuous phase, respectively, and $\psi_0$ is the surface potential (formally as measured at the plane of slip, a small distance away from the charged droplet surface incorporating some entrained ions and solvent). The parameter $\kappa$ is the reciprocal Debye length, a measure of the thickness of the double layer, which is inversely related to the ionic strength of the medium. According to this formulation the magnitude of the potential therefore decays with separation distance, the rate of decay increases with ionic strength and the magnitude of the potential depends on both the surface potential and the dielectric constant of the continuous phase.

*van der Waals forces*
Even neutral molecules will attract one another via the van der Waals or dispersion forces. Although a nonpolar molecule by definition has zero average electrical dipole, the instantaneous distribution of electrons usually leads to some polarity. The electrical field generated polarizes adjacent groups and sets up a weak attraction between the transient dipole and the resultant induced dipole. van der Waals forces are the dominant interaction in nonpolar fluids but, importantly for emulsion stability, the net action of many intermolecular forces leads to an interdroplet attractive force. A simple formulation for the van der Waals potential between two similar droplets of radius $r$ at separation $h$ is:[4]

$$w_{van\ der\ Waals} = -\frac{Ar}{12h} \tag{1.4}$$

The proportionality constant $A$ is the Hamaker function ($\sim 0.75 \times 10^{-20}$ J in food emulsions[4]). The precise value of the Hamaker function depends on the frequency-dependent dielectric properties of the component phases. According to this formulation the force is relatively long range (decays as the reciprocal of distance and often effective out to $\sim 10$ nm).

*Steric interactions*
Adsorbed material acts as a barrier to coalescence at very short range (i.e. when the surface layers begin to overlap). The two main contributions to this mechanism are compression (volume exclusion) and mixing (osmotic) effects. Compression is always strongly repulsive but mixing can be attractive at slightly longer ranges depending on the solvent–polymer interactions.

*Hydrophobic forces*
Water molecules strongly interact with one another to form a dynamic hydrogen-bonded structure. Water molecules adjacent to a hydrophobic (nonpolar) surface arrange their interactions to minimize the water–surface contacts and in doing so become more ordered and lose entropy. Bringing two hydrophobic surfaces together reduces the volume of low-entropy water and is the basis for the hydrophobic attraction between nonpolar surfaces in water. Partially covered surfaces or surfaces covered with certain (denatured) polymers will experience an attractive hydrophobic force. Hydrophobicity is only recently becoming understood as a force, but it is believed to decay exponentially with distance and to be relatively long range.

## 1.3 Emulsion dynamics

As seen above, the large interfacial free energy of emulsions means that they are thermodynamically unstable structures and will tend to phase-separate. There are various mechanisms that are important in emulsion destabilization, and although in reality they often occur simultaneously and catalyse one another, here we will start by examining them in isolation.

## 1.3.1 Creaming

The density of water is approximately 1000 kg m$^{-3}$ while food oils are about 970 kg m$^{-3}$. The 30 kg m$^{-3}$ difference provides buoyancy for oil droplets in water and they will eventually float to the surface and form a separate cream layer. (In some cases, e.g. solid fat droplets, the particles may be denser than the continuous phase and will sediment. The rates of both processes can be described in a similar manner.) A creamed layer has a higher fat content, and extensive creaming can lead to visible and textural differences in the product. A cream layer can initially be redispersed by mixing but, if left, the closely packed droplets may begin to flocculate and coalesce, in which case reversing the destabilization is less easy.

The stability of an emulsion to creaming can be calculated from the creaming velocity of a single particle.[24] The movement of a buoyant droplet is opposed by a frictional force from the continuous phase and counterbalancing these forces yields a simple Stokes expression for the terminal velocity ($v_{Stokes}$) for a creaming droplet:

$$v_{Stokes} = \frac{-2gr^2 \Delta \rho}{9\eta} \tag{1.5}$$

where $r$ is particle radius (in fact $r_{54}$), $\Delta\rho$ is the density of the dispersed phase minus the difference of the continuous phase, $\eta$ is the continuous phase viscosity and $g$ the acceleration due to gravity. Stokes' law is only true for isolated, noninteracting spheres, and several approaches have been used to modify it to account for finite droplet concentrations. Even in dilute systems, the movement of particles upwards is retarded by the countermovement of continuous phase downwards. This can be accounted for by modifying Stokes velocity in a volume fraction-dependent manner (e.g. $v = v_{Stokes}(1 - 6.5\phi)$; however, this approach will only work in dilute systems (roughly <2%) beyond which a semiempirical function must be applied (cf. Krieger–Dougherty equation below):

$$v = v_{Stokes}\left(1 - \frac{\phi}{\phi_c}\right)^{k\phi_c} \tag{1.6}$$

where $\phi_c$ is introduced as a critical concentration (often approximately close packing) where creaming is drastically retarded, and $k$ is a constant. Even this formulation deals poorly with polydisperse emulsions where there is a range of creaming rates. Large particles that cream rapidly can be particularly destabilizing if they flocculate with smaller particles as they move upwards.

Moderate flocculation increases creaming rate because the larger effective radius of the floc is weighted more heavily in Equation 1.5 than the loss of density difference due to the presence of entrained continuous phase. More extensive flocculation has the effect of lowering $\phi_c$.

### 1.3.2 Flocculation

Flocculation is the permanent or semipermanent association of droplets without mixing of their contents. Flocculation requires first that two particles collide, and second that they react in some way to bond together. Collision rates ($F_b$, collisions per unit volume per second) under Brownian motion can be calculated from Smoluchowski kinetics:[25]

$$F_b = k_b n^2 = \frac{3kT\phi^2}{2\eta\pi r^6} \tag{1.7}$$

where $k_b$ is a second-order rate constant for the process and $n$ is the number density of the particles. In fact, only about one collision in a million leads to reaction, and the Smoluchowski expressions given above proportionally overestimate the flocculation rate. In principle the proportion of collisions leading to interaction can be calculated from the interdroplet pair potential but in practice flocculation is usually considered more qualitatively. For example, electrostatic repulsive forces are reduced when the net charge is zero or their effects are shielded by high ionic strengths. Consequently, a whey protein-stabilized emulsion may flocculate when pH > pI or in the presence of salts.[26] Other forces are affected in other cases; for example, the surface proteins in similar emulsions were denatured by a heating step.[27] The denatured protein is more hydrophobic and its presence at the interface leads to increased hydrophobic attraction between droplets and flocculation. In all cases the balance of forces acting is important rather than the absolute magnitude of an individual contributor (Fig. 1.6a).

In other cases it is more profitable to consider the mechanism of flocculation as the simultaneous binding of two (or more) droplets to a third structural element (i.e. bridging flocculation, Fig. 1.6b). For example, emulsions prepared with a low protein:oil ratio may have the same protein chain adsorbed onto different droplets and therefore be flocculated. Similarly an oppositely charged added aqueous polymer may adsorb to multiple droplets and bind them together (Fig. 1.6b).[15]

The formation of a floc represents another level of emulsion structure not examined in Section 1.2. Floc structure has a strong effect on the properties of emulsions, particularly rheology (see Section 1.4.1). Figure 1.7 shows two simulated flocs formed from the same number of particles. Figure 1.7a is a more open structure than Fig. 1.7b and the effective volume entrained within the floc (i.e. the radius of gyration around the centre of mass) is greater. In many cases it is possible to ascribe a fractal dimension to the flocculated structure. All objects show some relation between their length and mass. For Euclidean solids, planes and lines it is a cubic, quadratic or linear relationship, respectively (e.g. an emulsion droplet is a three-dimensional Euclidean object – its mass is proportional to the cube of its radius). For fractal objects (e.g. Fig. 1.7) the relationship between length and volume (or mass) is noninteger. For a flat fractal the closer the fractal dimension to two, the more efficiently the object fills the space and the denser the aggregate (e.g. Fig. 1.7b is denser than 1.7a). Similarly, for a realistic three-dimensional aggregate, the closer the fractal dimensionality to three, the more dense the object and the lower its effective volume. Fast aggregation has been shown to give a more open structure than slower aggregation. Flocs can also rearrange after formation to maximize attractive interparticle interactions.

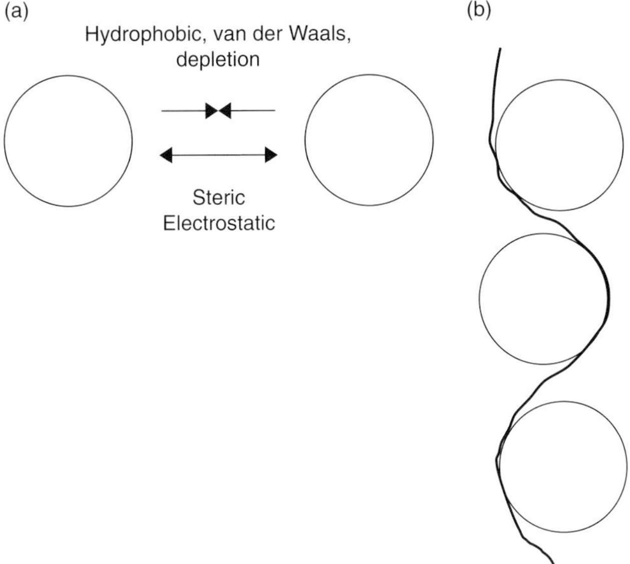


**Fig. 1.6**   Schematic diagram showing emulsion flocculation caused by (a) dominance of attractive over repulsive forces and (b) complex formation.

### 1.3.3 Coalescence

Coalescence occurs when the contents of two oil droplets merge to form a single larger droplet. Whereas flocculation can often be reversed by vigorous mixing or by changing the solvent conditions, coalescence can only be reversed by rehomogenization. Extensive coalescence will lead to oiling-off from the emulsion whereas limited coalescence in the mouth is believed to be responsible for some of the lubricity of food emulsions. The process

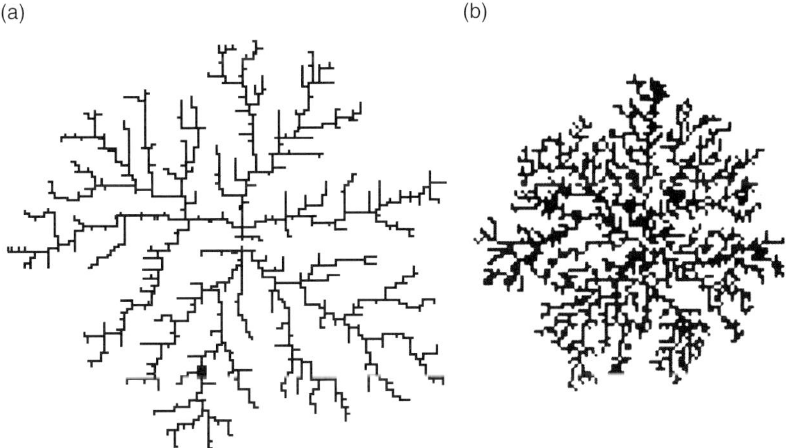


**Fig. 1.7**   Computer simulation of two-dimensional fractal aggregates formed from the same number of primary particles. The fractal dimension of (b) is greater than that of (a). Images generated using a JAVA simulation provided on-line by the Boston University Center for Polymer Studies.

is typically quite slow and so is most significant when droplets are in close proximity due to creaming or flocculation.

Coalescence requires the rupture of the lamella separating two adjacent droplets.[28] The formation of a channel is a nucleation event, and growth and coalescence follow very rapidly as the Laplace pressure reshapes the now-contiguous oil into a single sphere.

A large enough local deformation of one surface will lead to the formation of a channel between the two droplets (Fig. 1.8). The thinning of a lamella is opposed by the Gibbs–Marangoni effect. Adsorbed surfactant will laterally diffuse along the surface concentration gradient to the low-surfactant region of the incipient hole. The lateral movement of the surfactant sets up a flow of more continuous phase into the thinning point, which forces the approaching surfaces apart. The formation of a channel in the interfacial layer requires a

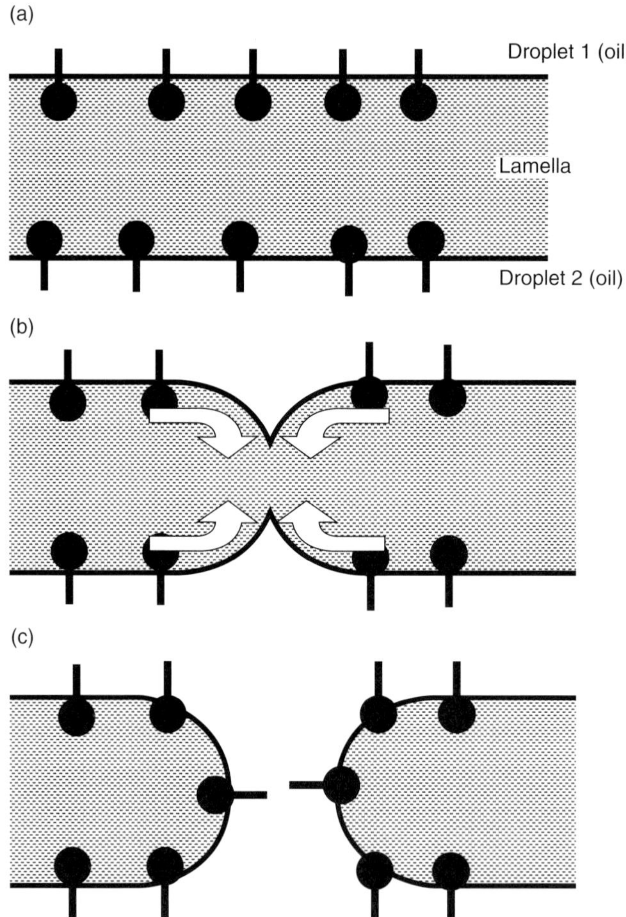


**Fig. 1.8**  Diagrammatic illustration of important steps during the formation of a pore in the lamella between two adjacent droplets, which is the first step of coalescence. (a) The thin lamella represents the aqueous layer separating two adjacent droplets. (b) Spontaneous deformations in the surface can lead to pore formation but this is opposed by lateral diffusion of the surfactant layer sweeping more water into the thinning lamella (i.e. Gibbs–Marangoni effect). (c) In some cases the spontaneous curvature favoured by the packing geometry of the surfactant layer favours pore formation and coalescence (i.e. the orientated wedge theory). See text for details.

dramatic change in surface curvature of the interfaces. Small-scale curvatures are affected by the packing of the surfactant molecules, and in some cases particular molecular geometries may ease channel formation (i.e. the orientated wedge theory[29]). The surface layer can also retard coalescence by mechanically damping surface deformations that would lead to hole formation. Interfacial lamellae can be mechanically ruptured by applied forces when the adjacent droplets cannot readily slip past one another, for example when they are constrained in a floc or in a highly concentrated emulsion. High shear can also provide the energy for two colliding droplets to coalesce.[28]

The effect of shear is particularly pronounced when the droplets are partially crystalline and the mechanical forces aid the penetration of the lamella by a fat crystal protruding from one of the droplets.[30] The liquid oil in the second droplet then preferentially wets the crystalline fat and flows out to reinforce the link (Fig. 1.9). Partially crystalline droplets do not completely merge but maintain a double shape because the mechanical strength of the solid fat network maintains the shape against the Laplace pressure; consequently this process is known as partial coalescence.

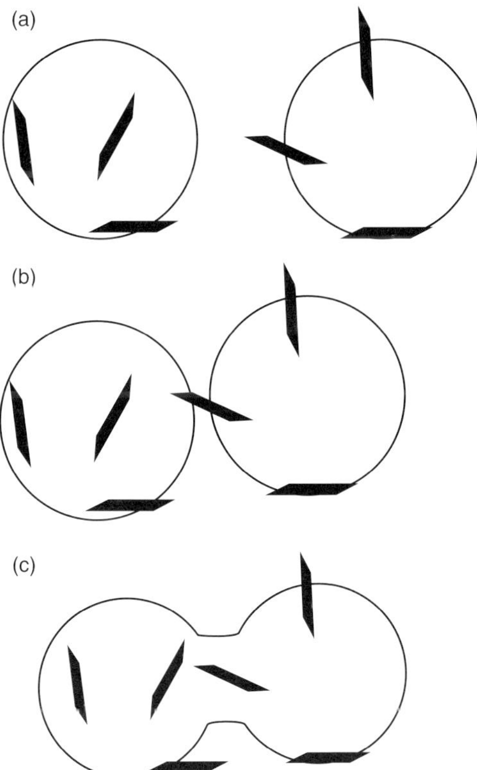

**Fig. 1.9**   Schematic diagram showing the mechanism of partial coalescence. (a) Fat crystals protrude from the surface of a semisolid droplet. (b) On collision the crystals can penetrate a second droplet. The rate of this process is highly affected by shear. (c) Liquid oil flows out to reinforce the link making the process irreversible.

## 1.4 Emulsion functionality

Eads defined functionality as 'the important set of specific materials to specific forces applied under particular circumstances' and took this to include processability, sensory quality and stability.[31] The functional properties of emulsions can be related to emulsion structure in a reasonably quantitative manner, at least for model systems, and provide some useful qualitative guidance in real foods.

### *1.4.1 Rheology*

The viscosity of a liquid increases upon the addition of rigid spherical particles due to the increased friction between the particle and the liquid layers causing greater energy dissipation. In a highly dilute, uncharged suspension containing noninteracting solid particles (volume fraction $\phi$) the relative viscosity ($\eta'$, apparent viscosity normalized to that of the dispersed phase) is given by the relationship:

$$\eta' = \sum_i k_i \phi^i \qquad (1.8)$$

where $k_i$ are constants (1, 2.5, 6.2, ...). Allowing only first-order terms in $\phi$, Equation 1.8 reduces to the well-known Einstein equation. However, the Einstein relationship is rarely quantitatively reliable in realistically concentrated emulsions where droplet–droplet interactions dominate the hydrodynamic drag. The second-order term allows for two-body interactions but is only itself quantitatively reliable at very low concentrations. At high concentrations it is common to resort to empirical and semiempirical expressions such as the Krieger–Dougherty relation:

$$\eta_r = \left(1 - \frac{\phi}{\phi_{max}}\right)^{-2.5\phi_{max}} \qquad (1.9)$$

where $\phi_{max}$ is a volume fraction related to close-packing (cf. Equation 1.6). At low concentrations all of these expressions reduce to the Einstein equation.

For a given volume fraction, droplet size is relatively unimportant whereas droplet flocculation can have a huge effect on the rheology of an emulsion.[24] Fractal dimension can be used to calculate an effective volume fraction $\phi_{eff}$:

$$\phi_{eff} = \phi \left(\frac{R}{r}\right)^{3-D_f} \qquad (1.10)$$

where $R$ is the floc radius, $r$ is the primary particle radius, $\phi$ is the particle volume fraction and $D_f$ is the fractal dimension (1–3). The effective volume fraction can be used in the Krieger–Dougherty type functions (Equation 1.9) to link microstructure and viscosity. Extensive flocculation can lead to the formation of a percolating network and the appearance of elastic properties (i.e. a particle gel). Because flocs are vulnerable to breakdown under shear, the rheological properties of flocculated emulsions are highly time- and shear-rate-dependent.

## *1.4.2 Chemical reactivity*

The presence of polar and nonpolar domains in a food emulsion allows reactive materials to interact at concentrations distinct from their bulk average and so the reaction kinetics may proceed in an unexpected manner. The partition coefficient (i.e. the ratio of activity of a component in the oil phase to the activity in the aqueous phase) defines the tendency of an ingredient to accumulate in one phase or another. A striking example of this is butter (a water-in-oil emulsion), which contains only about 2% salt yet is relatively microbiologically stable. A 2% salt concentration would only be expected to reduce the water activity of a solution to about 0.99, which would still be hazardous (S. Doores, pers. commun.). However, as the butter is only about 20% water and the salt is exclusively water soluble the aqueous salt concentration would be 10% with $a_w = 0.94$, which is responsible for the observed stability of the products. In an aqueous system Wedzicha and co-workers showed how benzoic acid would partition between the aqueous and oil components of an emulsion and also significantly accumulate at the interface.[32] This observation has subsequently been extended to a wider range of compounds,[33] and the rapid dynamics of interfacial exchange has been elucidated.[34]

The thermodynamics and kinetics of aroma partitioning between a food and the gas surrounding it are central to understanding the physical basis of flavour perception in foods. Aroma molecules typically have a significant lipid solubility so the presence of emulsion droplets can affect the perception of food flavour (e.g. the same amount of rancidity is often perceived less in a high-fat food compared with a low-fat food as the lipid acts as a reservoir to bind up the rancid volatiles). An effective partition coefficient between an emulsion and the headspace ($K_{ge}$) can be calculated from the oil–water ($K_{ow}$) and water–headspace ($K_{gw}$) partition coefficients and the volume fraction of the emulsion ($\phi$):[35]

$$K_{ge} = \frac{K_{gw}}{1+(K_{ow}-1)\phi} \tag{1.11}$$

According to this relation, the amount of headspace volatiles will increase with lipid content for more water-soluble flavours and decrease for more lipid-soluble flavours. In a real, complex food flavour this will affect the relative impact of different flavour notes and may subtly change the character of the aroma.

Emulsified lipids are vulnerable to oxidation but the kinetics of the reaction are distinct. Frankel *et al.*[36] hinted at the importance of interfacial phenomena in controlling reactivity when they pointed out the 'polar paradox' in that water-soluble surfactants are more effective in stabilizing bulk lipids whereas lipid-soluble antioxidants work better for the same lipid in an emulsified form. They suggested that the antioxidant concentrated in the most vulnerable portion of the sample in each case (i.e. at the air–lipid surface in bulk and in the lipid droplets in the emulsion). Since then the role of emulsion structure in controlling the location (and hence reactivity) of lipids, antioxidants and catalysts has been well established.[37]

# 1.5 References


1 Dickinson, E. & Stainsby, G. (1982) *Colloids in Food*. Applied Science Publishers, London.
2 Dickinson, E. & McClements, D.J. (1995) *Advances in Food Colloids*. Blackie Academic & Professional, Glasgow.
3 Hunter, R.J. (1986) *Foundations of Colloid Science*. Oxford University Press, Oxford.
4 McClements, D.J. (1999) *Food Emulsions: Principles, Practice, and Techniques*. CRC Press, Boca Raton, FL.
5 Evans, D.F. & Wennerstrom, H. (1994) *The Colloidal Domain*. Wiley-VCH, New York.
6 Walstra, P. (2003) *Physical Chemistry of Foods*. Marcel Dekker, New York.
7 Princen, H.M. (2001) The structure, mechanics, and rheology of concentrated emulsions and fluid foams. In: Sjoblom, J. (ed.) *Encyclopedic Handbook of Emulsion Technology*, pp. 241–278. Marcel Dekker, New York.
8 Weiss, J. (2002) Key concepts of interfacial properties in food chemistry. In: Wroldstad, R., Acree, E., Decker, T.E. *et al.* (eds) *Current Protocols in Food Analytical Chemistry*, pp. D3.5.1–22. John Wiley & Sons, New York.
9 Weiss, J., Herrmann, N. & McClements, D.J. (1999) Ostwald ripening of hydrocarbon emulsion droplets in surfactant solutions. *Langmuir* **15**, 6652–6657.
10 Walstra, P. (1993) Principles of emulsion formation. *Chem. Eng. Sci.* **48**, 333–349.
11 Euston, S.E., Singh, H., Munro, P.A. & Dalgleish, D.G. (1995) Competitive adsorption between sodium caseinate and oil-soluble and water-soluble emulsions. *J. Food Sci.* **60**, 1124–1131.
12 Damodaran, S. & Sengupta, T. (2003) Dynamics of competitive adsorption of alpha(s)-casein and beta-casein at planar triolein-water interface: Evidence for incompatibility of mixing in the interfacial film. *J. Agric. Food Chem.* **51**, 1658–1665.
13 Pugnaloni, L.A., Ettelaie, R. & Dickinson, E. (2003) Do mixtures of proteins phase separate at interfaces? *Langmuir* **19**, 1923–1926.
14 Dickinson, E. (1992) Structure and composition of adsorbed protein layers and the relationship to emulsion stability. *J. Chem. Soc. Faraday Trans.* **88**, 2973–2983.
15 Dickinson, E. (2003) Hydrocolloids at interfaces and the influence on the properties of dispersed systems. *Food Hydrocolloids* **17**, 25–39.
16 Dickinson, E. (1999) Caseins in emulsions: interfacial properties and interactions. *Int. Dairy J.* **9**, 305–312.
17 Douillard, R., Daoud, M. & Aguie-Beghin, V. (2003) Polymer thermodynamics of adsorbed protein layers. *Curr. Opin. Colloid Interface Sci.* **8**, 380–386.
18 Dickinson, E. & Matsumura, Y. (1991) Time-dependent polymerization of beta-lactoglobulin through disulfide bonds at the oil-water interface in emulsions. *Int. J. Biol. Macromol.* **13**, 26–30.
19 Chanyongvorakul, Y., Matsumura, Y., Sawa, A. *et al.* (1997) Polymerization of beta-lactoglobulin and bovine serum albumin at oil-water interfaces in emulsions by transglutaminase. *Food Hydrocolloids* **11**, 449–455.
20 Damodaran, S. & Rammovsky, L. (2003) Competitive adsorption and thermodynamic incompatibility of mixing of beta-casein and gum arabic at the air-water interface. *Food Hydrocolloids* **17**, 355–363.
21 Aveyard, R., Binks, B.P. & Clint, J.H. (2003) Emulsions stabilised solely by colloidal particles. *Adv. Colloid Interfac.* **100**, 503–546.
22 Israelachvili, J. (1992) *Intermolecular and Surface Forces*. Academic Press, London.
23 Claesson, P., Blomberg, E. & Poptoshev, E. (2004) Surface forces and emulsion stability. In: Friberg, S.E., Larsson, K. & Sjoblom, J. (eds) *Food Emulsions*, pp. 257–298. Marcel Dekker, New York.
24 Dickinson, E. (1998) Structure, stability and rheology of flocculated emulsions. *Curr. Opin. Colloid Interface Sci.* **3**, 633–638.
25 Vanapalli, S.A. & Coupland, J.N. (2004) Orthokinetic stability of food emulsions. In: Friberg, S.E., Larsson, K. & Sjoblom, J. (eds) *Food Emulsions*, pp. 327–352. Marcel Dekker, New York.

26 Demetriades, K., Coupland, J.N. & McClements, D.J. (1997) Physical properties of whey protein stabilized emulsions as related to pH and NaCl. *J. Food Sci.* **62**, 1–6.

27 Demetriades, K., Coupland, J.N. & McClements, D.J. (1997) The effect of temperature on the stability of whey protein stabilized emulsions. *J. Food Sci.* **62**, 462–467.

28 vanTan, G. (2004) Coalescence mechanisms in protein-stabilized food emulsions. In: Friberg, S.E., Larsson, K. & Sjoblom, J. (eds) *Food Emulsions*, pp. 299–326. Marcel Dekker, New York.

29 Kabalnov, A. & Wennerstrom, H. (1996) Macroemulsion stability: the oriented wedge theory revisited. *Langmuir* **12**, 276–292.

30 Boode, P. & Walstra, W. (1993) Partial coalescence in oil-in-water emulsions. *Colloid Surface A* **81**, 121–137.

31 Eads, T. (1994) Molecular origins of structure and functionality in foods. *Trends Food Sci. Technol.* **5**, 147–159.

32 Wedzicha, B.L. & Ahmed, S. (1994) Distribution of benzoic acid in an emulsion. *Food Chem.* **50**, 9–11.

33 Stockmann, K. & Schwarz, K. (1999) Partitioning of low molecular weight compounds in oil-in-water emulsions. *Langmuir* **15**, 6142–6149.

34 Wedzicha, B. & Couet, C. (1996) Kinetics of transport of benzoic acid in emulsions. *Food Chem.* **55**, 1–6.

35 Harrison, M., Hills, B.P., Bakker, J. & Clothier, T. (1997) Mathematical models of flavor release from liquid emulsions. *J. Food Sci.* **62**, 653–658, 664.

36 Frankel, E., Huang, S.W., Kanner, J. & German, B. (1994) Interfacial phenomena in the evaluation of antioxidants: bulk oils vs. emulsions. *J. Agric. Food Chem.* **42**, 1054–1059.

37 McClements, D.J. & Decker, E.A. (2000) Lipid oxidation in oil-in-water emulsions: Impact of molecular environment on chemical reactions in heterogeneous food systems. *J. Food Sci.* **65**, 1270–1282.

Chapter 2

# Physicochemical Behaviour of Starch in Food Applications

*Alain Buleon and Paul Colonna*

## 2.1 Introduction

Starch represents the major reserve polysaccharide of photosynthetic tissues and of many types of storage organs such as seeds, swollen stems, tubers and roots. Starch is contained in a large variety of plant crops such as cereals (50–80%), legumes (25–50%) and tubers (60–95%). Native starch granules are used in several applications as solid particulate materials for cleaning or drying.

But the main uses of starch follow disruption of the starch granules, as ingredients and foods, either home-made or manufactured. These modifications affect sensory properties (texture, visual aspect and flavour) as well as nutritional ones.

Starch, predominantly composed of amylose and amylopectin, is one of the major polysaccharides of many higher plants. It provides an essential carbohydrate food energy source for the human population. In 1992 the US Department of Agriculture emphasized the importance of cereal-based foods in the human diet by introducing the Food Guide Pyramid.

An understanding of the physical chemistry of starch in foods is now possible thanks to inputs from polymer science. Unfortunately, the application of these approaches to biopolymers and foods is problematic, for the following reasons.

- In contrast to synthetic polymers, biopolymers are variable products, the suitability of which for a given process/product goal is affected by the genotype, environmental conditions during grain development and the milling process. These difficulties have been reinforced by progress in plant breeding and by the availability of new botanical sources thanks to the global market.
- Very often food products have a complex formulation, with several components (starch, proteins, water, sugars, lipids) that can interact by complexation or phase separation, and lead to more or less organized structures.
- Starch itself is made of two types of macromolecules, the linear amylose and the branched amylopectin: it can thus be considered a complex material.
- Doughs, pastes and melts from cereal products are non-Newtonian, with a high level of elasticity, and are very sensitive to temperature, water content and, more generally, to composition (starch origin, presence of lipids, etc.).

- Biopolymers are not as thermally stable as standard polyolefins; they start to decompose at 200–220°C. An important emerging issue for starch-based processed foods is acrylamide, which has been found as a by-product in starchy foods baked or fried at very high temperatures, such as potato crisps, crisp breads and cereals. Acrylamide has been shown to be a carcinogen in *in vivo* studies using mice. In April 2002 the Swedish Food Administration identified levels of up to 1200 ppm of acrylamide in potato crisps – the EU safety threshold is 0.5 ppm. This study led to a burst of studies in the West devoted to the effect of high-temperature processing of starch on acrylamide formation. However, before this critical event, it must be remembered that chemical interactions can occur, leading to intra- and intermolecular covalent cross-linkings; the kinetics of these reactions are determined by temperature and also water content.

All these reasons demonstrate that any investigation should use samples whose biological origin and conditions of preparation are not only known, but also reproducible. The variability caused by these various factors leads to a broad range of behaviours during the three basic operations in food technology: mixing, pasting-baking-cooking and storage. To characterize the complexity of these changes, laboratory tools such as the Rapid Visco analyzer (Newport Scientific Products, http://www.newport.com.au/products/) or Brabender Amylograph (Brabender, http://www.brabender.com/) have been used, in order to describe food component performance. Their use does not require great knowledge of physical chemistry and rheology since their goal is to provide qualitative information related to process adequacy. More objective experimental approaches with necessary adaptations are now available, such as differential scanning calorimetry (DSC), high-performance liquid chromatography (HPLC), infrared (IR), nuclear magnetic resonance (NMR), magnetic resonance imaging (MRI), transmission electron microscopy (TEM) and confocal scanning laser microscopy (SLM). The variety of technologies and methods and their rapid evolution explains why no literature synthesis is available on this topic.

This chapter summarizes current knowledge of the physical chemistry of starches in foods, up to the lastest improvements in physicochemical methods, to the point at which they can actually be useful for food technology.


## 2.2 Starch composition and chemical structure

Starch granules are semicrystalline particles ranging from 1 to 100 µm in size. The basic building block is a glycosyl monomer, measuring 0.3 nm. Therefore any description of the native granules must encompass the different scales (Fig. 2.1). When viewed under the polarizing microscope, native starch granules show a dark birefringent cross ('Maltese cross'), which is characteristic of the spherulitic organization of the macromolecules inside each granule.


### 2.2.1 Granular structure

Native starch granules exhibit different sizes and shapes depending on the botanical source[1,2] (Table 2.1); both shape (round, oval, polyhedral) and particle size distribution (unimodal,

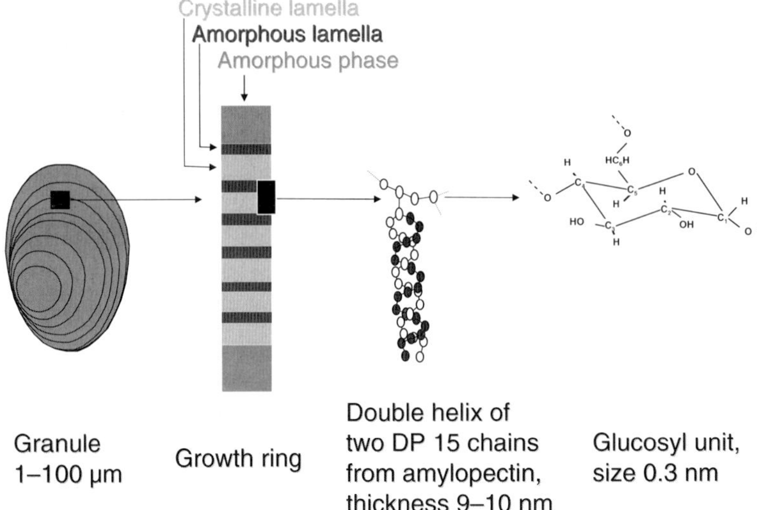


**Fig. 2.1**  Schematic representation of the different structural levels of the starch granule and the involvement of amylopectin.

bimodal) are typical of the botanical origin. Starches from *Triticicae* have a bimodal distribution comprising large A-type (lenticular, 10–35 μm) and small B-type (spherical, 1–8 μm) granules. Despite several decades of research into the crystalline ultrastructure of starch, many questions remain, such as the respective contributions of amylose and amylopectin to crystallinity, the distribution of ordered and unordered areas in the granule, the size distribution of crystalline areas or the organization of mixed A- and B-type granules.

**Table 2.1**  Morphological features of starch granules and amylose content in major plant sources.

| Source | Amylose content (% total starch) | Granule type | Average size (μm) | Shape |
| --- | --- | --- | --- | --- |
| Barley normal (wild) | 21–24 | A-granule | 20 | |
| | | B-granule | 2-3 | |
| Wheat normal (wild) | 25–29 | A-granule | 30 | Discs |
| | | B-granule | 2–3 | Perfect spheres |
| Wheat waxy | 1.2–2.0 | One type | | |
| Maize normal (wild) | 25–28 | One type | 30 | Polyhedral and rounded |
| Maize waxy | 0.5 | One type | 15 | |
| Maize high amylose | 60–73 | One type | 5–25 | Highly elongated irregular filament |
| Oat | | A-granule | 15 | Compound |
| | | B-granule | 2–3 | Oval |
| Potato normal (wild) | 18–21 | One type | 40 | Large oval |
| Potato amylose-free | 1 | One type | 40 | Large oval |
| Pea *RR* (wild) | 33–36 | One type | 30 | Oval |
| Pea *rr* | 66–72 | One type | 50 | Compound |
| Pea *rbrb* | 23–32 | One type | 20 | Round |
| Pea *rr rbrb* | 49 | One type | | Compound |

The internal architecture of native starch granules (Fig. 2.1) is characterized by 'growth rings' that represent concentric semicrystalline shells (thickness 120–400 nm) separated by amorphous regions.[3–5] There is much evidence that the crystalline shells consist of regular alternating amorphous and crystalline lamellae repeating at 9–10 nm.[6] In this structural organization, parallel double helices of amylopectin side chains are assembled into radially oriented clusters. Nevertheless, knowledge is still limited regarding the detailed structure, organization and arrangement of the lamellae. From electron diffraction data and silver staining of potato starch fragments, Oostergetel and van Bruggen[7] concluded that the semicrystalline domains form a network of left-handed superhelices (diameter 18 nm, pitch 10 nm), which could be a well-ordered skeleton for the starch granule. On the basis of electron and atomic force microscopy observations, Gallant *et al.*[8] proposed that lamellae are organized in spherical blocklets, whose diameters range from 20 to 500 nm, depending on the botanical origin of starch and its location within the granule. Lastly, Waigh *et al.*[9,10] proposed that amylopectin is structurally analogous to a synthetic side chain liquid crystalline polymer (Fig. 2.2) with three distinct components: rigid units (mesogens) corresponding to double helices, flexible spacers and a flexible backbone. In the nematic state, helices are not aligned into lamellae whereas in the smectic state, as in the granules under normal conditions, the mesogens are aligned, creating a 9-nm repeat between the lamellar lengths (Fig. 2.2).

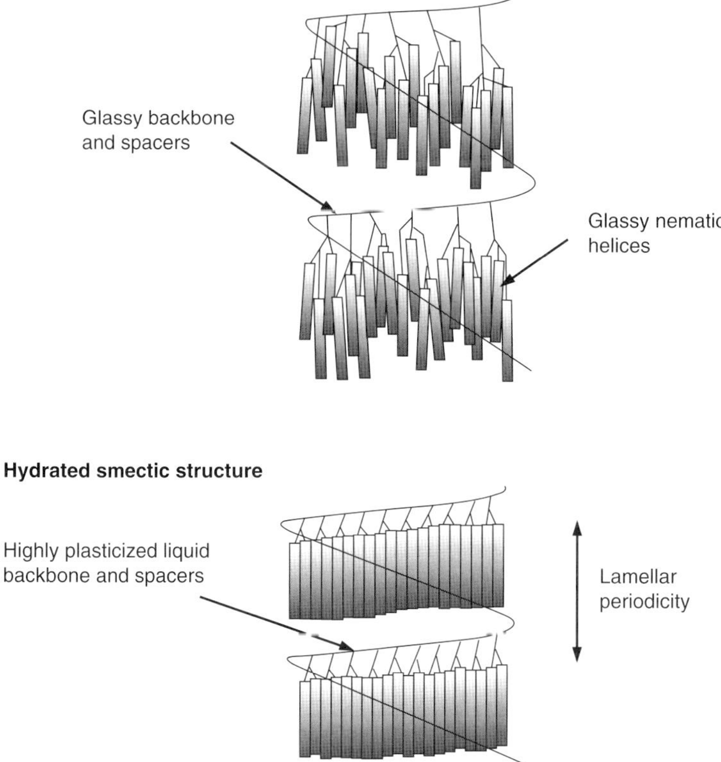


**Fig. 2.2**    Side chain liquid crystal model for amylopectin and nematic–smectic transition.[10]

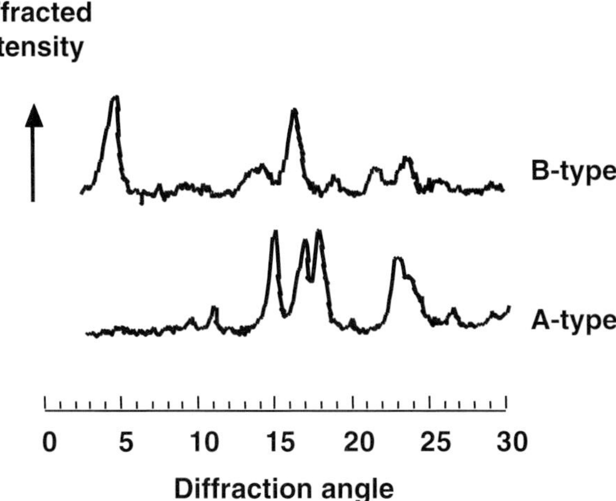



**Fig. 2.3**   X-ray diffraction diagrams of A- and B-type starches.

The crystallinity of starch is due to the packing of double helices. Native starch granules exhibit two main types of X-ray diffraction diagrams (Fig. 2.3), A-type and B-type; mixtures of A- and B-types designated C-type also occur in most legume starches and some mutant starches: B-type occurs at the centre of the granule and A-type around the periphery. Mild hydrolysis is a way to demonstrate such heterogeneities inside complex starch granules.[11]

Such crystalline structures may also be encountered in food applications of starch depending on the processing conditions. The structural models established from A and B amylose crystals were transposed to crystalline regions of starch because the main reflections contained in the powder diffraction diagrams of native starch were present in the diffractograms of these crystals. The most recent models for A and B amylose structures are based upon sixfold left-handed double helices with a pitch height of 2.08–2.38 nm.[12,13] In the A structure, these double helices are packed with the space group B2 in a monoclinic unit cell ($a = 2.124$ nm, $b = 1.172$ nm, $c = 1.069$ nm, $\gamma = 123.5°$) with eight water molecules per unit cell (Fig. 2.4). In the B-type structure, double helices are packed with the space group $P6_1$ in a hexagonal unit cell ($a = b = 1.85$ nm, $c = 1.04$ nm) with 36 water molecules per unit cell (Fig. 2.4). The symmetry of the double helices differs from A to B structures, since the repeated unit is a maltotriosyl unit in the A form and a maltosyl unit in the B form.[13] Independent evidence for the individuality of each glucosyl residue in maltosyl and maltotriosyl units comes from solid state $^{13}$C nuclear magnetic resonance (NMR). The C1 peak in the A-form spectra is a triplet while it is a doublet in spectra of B-form.[14]

## *2.2.2 Molecular composition*

In most common starches, the relative weight percentages of amylose and amylopectin range between 65 and 82% for amylopectin and 18 and 35% for amylose (Table 2.1). However, some mutant genotypes of maize, barley, oat and rice contain as much as 70% amylose whereas other genotypes, called waxy, contain less than 15% (maize, barley, rice, sorghum,

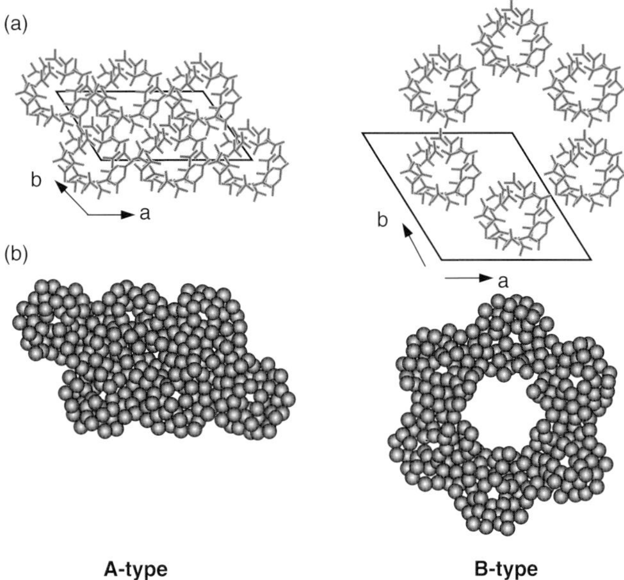



**Fig. 2.4**   (a) Crystalline packing of double helices in A-type and B-type amylose. (b) Projection of the structures onto the $(a,b)$ plane.

wheat). Diploid species of maize, rice or barley have been intensively investigated in contrast with tetraploid and hexaploid species such as oats.[15]

### 2.2.2.1 Amylose

Amylose is defined as a linear molecule of $(1{\rightarrow}4)$-linked $\alpha$-D-glucopyranosyl units; however, some molecules are slightly branched by $(1{\rightarrow}6)$-$\alpha$-linkages.[16] No effective methods for the separation of linear and branched amyloses are known, so all results concerning amylose branching have been obtained on the assumption that two quite distinct populations exist, one strictly linear and the other characterized by a 40% $\beta$-amylolysis limit.[17] This low value suggests that the branch linkages are frequently located near the reducing terminal end and/or they have multiple branched side chains.  The branched molecule amount ranges from 25 to 55% on a molecular basis,[17,18] and for one starch was shown to increase continuously as a function of the molecular weight; the presence of 9–20 branch points equivalent to 3–11 chains per molecule[19] does not alter significantly the solution behaviour of amylose chains as evidenced by $R_i \sim M_i^n$ relations.[20] $M_i$ and $R_i$ are, respectively, the molecular weight and the radius of the component $i$ of a series of particles of the same architecture but different molecular weights.

Molecular weight distributions and average molecular weights have been extensively measured for a large number of starches. In contrast with proteins, which are genetically coded, polysaccharides have a molecular weight distribution, usually represented by the average molecular weights: number-average MW ($M_n$), weight-average MW ($M_w$) or $z$-average MW ($M_z$). Important discrepancies in molecular weight and polydispersity of amylose are observed in the literature due to (i) the biological origin of amylose, leading to uncontrolled

**Table 2.2**    Number-average ($M_n$) and weight-average ($M_w$) molecular weights of potato and wheat amyloses.

| Starch | $M_n \times 10^{-5}$ | $M_w \times 10^{-5}$ | Reference |
|---|---|---|---|
| Potato | 10.3 | – | 16 |
| | – | 7.0 | 20 |
| | 1.3 | 9.0 | 21 |
| Wheat | 2.1 | – | 16 |
| | – | 5.1 | 20 |
| | 0.6 | 3.9 | 21 |

variations in the biosynthesis mechanisms; and (ii) molecular degradation occurring during amylose fractionation. The different values (Table 2.2) quoted in the literature for potato and wheat amyloses illustrate these variations. Generally, values of average molecular weight range from $10^5$ to $10^6$. Using a specific optimization algorithm, experimental molecular weight distribution (MWD) obtained by coupling between size-exclusion-chromatography and multi-angle-laser-light-scattering (SEC–MALLS) creates the possibility to fit them to mathematical MWD such as normal distribution, 'most probable' (MP) distribution and 'log-normal' (LN) distribution. Good agreements were obtained using either a sum of overlapping Gaussian curve[21] or a 'most probable' model.[20]

The flexibility of polymeric chains is measured by the dimensionless quantity $C_\infty$ (the characteristic ratio), generally defined as the ratio between the dimensions of real and freely-jointed chains. An equivalent characteristic is the persistence length $a$, defined as the average projection of an infinitely long chain on the initial tangent of the chain. Amylose chains in solution ($a = 1.71$ nm) are more flexible than those of modified cellulose (cellulose diacetate $a = 4.8$–$7.2$ nm; carboxymethylcellulose $a = 8.0$–$12.0$ nm) but stiffer than those of pullulan ($a = 1.2$–$1.9$ nm). Therefore, all these polysaccharides belong to the class of loosely jointed polysaccharides, in contrast with stiff polysaccharides,[22] such as xanthan ($a = 310 \pm 40$ nm) and scleroglucan ($a = 180 \pm 30$ nm). This conformational feature explains why amylose has a low intrinsic viscosity value compared with other polysaccharides.

Another specific feature of interest concerning amylose is its capacity to bind iodine. The existence of $I_3^-$ and $I_5^-$ ions was checked using Raman spectral measurements and ultraviolet/visible (UV/VIS), coupled to theoretical analyses. The four dominant polyiodide chains that coexist are longer species such as $I_9^{3-}$, $I_{11}^{3-}$, $I_{13}^{3-}$ and $I_{15}^{3-}$. An interesting result is the demonstration of the absence of participation of $I_2$ in the polyiodide chain.[23]

### 2.2.2.2 Amylopectin

Amylopectin is a highly branched component of starch: it comprises chains of $\alpha$-D-glucopyranosyl residues linked together mainly by $(1 \rightarrow 4)$-$\alpha$-linkages but with 5–6% of $(1 \rightarrow 6)$-$\alpha$-bonds at the branch points.

The basic chain organization is described in terms of the A, B and C chains as defined by Peat *et al.* in 1956.[24] Thus, the outer chains (A) are glycosidically linked at their potential reducing group through C-6 of a glucose residue to inner chains (B), which in turn can be linked to other B chains or to the backbone of the molecule, the single C chain; this C chain carries other chains as branches but contains the sole reducing terminal residue. When taking into account their relative sizes, amylopectin is built with three types of chains (Fig. 2.5): short chains (S) consisting of both outer (A, chain length (CL) 12–16) or inner (B1, CL 20–24) chains, inner (B2) long chains (L) of degree of polymerization (DP) 42–55, and a few B chains of DP >60.[25–27] Differences related to the botanical species concern, on the one hand, the L/S ratio expressed on a molar basis and, on the other hand, the length of A and B1 chains. The L/S ratio is estimated at 5 for amylopectins from B-crystalline type (potato) starch and at 8–10 for normal cereal amylopectins from A-crystalline type (normal genotypes) granules. Starches with A-type crystallinity have shorter chain lengths (12–16) on average than B-type starches.

Amylopectin has a high degree of structural organization with the nonrandom distribution of linear chains and the clustering positioning of branch linkages. Each amylopectin cluster, which is the smallest unit of amylopectin, is composed of a different ratio of chains depending on plant origin. A and B1 chains form one cluster whereas B2, B3 and B4 extend into two, three and more clusters. The C chain of amylopectin is the only chain having the reducing terminal residue per molecule: it has a similar size to other unit chains of amylopectin. Regions of high branching frequency alternate with regions that are devoid of branches enabling intervening linear chains to align in parallel arrays of double helices. The general rule is that amylopectins have more A chains than B chains, with the chain ratio ranging from 1.0:0 to 1.5:1. These values are consistent with the cluster and Meyer's structures, but not with those of Haworth and Staudinger.[28]

Without doubt, the most important feature of this branched molecule is that S chains are found in discrete clusters.[29,30] Two different two-dimensional representations exist in the literature. In that of French,[29] the (1→4)-α-chains linking the cluster units have relatively long sections free of any (1→6)-α-linked branches whereas in Robin *et al.*[30] these longer

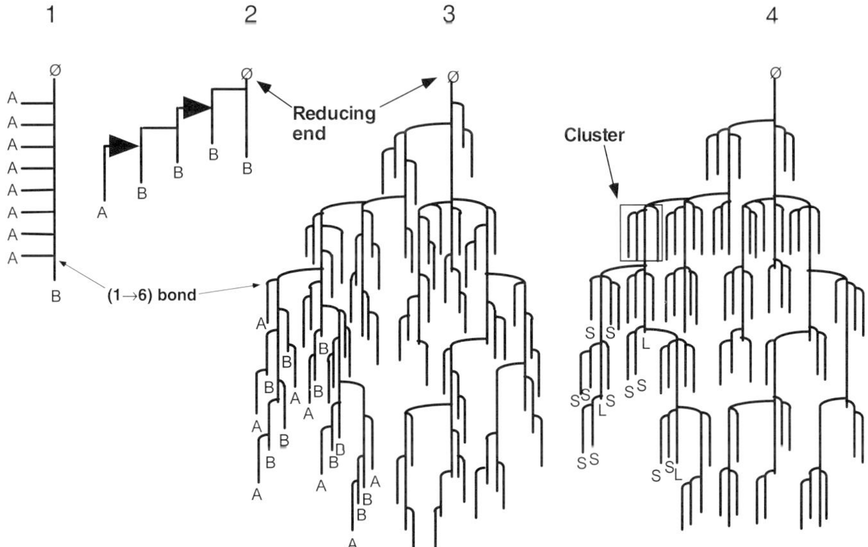


**Fig. 2.5** Schematic representation of the molecular structure of amylopectin as proposed by (1) Haworth, (2) Staudinger, (3) Meyer and (4) Meyer redrawn as a cluster-type architecture.[30]

linking chains have (1→6)-α branches attached with a significant proportion of A chains in the internal sections of the molecules. Another difference lies in the location where the longer (1→4)-α-chains linking cluster units arise within a cluster unit, from any tier in the lower stratum or by an extension of the zero tier. No particular data enable the determination of which pattern occurs (Fig. 2.6). The interesting new fact at this stage is that the fine structure of an amylopectin cluster can be related to the crystalline type observed inside the native crystallites[31] (Fig. 2.7).

The two aforementioned L chain populations represent the backbone supporting the S chains bearing clusters. The success of the cluster-type model is also explained by its ability to account for the higher viscosity of amylopectin when compared with glycogen, the involvement of the amylopectin chain in crystallinity and the degradation pattern displayed by α-amylases. S chains can be considered as the limiting factor defining crystallite thickness. Assuming classical models for A- and B-crystallites, involvement of S chains in crystallite thickness leads to a helicoidal length of about 5.7 nm, with 16 glucosyl units, each one giving a repeating distance of 0.35 nm per glucose. For waxy maize and amylose-free potato starch, clustered chains ranged in DP from 9 to 34, while short chains with DP 6–8 and all long chains DP >35 are amorphous. According to the model of Bertoft,[32] the clusters are connected to a backbone that extends in almost a perpendicular direction and is formed by the amorphous chain. Thereby a superhelical structure can be built from a single amylopectin macromolecule (Fig. 2.8).

This branched character based upon short chains also explains the low binding capacity; 100 g of amylopectin bind less than 1 mg of iodine, giving a $\lambda_{max}$ around 540–550 nm. The low iodine-binding capacity is based upon the formation of an arrangement of four iodine atoms more or less arranged linearly within the cavity of the helix structure of 11 glucosyl units.[33] This fits nicely with the abundant S-chain populations.

Using existing chromatographic profiles of debranched (1→4)-α-chains, Caldwell and Matheson[34] have successfully developed an *in silico* random generated dendrimer model where chains are located on a three-dimensional cubic grid. Extension or branching of A and B chains were chosen randomly. One construct generated on a three-dimensional cubic grid, in which the positions occupied by the chains were exclusive, with random extension and branching, was compared with a spatially unrestricted model with the same average chain length and fraction of A chains. In all cases amylopectin had an oblate ellipsoidal shape. This opens the possibility of relating the macromolecular level to the crystalline level as depicted by the semicrystalline lamellae composed of stacks of amorphous and crystalline lamellar structures (9 nm).

Another way is to study native amylopectin using dynamic and static light scattering. Amylopectin has one of the largest molecular weights ($10^7$–$10^9$), mostly $>10^8$. Takeda *et al.*[35] observed three populations after chromatography of fluorescent labelled molecules, with number-average molecular weights of 13 400–26 500, 4400–8400 and 700–2100. Different models for amylopectin were examined through comparison of calculated and experimentally determined particle scattering functions from combined static and dynamic light scattering studies. For maize amylopectin, these authors concluded that each L chain had 1.4 clusters made of 3.22 S chains on average, whereas Robin *et al.*[30] assumed exactly 2 clusters per L chain. The distance between two clusters on the same B chain is 22 glucosyl units on average. This modelling, based upon the cascade branching theory, should be renewed in the light of the last refinements of Hizukuri and colleagues.[25–27]

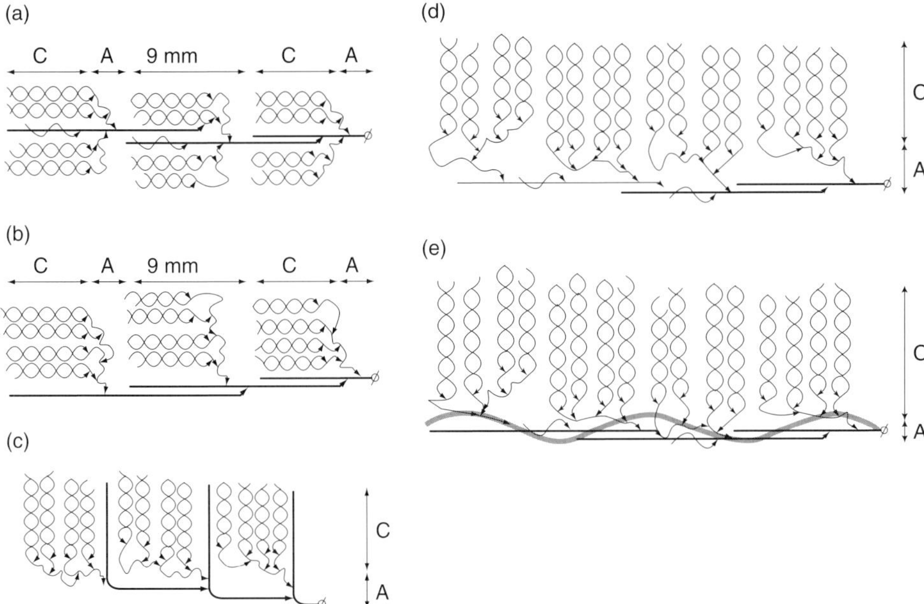

Fig. 2.6   Possible cluster models that cannot be distinguished on the basis of the unit chain distribution of native or acid-treated (lintnerized) starches.[32] The crystalline and amorphous lamellae found inside starch granules are indicated by 'A'. External segments of clustered chains form double helices. Long B-chains (drawn as B2 chains) are shown as bold lines and the reducing end is indicated. (a) The cluster model of Hizukuri:[26] the B2 chains are part of the clusters and partly crystalline. (b) Unidirectional backbone model: the B2 chains form a backbone to which the clusters are anchored. (c) A partial structure of Robin *et al.*[30] in which the internal parts of the B2-chains are amorphous. (d) Two-directional backbone model. The entire B2-chains are amorphous and extend in a direction perpendicular to the clustered chains. Some fingerprint A (Afp)-chains are attached in a perpendicular direction to the clustered chains. Some Afp-chains are attached to the B2-chains. B3 and long A-chains will also be found in the amorphous lamella when present. (e) The two-directional backbone model in a normal starch, where the amylose (thick wavy line) is found together with the amorphous chains of amylopectin.

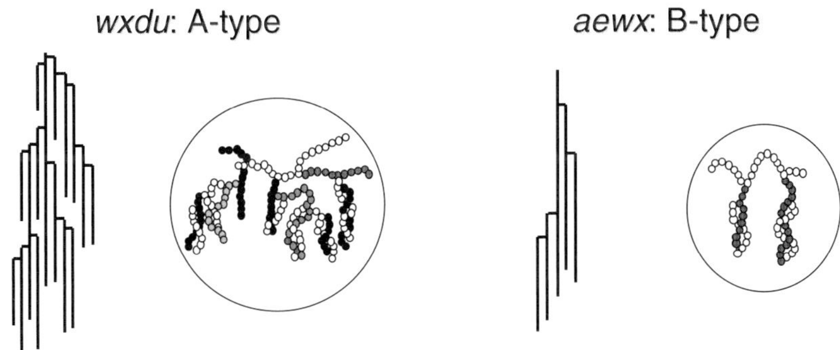



Fig. 2.7   Schematic representation of amylopectin clusters from maize mutants.[31] A cluster of amylopectin from *wxdu* starch (left), which shows A-type crystallinity, is represented in symbolic form and as an organized structure (circled). These clusters are large and comprise numerous short chains with close branch points. This type of organization leads to a high branching density: 0.18 per branching zone of cluster (BZC). For amylopectin from *aewx* starch (right), which shows a typical B-type crystallinity, clusters are composed of fewer, longer chains, with a greater distance between branch points and a low branching density (0.13).

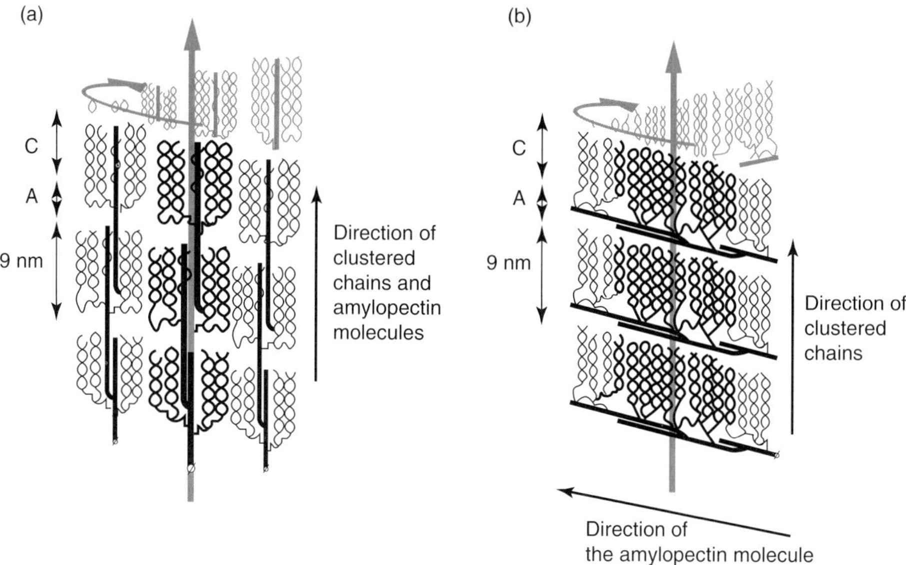



**Fig. 2.8**    The superhelix model of amylopectin schematically redrawn from Oostergetel and van Bruggen[7] and Waigh *et al.*[9]. The axis and its turns of the superhelix are indicated by grey arrows. (a) The superhelix based on the cluster model of Hizukuri[26] is a cooperative structure formed by several individual amylopectin molecules. (b) The superhelix based on the two-directional backbone model is formed by a single amylopectin molecule.[32]

Amylopectin has been investigated by size-exclusion chromatography (SEC)-HPLC and asymmetrical field flow fractionation. Values for $M_w$ and radius of gyration ($R_G$) of amylopectin are around $1.07$–$2.98 \times 10^8$ g mol$^{-1}$ and 165–220 nm respectively. Maize amylopectins gave the higher values of $M_w$ and $R_G$ and potato amylopectin the lower values. Kratcky diagrams have been determined by superimposition of $P(\theta)$ (particle scattering factor) for radii of gyration of 150, 160, 180, 200 and 220 nm for every amylopectin. All amylopectins followed the polycondensation model ABC[36] with C = 0. The slope of the log log plot of $R_G$ vs $M_w$ ($\nu_G$) ranged from 0.36 to 0.44 for the studied amylopectins, demonstrating a high degree of branching. Fractal dimensions ($d_{fl}$) can calculated for the log $P(\theta)$ linear zone. Values of $d_{fl}$ were between those for a statistically branched molecule swollen in a good solvent ($d_f = 2$) and those specific for a nonswollen branched coil ($d_f = 2.5$).

### 2.2.2.3 Intermediate materials

Branched macromolecules have been observed in high-amylose mutants of maize, pea and oat, where they can represent up to 30%, with intermediate iodine-binding capacities and molecular weights.[16,37,38] More robust evidence[39–41] has come from the presence of an intermediate fraction at intermediate elution volume in low-pressure size exclusion chromatography or density-gradient ultracentrifugation, between amylose and amylopectin. This material cannot be confused with phytoglycogen, a soluble biopolymer that is washed from the endosperm or cotyledon during granule purification. This phytoglycogen is not included in the starch granule and cannot interfere with the intermediate material. Intermediate material represents 11.5% and 7.7% for normal corn and potato starches, respectively.[40] Higher values are obtained for mutants.[41]

### 2.2.2.4 Minor components

Lipids represent the most important fraction associated with the starch granules, with values up to 0.8–1.2% and 0.6–0.8% for wheat and normal maize, respectively.

The main constituents of surface components removable by extraction procedures are proteins, enzymes, amino acids and nucleic acids. Some components can be extracted without granule disruption, amounting to about 10% of proteins and 10–15% of lipids. Triglycerides represent a major fraction of surface lipids of maize and wheat. Glycolipids and phospholipids correspond to amyloplast membrane remains. The location of the lipids at the surface of starch granules is still unknown.

A number of proteins[42] associated with wheat starch granules isolated by aqueous extraction from grain or flour have received special attention. Some seem to be integral components of the granule structure whereas others appear to be associated with the granule. One of the starch granule proteins, friabilin, was studied in detail because of its association with changes in wheat grain endosperm texture, from soft wheat to hard wheat.[43,44] The association of friabilin with starch could be considered artefactual. Being located at the starch granule peripheries, the puroindolin-β-polypeptides would presumably become accessible for adsorption to the granule surface immediately after flour wetting. They are present long before the development of grain hardness in both soft and hard varieties.[45]

By contrast, internal components are composed mainly of lipids. Proteins, including granule-bound starch synthase, are in the minority. Extraction procedures have been optimized by Morrison:[46,47] the presence of internal lipids is a characteristic of cereal starches (Table 2.3).

Cereal starches are characterized by the presence of monoacyl lipids – free fatty acids (FFA) and lysophospholipids (LPL) – in amounts positively correlated to amylose content.[48] Wheat, barley, rye and other triticale starches contain almost exclusively LPL, whereas

**Table 2.3**  Free fatty acids and lysophospholipids present in cereal starches.

| Source | Free fatty acid content range | Lysophospholipid content range |
|---|---|---|
| *Barley* | | |
| Waxy | 0.03–0.04 | 0.12–0.75 |
| Normal | 0.03–0.05 | 0.47–1.14 |
| High amylose | 0.05–0.09 | 0.86–1.36 |
| *Maize* | | |
| Waxy | 0.01–0.05 | 0.01–0.03 |
| Normal | 0.30–0.53 | 0.16–0.35 |
| High amylose | 0.38–0.67 | 0.26–0.61 |
| *Rice* | | |
| Normal | 0.22–0.50 | 0.41–0.86 |
| *Wheat* | | |
| Normal | 0 | 0.78–1.19 |
| Waxy | 0 | 0.07–0.17 |

other cereals contain mainly FFA together with a minority of lysophospholipids (Table 2.3). Lysophosphatidylcholine is the major lipid found for both wheat and maize, with palmitic and linolenic acids. For barley starches, the fatty acid composition of the lipids becomes progressively more unsaturated as lipid content increases, but this pattern is less consistent in starches from maize. Most waxy starches have negligible lipids. In the family of maize starches, by comparing different mutants, lipid content appears most directly correlated with long-chain linear $\alpha$-1,4-glucans (i.e. the backbone of amylose) revealed by enzymatic debranching. In starch from wheat and barley harvested at various stages of grain development, both amylose and LPL contents increase with maturity. The picture thus emerging is that of a large A-type starch granule displaying a gradient (from the hilum to the periphery) of increasing amylose and LPL.

Numerous studies have observed correlation between these monoacyl lipids and the functional properties of barley,[49] oat and wheat[50] starches. Monoacyl lipids will induce the formation of amylose–lipid complexes during gelatinization. They restrict swelling, dispersion of the starch granules and solubilization of amylose, thereby generating opaque pastes with reduced viscosity and increased pasting temperatures.

Mineral fractions ($<0.4\%$) are negligible in cereal starches in contrast with tuber starches. Phosphorus is the most important one, which can easily be tracked using $^{31}$P-NMR.[51] Cereal starches contain phosphorus mainly in the form of phospholipids. Root and tuber starches are unique because they contain phosphate monoesters, with an exceptionally high level in potato ($0.1\%$). Phosphate monoesters of all these starches are located mainly on the primary C-6 alcohol function and less on the secondary function C-3 of the glucosyl unit. Potato starch has the highest level of phosphate among commercial starches: this feature is responsible for the high swelling power and stable-paste properties of this starch.


## 2.3 Modifications of starch by hydrothermal treatments and shearing

This part focuses on the impact of water, heat and shearing on starch transformation including melting/gelatinization, and the formation of starch glasses or gels. The resulting structures strongly depend upon the hydrothermal history and kinetics. The associated properties are related to phase transition, glass transition, physical ageing and the plasticizing effect of water.

Food technologists must keep in mind that the 'native state' can be considered more or less as a theoretical state. During milling of cereal grains, some starch granules sustain mechanical and thermal damage, governed by process intensity and hardness of the grain. When processed under the same conditions, hard wheat is more susceptible to mechanical damage than soft wheat. This starch damage is defined as any change in granule structure, giving reduced resistance to action of amylases. It leads to a quicker hydration and a higher amylase susceptibility, factors that play during dough formation and fermentation, respectively. Depending on the technology, there is an optimum level of starch damage otherwise dough stickiness and gas overproduction occur.

## 2.3.1 Gelatinization, pasting and melting

### 2.3.1.1 Structural changes

Gelatinization corresponds to irreversible disruption of molecular order within the starch granules[52] observed when aqueous dilute suspensions of starch granules are heated above 60°C (Fig. 2.9). A loss of birefringence occurs in a temperature range (10–15°C) characteristic of each type of starch: potato 56–66°C, wheat 52–63°C, maize 62–72°C and rice 66–77°C.

With excessive moisture (water:starch ratio >1.5), several events take place simultaneously: diffusion of water inside the starch granule with a limited swelling,[53] disappearance of birefringence, loss of crystallinity of the granule, endothermal phase transitions, predominant swelling of the granule after the loss of birefringence, and decrease in the relaxation times of the water molecules. Amylose-rich starches (maize, wrinkled pea) yield very broad endotherms with higher melting temperatures ranging from 80 to 130°C.[54,55] In addition, a second reversible endothermic transition is observed near 100°C for lipid-containing cereal starches. This transition is usually assigned to the melting of the amylose–lipid complex (cf. Section 2.4.3.1 below).

The breakage of the hydrogen bonds of crystalline zones produces initially a huge water absorption followed by leaching of macromolecules of lower molecular weight, mainly amylose. This event would be mainly explained by the difference in diffusion coefficients: values

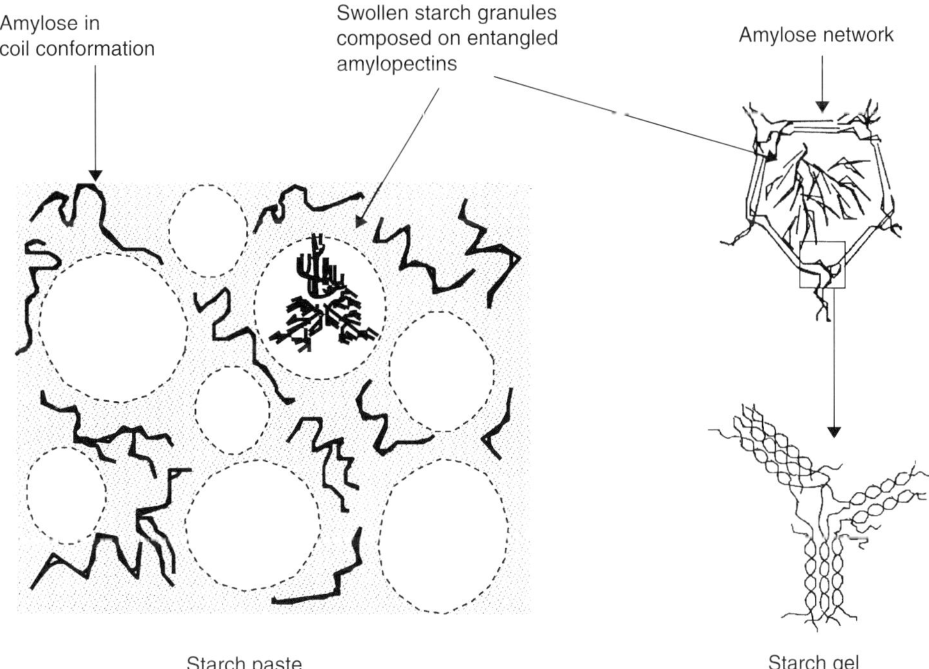


**Fig. 2.9**   Starch gelatinization and gelation.

of the translational diffusion coefficient ($D_T$) are $1.2 \times 10^{-7}$, $8$–$15 \times 10^{-12}$ and $1.7 \times 10^{-12}$ m$^2$s$^{-1}$ for water, amylose and amylopectin, respectively. Pasting is the rheological description of this event. Starch pastes correspond to suspension of swollen starch granules (ghosts) and dissolved macromolecules outside the ghosts.

On swelling and gelatinization, the starch granules may undergo changes into various shapes. A tangential swelling has been observed for normal cereal starches, supposed to be due to the hydration and lateral expansion of lateral crystallites. Gelatinization is a semico-operative process, due to the relationship between the hydration-facilitated melting of the crystalline regions and the swelling owing to further hydration of disordered polymer chains in the amorphous zones.[56] The undispersed fraction of the granules swells by absorbing from 10 to 30 times its weight of water as a function of temperature according to the type of starch. The dissolved fraction increases continuously, with great differences between genotypes. The preparative procedure determines the extent of swelling and solubility, without changing the ranking between different genotypes.

All cereal starches display two-stage behaviours. The first step is a limited swelling and a low level of solubilization, around the gelatinization temperature at 60–75°C. The second step occurs above 90°C: granules swell to a large extent and disrupt, leading to incomplete solubilization (30–60%). All laboratory procedures are far from industrial conditions of gelatinization, where shearing and high heating rate play an important function. Tester and Morrison[50] stated that swelling is primarily the property of amylopectin and that amylose and lipids inhibit swelling of starches.

Melting occurs at low moisture (as low as 11%) and high temperature (as high as 180°C), where shear and heat enable the formation of a viscoelastic melt.[57] By submitting starchy products to high shear stresses and temperatures, a macroscopic homogeneous molten phase is obtained, due to starch melting. The term 'starch melting' includes the loss of crystallinity and native granular structure. Such phenomena occur on extruders and their extent may vary according to the numerous parameters governing the versatility of this process: screw and die geometries and arrangement, barrel temperature, screw speed, water addition, etc. The viscous behaviour of starches with varying amylose content can be adjusted to an empirical model[58] with two parameters: moisture content and specific mechanical energy. Viscosity ($\eta$) shows chiefly an increasing sensitivity to water content and mechanical treatment when amylopectin content increases; increasing amylose content leads to higher values of viscosity and more pronounced shear thinning behaviour. The first trend is in agreement with the most significant plasticizing action on highly branched macromolecules noticed by Lourdin *et al.*[59] The second trend may be related to the larger sensitivity of amylopectin to macromolecular degradation due to its higher molecular weight.


*2.3.1.2 Mechanisms of gelatinization-melting*

These changes at a microscopic scale occur simultaneously with changes at a nanoscale. On X-ray diffractometry, A- and B-patterns disappear, and a V-pattern is observed except for starches that are either waxy or do not contain lipids such as legume ones. The loss of order can be detected as an endothermic event by differential scanning calorimetry. The use of synchrotron radiation allows the diffraction diagrams of starch to be picked up at very short intervals upon heating and thus the kinetics of starch melting and amylose complexation to be followed. Jenkins and Donald[60] and Garcia *et al.*[61] have already used synchrotron radiation to study the gelatinization/melting behaviour of wheat and cassava starches, respectively. Each type of starch gives an endotherm characterized by its enthalpic change (10–20 J g$^{-1}$) at the characteristic temperatures (beginning, midpoint, end). For most cereal starches containing complexable lipids, an additional reversible endothermic transition is observed at 95–105°C, which represents a disordering transition of amylose–lipid complexes. These complexes are formed when amylose chains are released during gelatinization. X-ray diffractometry[62] gives quantitative information about the residual crystallinity.

When studying gelatinization by IR, the intensity of the band at 1047 cm$^{-1}$ decreases mainly due to line broadening of the bands at 1047 and 1022 cm$^{-1}$. The reverse process involving the reappearance of the 1047 cm$^{-1}$ band is usually observed during gelation. Therefore the ratio 1047/1022 cm$^{-1}$ is often interpreted as being related to the amount of short-range order (double-helix content) relative to the amorphous content.

Starch gelatinization is generally considered to occur with excess water (more than 100% added water on dry basis), whereas melting corresponds to the disappearance of native starch crystallinity at low hydration. When smaller amounts of water are present, multiple endothermic events are observed, which are more relevant to melting transitions as in extrusion-cooking (Fig. 2.10). During heating at low and intermediate water contents, various successive endothermic transitions are observed depending on the water content

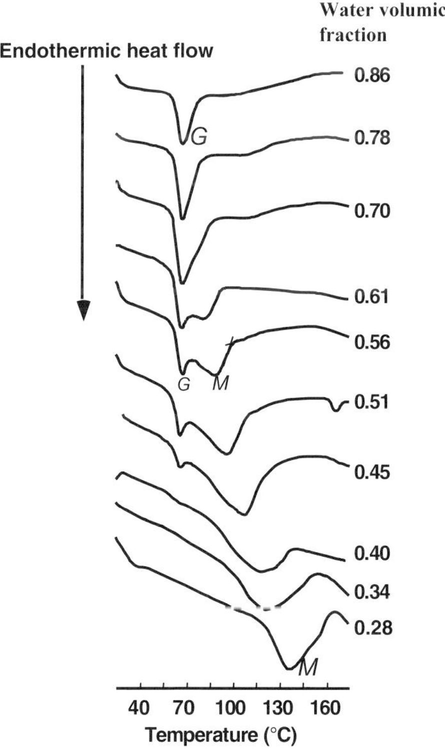


**Fig. 2.10**   Melting curves of cassava starch as a function of the water volume fraction.

and the nature of the substrate studied.[63–65] The first peak, termed the gelatinization peak and observed above a water threshold of 66% (w/w), disappears progressively whereas further endotherms are progressively shifted towards higher temperatures as the water content decreases. However, melting temperature is often mistaken for the gelatinization temperature, as DSC studies have shown that there is no discontinuity differentiating gelatinization and melting. Experimental melting temperatures are known to vary as a function of water content according to the Flory equation[66] for volume fractions of water between 0.1 and 0.7, where the inverse of melting temperature was a positive linear relation to volume fraction of water. The melting enthalpy depends on water content and starch origin. The theoretical melting temperature of the most perfect crystallites with no water, $T_m^0$, is between 160°C and 210°C with the corresponding enthalpy between 160 and 875 J mg$^{-1}$ depending on the botanical origin of the starch. Using spherulites of DP 15, Whittam *et al.*[67] observed a higher melting temperature for A-type than B-type crystallites at the same water content, over a large range of volume fraction of water (0.4 to 0.95) (Fig. 2.11). Surprisingly, $\Delta H_m$ values for the melting transition appear to be similar, ≈35 J g$^{-1}$. In many starch-based foods, water content depends upon the nature of preparation: for example, 50–60% in bakery products, <30% in biscuits. Therefore, incomplete melting explains the persistence of native forms of starch observed in biscuits and bread crust in contrast with bread crumb, where all the starch is gelatinized.

All these interpretations are based upon a direct interaction between starch crystallites and water. Although the Flory equation fits well with the experimental results, true equilibrium between crystallites and solvent is not present during gelatinization, rendering the application of this equilibrium thermodynamics questionable. Furthermore, Donovan[63] has proposed that the coupling between the crystallites and the amorphous zones would be involved in the multiple peaks observed in low-moisture conditions. However, for A- and B-type spherulites of short-chain amylose, the melting behaviour follows perfectly a behaviour typical for a poly-

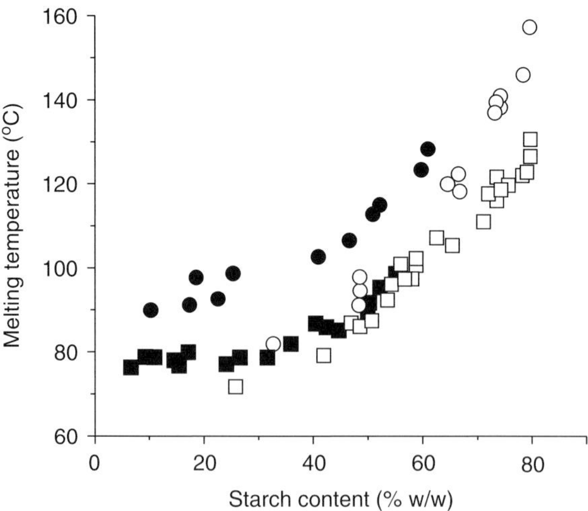

**Fig. 2.11**   A comparison of the effect of water content on the melting temperatures of A- and B-type spherulites (●, ■, respectively), waxy maize starch (○) and potato starch (□).[69]

mer-diluent system:[68] on reducing water content, the melting temperature of both allomorphs increases. A-type spherulites melt at higher temperature than B-type ones. By comparing DSC and solubilization at low moisture contents, Crochet *et al.*[69] come to the conclusion that the amorphous parts of the granule exert little influence on the gelatinization process, which is to a large extent a thermodynamically controlled dissolution process. By contrast, at higher water contents, the cooperative nature of granule gelatinization is evidenced, with the swapping of the crystalline/double-helical for a coiled conformation.

Another approach was developed by Slade and Levine[70] by introducing glass transition as a key event in determining the change in polymer mobility (Fig. 2.12). Wheat and waxy maize starches present a melting process that is irreversible, kinetically controlled and mediated by water plasticization. Below the glass transition temperature, $T_g$, the segmental mobility of polymer chains is frozen in a random conformation, rendering starch phases solid and glassy. When temperature increases, molecular motion is initiated, enabling molecules to slide past one another. At this point, the polymer becomes rubbery and flexible and presents a viscosity of around $10^{12}$ Pa s: this physical event reflects the increase in the motion of short segments (3–20 monomers) of the polymer backbone and is called the glass transition temperature ($T_g$). Above the $T_g$ the material becomes rubber-like (mobile): structural transformations are allowed in the amorphous zones. The melting of interconnected microcrystallites depends on a glass-to-rubber transition of the amorphous zones on which water exerts a plasticizing effect. Observation of more than one melting endotherm at limited moisture contents could be attributed to nonuniform moisture distribution or the

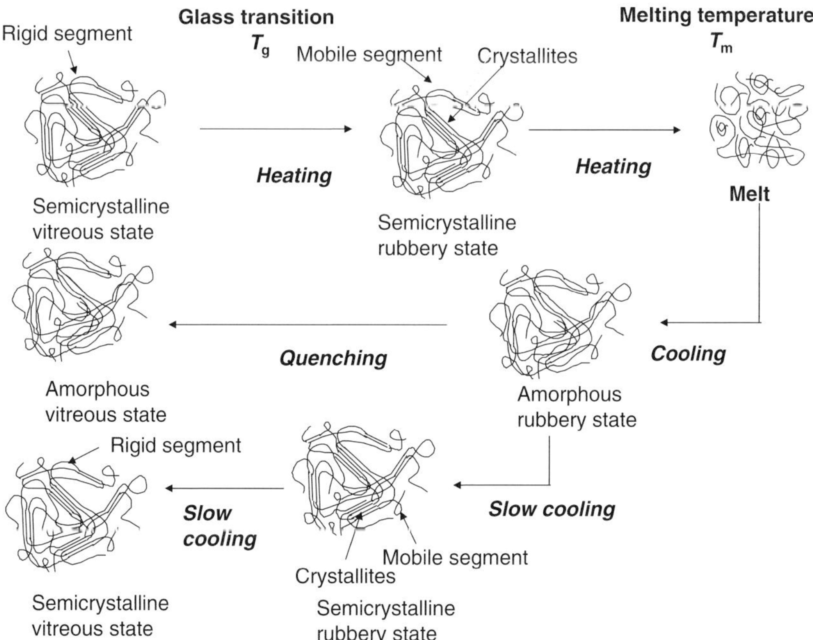


**Fig. 2.12** Physical state of a semicrystalline solid as a function of glass transition ($T_g$) and melting ($T_m$) temperatures.

presence of microcrystalline domains with different thermal stabilities. However, the melting of isolated crystallites remains explainable by the Flory law. Slade and Levine[70] came to the conclusion that a range of values of $T_g$ should be observed in the amorphous zones as a function of the constituents constrained by the crystalline zone. This rule does not take into account that gelatinization is a broad event for a population of starch granules (10–20°C), but a narrow event (1–1.5°C) when considering one granule. Unfortunately, Tan *et al.*[71] were unable to observe a plasticization of starch granules prior to gelatinization despite the use of modulated temperature differential scanning calorimetry.

$T_g$ and $T_m$ define the boundaries within which crystallization of amylose–lipid complexes, annealing or melting of metastable crystallites can occur without loss of granular integrity. Heating starch with water to a temperature slightly lower than the gelatinization-melting temperature is a well-known method of increasing the gelatinization temperature (annealing) and reducing the gelatinization range of populations of granules. This annealing produces an ordering of the crystallites by propagation. Solvents (i.e. water) or low molecular weight compounds (called plasticizers) depress the $T_g$ of polymers according to the Couchman and Karasz[72] equation.[73] The magnitude of $T_g$ depression depends on the amount and the nature of these molecules. Zeleznak and Hoseney[74] published $T_g$ data on native and pregelatinized wheat starches. With increasing water content $T_g$ ($\approx 151$°C, dry starch) is depressed by 137°C upon addition of 10% water to the dry starch. Above 30% water $T_g$ remains stable (63–65°C), which corresponds to the minimum water requirement for the plasticization. It must be recalled that $T_g$ for water is $-134$°C. Whittam *et al.*[75] observed a disagreement within 25 K between experimental $T_g$ determination and theoretical $T_g$ values.

Recently, the side chain liquid crystal model was used to investigate the melting/gelatinization process.[10,76] At low water contents (<5% w/w), the amylopectin helices are in a glassy nematic state, the rigid crystalline parts (mesogens) are somewhat disordered (Fig. 2.13a) and the single (M) peak corresponds to helix–coil transition. At intermediate water contents, the first DSC endotherm is attributed to dislocations between double helices leading to a smectic–nematic transition (Fig. 2.13b). The second endotherm is the helix–coil transition with irreversible disentanglement of double helices. Lastly, in excess water lamellar break-up and disentanglement of double helices occur simultaneously (Fig. 2.13c). The gelatinization appears as the result of four processes: (i) cleavage of existing starch–starch–OH bonds (endothermic); (ii) formation of starch–solvent–OH bonds (exothermic); (iii) unwinding helix–coil transition of amylopectin helices (endothermic); and (iv) formation of amylose–lipid complexes. Therefore the magnitude of the $\Delta H$ reflects more the level of helices than the crystallinity level. The contribution of the different processes can be determined by using different solvents.[71]

One output of this new approach is to focus attention on solvent mobility, which in local conditions will determine the possibility of molecular rearrangements on a non-equilibrium basis. The main physical properties in the range $T_m$–$T_g$ are governed by the Williams–Landel–Ferry theory. Consequently rates of all physical changes will increase exponentially as a function of $T_{\text{storage}} - T_g$ ($T_{\text{storage}}$ being the storage temperature of the studied sample).[77] Management of solvent and temperature becomes a main parameter controlling long-term changes of starchy products. There is still a need to determine the complete state diagram of starch-water, as a function of temperature and time.

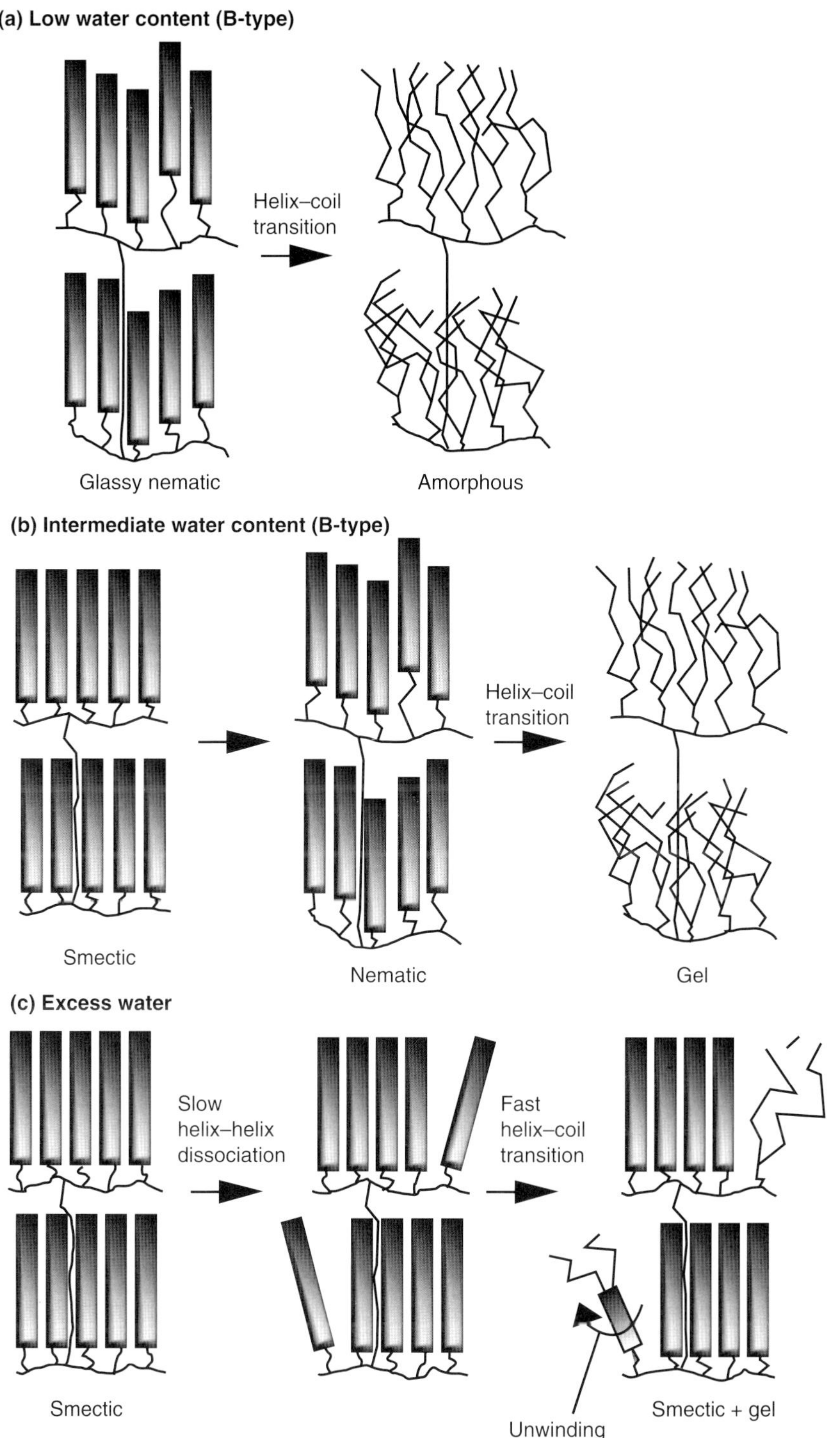


**Fig. 2.13** Starch gelatinization and side chain liquid crystal approach: (a) low water content, (b) intermediate water content and (c) excess water.[76]

### 2.3.1.3 Functional properties

Functional properties correspond to the appearance of thickening behaviour and increased enzyme susceptibility. They are fundamental to starch usage and therefore have received much attention. Important differences are observed between starches of different botanical origins. Whatever the gelatinization procedure, starch pastes must be considered as mixtures, in varying proportions, of amylose and amylopectin solutions and swollen particles. The rheological behaviour will depend upon the relative proportions of these two different systems, which are determined by concentration, heating rate, shearing and starch damage. Due to their complex composition, starch pastes are difficult to model.

For low starch concentrations (<10%) pastes made from various starches exhibit shear-thinning thixotropic behaviour. For starch pastes, the shear stress ($\tau$) vs shear rate ($\gamma$) curves are linear over the range of 10 to 1000 s$^{-1}$ and the viscosity ($\eta$) can be represented by a power law with $n$ the flow behaviour index <1.

At low shear rates below 10 s$^{-1}$, a yield stress has been observed, which is typical of suspensions.[78] The consistency index, $K$, follows an Arrhenius-type equation whereas $n$ is almost independent of temperature. A similar power law equation can be used to follow the influence of concentration with a power $p$ dependence on concentration, with $p \approx 3.5$–4. For highly concentrated starch pastes (>10%), flow behaviours depend upon cooking conditions, giving either shear thickening or shear thinning. These starch pastes can be considered as concentrated suspensions of closely packed deformable gel particles.

Upon extrusion, the melt expands and cools rapidly due to vaporization of moisture, eventually settling into an expanded solid foam.[57] Starch with a $T_m/T_g$ <<1 has a high glass-forming tendency, which demonstrates a large free volume requirement and thus a large temperature increase required for mobility. Expansion, sometimes called puffing, is the phenomenon by which foods, mainly cereals, acquire a porous structure, like solid foam, due to transient heat and vapour transfers. Pore size and distribution, thickness and mechanical properties of the wall material define, in turn, the texture of the product. This texture is strongly influenced by how the pores are generated. In the last ten years, there has been more attention paid to the basic phenomena that rule expansion: nucleation, bubble growth and water evaporation, coalescence and shrinkage.[79] This solid foam is sensitive to moisture plasticization (moisture induces a lowering of $T_g$) especially in the 0–10% water content range; crispiness is lost when water activity shifts above a water activity range between 0.35 and 0.5. Katelunc and Breslauer[80] reported that for corn flour extrudates, an increase in $T_g$ is related to an increase in the sensory–textural attribute of crispiness.

## 2.3.2 Gelation

Pastes and molten materials are metastable nonequilibrium states and undergo structural transformations when cooling and during storage.

### 2.3.2.1 Structural changes

Amylose gel is an interconnected three-dimensional network, formed with interchain association over lengths of $\approx$13 nm.[81] These crystallites are composed of long constitutive chains (DP 35–40), are acid resistant and exhibit a high fusion temperature ($\approx$140–150°C) with an enthalpy change of 9–24 mJ mg$^{-1}$. Amylose gels are obtained in a few hours. When linear chains are elongated in solution, gelation occurs in a few minutes.[82] After several days and for up to 50 days, amylopectin forms crystallites that melt at $\approx$45–55°C (enthalpy transition 3–15 mJ mg$^{-1}$). These crystallites are thin, mainly composed of linear chains of DP 15 and acid labile. Crystallites in amylopectin gels consist of single or multiple branched molecules that originate from the S-chain clusters of the molecule. The X-ray diffraction pattern associated with this gelation is the B-type.

Amylose chains aggregate in an infinite three-dimensional network. The chain segments inside the crystallites are disposed obliquely to the microfibre axis.[83] This model enables the microfibrous structure observed by electronic microscopy to be related to the macromolecular characteristics of the associated regions determined by acid hydrolysis. The length of the crystallites ranges from 8 to 18 nm, which is in good agreement with the filament width determined by electron microscopy (20 $\pm$ 10 nm). The aggregation of these segments generates a three-dimensional network. Therefore a network strand would consist of contiguous associated blocks, aligned along the length axis of the microfibre. Double helices would then be linked to others by loops of amorphous amylose segments, dangling in the gel pores. This fraction would be responsible for the hydrodynamic behaviour of the amylose gels. This model suggests the occurrence of both parallel and antiparallel packings, which correspond to two local energetic minima of amylose double helices.[12] This packing of crystalline blocks is easier to relate to the coil conformation of amylose chains in solution.[84]

Amylopectin[85] at high concentrations gives gels that are fairly well degraded enzymically but have poor thermal stability, around 45°C in excess water, due to the short chains involved in crystalline domains (DP $\approx$ 15) and poor mechanical properties.

Water is entrapped in these three-dimensional networks, in which macromolecular probes (enzymes) can diffuse. The $D_T$ of bovine serum albumin in amylose and amylopectin gelled networks decreases with increasing polysaccharide concentration.[81] The value of $D_T$ decreased from $5.5 \times 10^{-11}$ to $1.5 \times 10^{-11}$ m$^2$ s$^{-1}$ over the concentration range 5–15% w/w of gelling polymer. Accessibility has been studied by diffusion until equilibrium: probe size and macromolecule concentration are the two main parameters. For probes where the hydrodynamic radius ($R_H$) < 1 nm, the volume of solvent trapped within the network is completely accessible to the probe. Accessibility is a reasonably continuous function of hydrodynamic radius.

Modelling has been carried out in considering a gel network as a collection of immobilized, randomly spaced, rigid rods. A particle diffuses through the network by taking directionally random steps; if collision with the network occurs, the random step is not completed and the diffusion is retarded. More recently, hydrodynamic screening has been thought to be a more appropriate description of the physical process responsible for the retardation of diffusion.[81] The relationship between the average mesh size ($\xi$) of the semidilute polymer solution and the hydrodynamic radius $R_H$ of the probe particle is important. When $R_H >> \xi$, the matrix appears as a continuum, and the Stokes–Einstein relationship may be used to describe the diffusion process, with the viscosity term being the macroscopic viscosity of the polymer solution.

### 2.3.2.2 Mechanisms

Studying amylose gelation, Miles *et al.*[86,87] and Doublier and Choplin[88] observed first an increase in turbidity ascribed to a phase separation process. In a subsequent phase, increase in elasticity was recorded, followed on a longer time scale by a crystallization process. In this gelation model, the two latter phases correspond to the establishment of the gel junction zones. By contrast, Gidley and Bulpin[89] and Clark *et al.*,[90] studying gelation of fairly monodisperse amylose (DP ~2500), showed an immediate increase in elasticity occurring before the turbidity development. They suggested that the formation of junction zones leads to a phase separation phenomenon.

Opaque gels are obtained upon cooling of concentrated starch aqueous dispersions as a result of a ternary phase separation occurring in the water–amylose–amylopectin system, which is followed by a nucleation leading to the development of a 'thin' three-dimensional amylose network following this phase separation.[88,89] At this stage, elementary junction zones are locally established between the macromolecules. These should adopt locally a left-handed, parallel-stranded double-helical conformation. On ageing, starch gels develop a B-type crystallinity resulting from an aggregation process in a parallel register of the elementary junction zones. Therefore by cooling, gelation occurs in two steps: (i) a phase separation that produces polymer-rich and polymer-deficient regions; and (ii) crystallization within these polymer-rich regions.

If the concentration is sufficiently high, the polymer-rich regions form an interconnected gel network. For amylose, gelation occurs only above C* (semidilute region), which corresponds to a concentration of ≈1.5%. For amylopectin, chain entanglements begin at 0.9%; however, the liquid–solid transition is gradual, with three defined stages. The 6–10% zone is a critical zone above which solid behaviour is evident.[91]

Incompatibility between gelling polymers (A and B) is a fairly common phenomenon and usually leads to mixed gels formed of two phases, each of these being essentially composed of a single polymer (A or B). The polymer compositions of the continuous and dispersed phases are dependent on the ratio A:B. For a particular value of the A:B ratio (called the phase inversion point), the continuous matrix becomes the discontinuous filler and vice versa. Starch gels obtained from a total dispersion of starch macromolecules can be considered as a particular case of two-phase mixed gels. Amylose and amylopectin are incompatible in solution[92] at 80°C. For concentrations greater than 3%, phase separation occurs below 90°C, giving two phases, each one being composed essentially of 70% of either amylose or amylopectin. Therefore mixed gels are composed of a continuous network of one polymer (amylose or amylopectin) entrapping droplets of the other polymer. Based upon rheological and thermal properties and hydrolytic resistance of amylose-rich mixed gels, Leloup *et al.*[93] and Ortega-Ojedo *et al.*[91] concluded that the transition in the gel behaviour observed for an amylose:amylopectin ratio of 30:70 (weight basis) corresponded to the inversion in the polymeric composition of the continuous and discontinuous phases. Mixed amylose–amylopectin gels have a continuous amylopectin matrix below this threshold but a continuous amylose matrix above. This behaviour is emphasized for gelatinized starch granules, where amylose leaching reinforces this exclusion. This threshold can be affected by the preparation method and the origin of starch macromolecules.

When cooling starch melts at high kinetic rates below $T_g$, gelation can be prevented and replaced by the formation of a glass. Figure 2.12 shows the behaviour of a semicrystalline solid in the different temperature domains around melting ($T_m$) and glass transition ($T_g$) temperatures. A rapid quench of a starch paste or a melt leads to the formation of a vitreous glass with brittle behaviour. A similar state is obtained by rapid removal of water as in extrusion-cooking at the expansion stage, or during the processing of pregelatinized starch, water being a plasticizer of starch. Starch retrogradation and gelation occur within the temperature range of $T_g < T < T_m$. In that case, the rate of crystallization increases exponentially, with $\Delta T = T - T_g$.[94] When the storage temperature is cycled between low temperature (6°C), facilitating the nucleation, and a higher temperature (40°C), facilitating the growth of crystallites, the quality of the crystallites is improved and they melt at higher temperatures.[95]

A rapid quench at $T < T_g$ prevents any type of large-scale molecular rearrangement. Such glasses are almost amorphous because only very restricted and slow molecular motions can occur. Recently, local order in such products, prepared by casting, extrusion or freeze-drying of starch, amylose and amylopectin, was assessed by solid state $^{13}$C cross-polarization (P) technique combined with magic angle spinning (MAS) NMR through the decomposition of the C1 signal.[96] Four to five types of $\alpha(1{\rightarrow}4)$ linkages were quantified for all samples studied. The influences of the intrinsic primary structure (linear or branched) and of the preparation procedure on conformational changes were interpreted in terms of distribution of average glycosidic linkage dihedral angles ($\Phi,\Psi$) and local order. Two resonances, at 103.3 and 94.4 ppm, were associated respectively with an angular conformation similar to single helices present in Va type structures (A line) and to more constrained conformations favoured by drastic methods of preparation such as freeze-drying. The 102.9 ppm resonance was associated with the helical conformation present in Vh and V-isopropanol structures. The two other resonances in the 100.4–101.4 and 97.0–98.6 ppm ranges are influenced by crystallinity, branching points and the techniques of preparation. The former line intensity is connected to conformational or crystalline defects while the chemical shift of the latter line is more linked to the presence of $\alpha(1{\rightarrow}6)$ linkages. The most frequently observed tendencies were a decrease in line width with water sorption, which expressed a more homogeneous distribution of conformations, and higher sensitivity to rehydration of more constrained conformations (freeze-dried samples).

### 2.3.2.3 Functional properties

The macromolecular reorganization occurring during gelation has major effects on texture and nutritional value of foods containing starch. Starch gels are elastic although the yield values are low so that the gels are easily broken down. Amylopectin gels behave as Hookean solids at strains <0.1, amylose gels at strains <0.2. In the concentration range 10–25% after storage for 6 weeks at 1°C, there is a linear relationship between modulus and concentration, in contrast with amylose gels, where a seventh power dependence of modulus on concentration is observed. For concentrated starch gels (10–20%), (loss modulus) storage modulus ($G'$) increases rapidly during the first 100 min and then the rate slows down; however, the final $G'$ value is only obtained 21 days later. Amylose gels exhibit a rapid rise in storage modulus whereas amylopectin gelation is a much slower process.

With amylose-amylopectin mixtures, the higher the starch concentration, the smaller is the amount of amylose needed to obtain $G' \gg G''$ and higher values of moduli.[91]

Most models used for the prediction of starch retrogradation kinetics are grounded on the Avrami equation, originally derived for polymer crystallization in melt. A good fit of the experimental values was obtained.[97] The equation has two parameters, the constant rate and the Avrami exponent, which depends on the type and nucleation and the dimensions in which growth takes place. Such data have been used to predict starch crystallization, taking into account both the presence of residual starch crystallites and the mechanism of crystal growth in freshly cooked durum wheat bread.[98]

On long time scale syneresis occurs, which is one of the disadvantages of unmodified starch as a gelling agent.

Technological control of texture is possible by manipulating the dispersion level of starch granules, the amylose and water contents, and the storage conditions. Whatever the macromolecular composition, gelation is more rapid at lower temperatures. Rigidity development is significantly enhanced at lower temperatures, mainly for amylopectin gelation. Crystallization presents a maximum in the range of 50–60% starch. Above 80% starch, crystallization is blocked by $T_g$. After pasting or drum-drying of maize starch, amylose crystallizes independently from amylopectin by complex formation and retrogradation. This phenomenon is explained by the leaching of amylose during the thermal treatment, the subsequent phase separation and then the crystallization within the amylose volume fraction. These crystallites, composed of linear chains (DP 35–40), are acid-resistant and exhibit a high fusion temperature (above 100°C). By contrast, after extrusion-cooking, amylose and amylopectin co-crystallize in the same manner as pure amylopectin.

Mungo starch vermicelli and rice flour noodles are based upon these starch networks.[99] In both types of foods, an amylopectin-based structure is reinforced by amylose-based structures that are present mainly in the complexed (V-type diffraction pattern) and retrograded (B-type diffraction pattern) forms for rice flour noodles and mungo starch vermicelli, respectively. These crystallites are more resistant towards acid hydrolysis and cooking (melting temperature 82 and 119°C). The cooking behaviour of these glutenless noodles can be explained by these amylose networks.

### *2.3.3 Glass transition and plasticization by water*

The ever increasing interest of food scientists in the concept of glass transition arises from an attempt to improve the description of complex solid systems out of thermodynamic equilibrium. Based on concepts of polymer science transferred to hydrated biomaterials, the applicability of the kinetic description of amorphous glassy/rubbery systems is being probed producing limited dynamic views of what might be called 'mobilities'. In amorphous and semicrystalline polymers, at sufficiently low temperatures, the mobility of polymer chains is strongly restricted, limited to very slow and local molecular motion (physical ageing). By applying thermal energy, molecular motion is initiated at a given temperature called the glass transition temperature, $T_g$; above $T_g$ the materials become thermoplastic, rubbery and flexible. At the glass transition there is a discontinuous change in heat capacity and in the thermal expansion coefficient. $T_g$ also decreases with increasing water content (water acts as a ubiquitous plasticizer of starchy products) and increases with increasing crystallinity.

This approach was applied to interpret food systems, and especially the mechanical and storage properties of low-moisture starchy foods, when water was shown to be a ubiquitous plasticizer of biopolymers.[100] It was subsequently developed and expanded by numerous authors.[94,101–105]

Dynamical mechanical thermal analysis (DMTA) and differential scanning calorimetry (DSC) are the most common techniques used to investigate glass transition and plasticization. The advantage of DSC is its use of a confined sample whose composition may be kept constant during temperature changes. The heat capacity change detected by this method mainly probes the mobilization of chiefly translational and rotational degrees of freedom allowing chain flow in the case of polymers, although specific motion in the structure cannot be precisely identified.

By comparing values of $T_g$ determined by DSC, Bizot *et al.*[106] illustrated (Fig. 2.14) in a structure–property relationship the influence of molecular weight, degree of branching and $(1{\rightarrow}4)$ versus $(1{\rightarrow}6)$ glycosidic linkage ratio on the depression of $T_g$ with water content, thus extending results of earlier studies.[107,108] The effect of water plasticization on amylose, amylopectin, phytoglycogen, pullulan and different low-molecular-weight products derived from hydrolysis was described by the Couchman and Karasz correlation.[72] Linear chains seem to favour chain–chain interactions and induce partial crystallinity; branched molecules

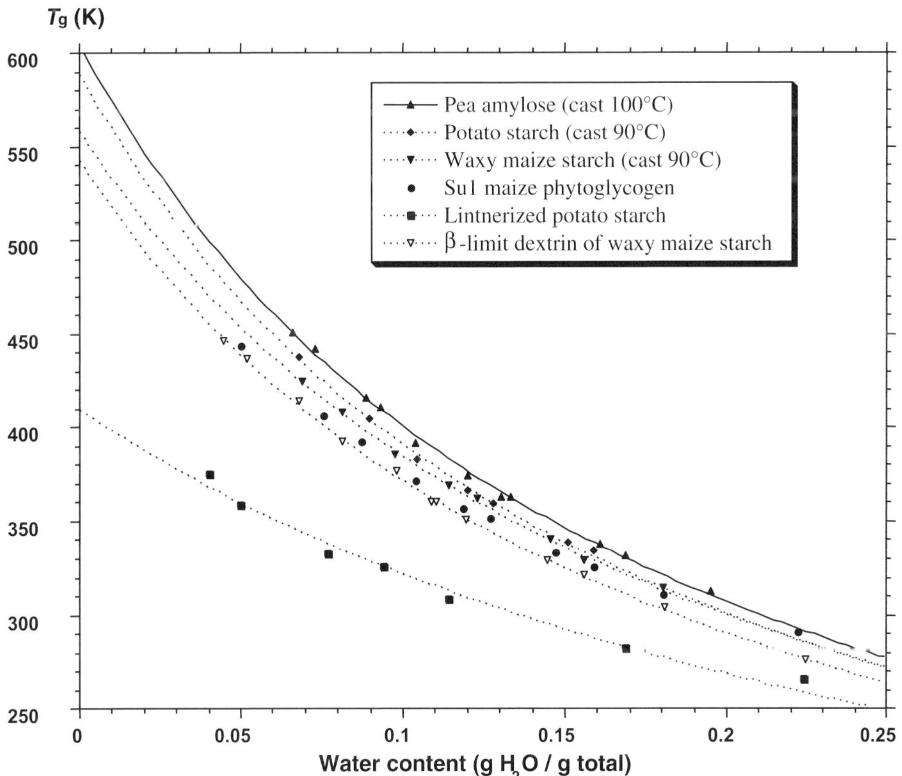


**Fig. 2.14**  Calorimetric glass transition (at 3°C/min) of starch and subfractions versus water content.[106]

display lower values of $T_g$ due to chain end effects and flexibility of branching points.[109] The three dihedrals present in $(1\rightarrow6)$-$\alpha$ linkages seem to depress $T_g$ in a similar fashion to internal plasticization. The case of linear $(1\rightarrow4)$-$\alpha$ amylose chains bearing $(1\rightarrow6)$ grafted fructoses was also examined as a first step towards tailored structures designed to optimize mechanical properties and internal plasticization, and inhibit recrystallization.[110]

### 2.3.4 Physical ageing

Being far from thermodynamic equilibrium, glassy materials are subject to the so-called 'structural relaxation', molecular rearrangements leading to lower states of energy. These alterations may be of significance for food products because they directly affect enthalpy, volume and mechanical as well as diffusion properties. The term 'physical ageing' was introduced by Struik[111] to distinguish these effects from other ageing processes such as chemical reactions, degradations or changes in crystallinity. Physical ageing has been widely studied in relation to the evolution of physical properties such as volume, enthalpy, creep compliance[112] and relaxation times evaluated from dielectric spectroscopy. Enthalpy relaxation is now clearly recognized as a significant phenomenon occurring during storage of food materials and a signature of physical ageing. The corresponding endothermic peaks were observed in the 50–60°C region on fresh samples and disappeared on second scans. Its importance depends on $\Delta T (= T_g - T_{ageing})$ and on the time of ageing.[113] This behaviour was observed on a wide variety of products, including wheat gluten,[114] breakfast cereals,[115] amylopectin,[107] hydrated starch lintners[116] and native rice starch.[117] Enthalpy relaxation was studied on a series of hydrated polysaccharides after ageing at different times and temperatures (15, 45, 65 and 100° below $T_g$),[118] in comparison with synthetic polymers, polymethylmethacrylate (PMMA) and polyvinylpyrrolidone (PVP). Structural relaxation during storage yielded either sub-$T_g$ enthalpy recovery peaks or overshoots in the glass transition region, demonstrating the width of the relaxation time spectrum. When the ageing time is increased, both amplitude and position on the temperature scale of the enthalpy recovery peak are observed to increase (Fig. 2.15). The enthalpy recovery peak develops less markedly (amplitude and area) for the lower ageing temperatures, but the shift on the temperature scale is larger. While extruded starch equilibrated apparently within the experimental time, amylopectin and phytoglycogen aged slightly faster than PMMA and hydrated PVP without completely reaching their equilibrium. Ageing effects detected after short ageing at very low temperatures ($T_g - 100°C$) revealed the width of the relaxation time spectrum.

## 2.4 Interactions with other molecules

Other components of foods affect kinetics and characteristic temperatures of phase transitions, as well as the inherent pasting and gelling of starch. So their contributions to the whole characteristics of complex foods may be high. The effects of salt (NaCl) can be observed at the concentrations used in foods.[119]

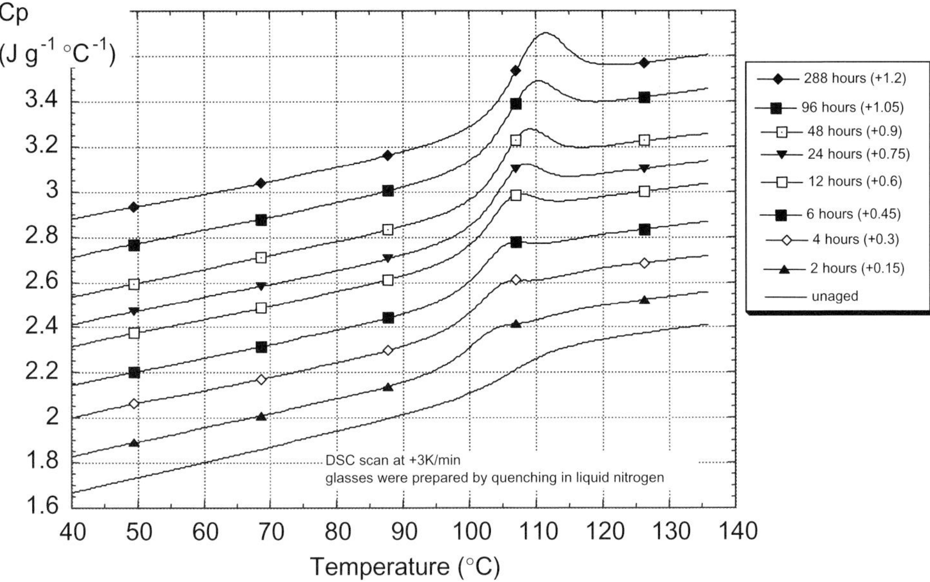


**Fig. 2.15**  Enthalpy relaxations observed for amylopectin glasses (10% $H_2O$) aged at 90°C for different times (Cp = heat capacity).

## 2.4.1 Hydrocolloids and proteins

Mixing starch pastes with other biopolymers results in thermodynamic incompatibility with all species tested to date.[120] After their preparation, blends tend to demix into polymer 1-rich areas and polymer 2-rich areas (Fig. 2.16), known as phase separation. The phase boundary (binodal) between the two final stable polymer 1-rich areas and polymer 2-rich

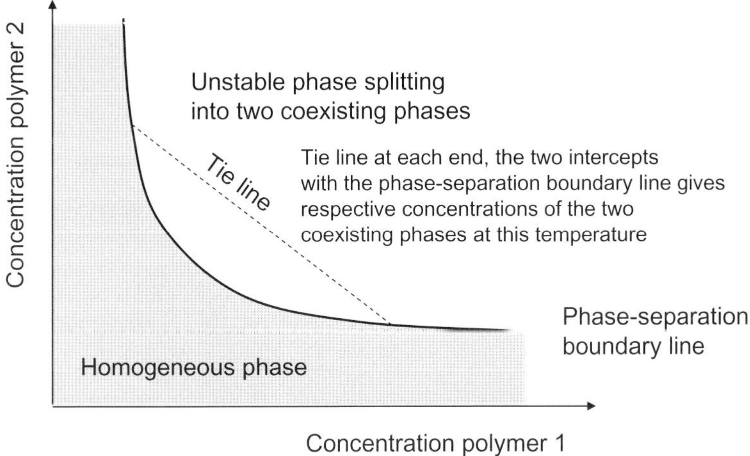


**Fig. 2.16**  Phase separation in biopolymer mixtures.

areas and the initial unstable polymer 1–polymer 2 is a key result, for which long and tedious observations are required.

The key functional parameter is the phase separation threshold, that is, the minimal bulk concentration of biopolymers at which phase separation occurs. Easily detected with potato starch and gelatin,[121] this threshold is high for mixtures of starch and proteins.[120]

Hydrocolloids modify the rheological properties of starch pastes and gels.[78,122,123] Addition of hydrocolloids enhances or modifies the gelatinization and retrogradation behaviour of starch and improves the water-holding capacity and the freeze–thaw stability of starch-based preparations. Incompatibility between leached amylose and hydrocolloid is responsible for a higher concentration of the hydrocolloid macromolecules in the suspension. Whereas galactomannans increase only the rate of gelation without modifying the final rigidity, xanthans increase $G'$ values towards a solid-like behaviour. These synergistic behaviours remain stable at room temperature. A starch-hydrocolloid system has to be considered as a suspension of starch ghosts dispersed in a solution of hydrocolloids.[124]

Gluten comprises an important group of proteins that interact with starch during cooking. In bread, gluten underwent a transformation resulting in the release of water, which became absorbed by retrograding starch.[125] Using a combination of DSC, X-ray diffraction and NMR, it has been found that the presence of gluten[126] has no effect on the kinetics, extent or polymorphism of amylopectin retrogradation. However, on a bigger scale, gluten retarded water loss in the starch granule remnants. This could explain the decrease of retrogradation enthalpy of starch when the level of gluten is increased.[127,128] Sevenou,[129] using Fourier transform infrared (FTIR), demonstrated that starch gelatinization did not lead to gluten dehydration during the heating of the dough. The question of water transfers during baking has to be considered at different scales in order to disconnect rheological behaviour from short-distance interactions.

Interactions with milk and its derivatives (mainly milk powder) have also been extensively investigated. The presence of minor amounts of starch in milk increased the viscosity. Although milk salts appeared to have no direct influence on the storage modulus, lactose and whey proteins exerted a minor effect on starch paste. By contrast, casein micelles greatly increased the modulus of the starch suspension, due certainly to the inaccessibility of swollen starch granules to the casein.[130] The addition of sugars elevated the apparent viscosity of the starch–milk paste, with the effect of fructose higher than that of sucrose and glucose.[131] When considering the effect of the addition of starch on milk behaviour,[132] phase separation (binodal) is observed as with guar and locust gums. However, it occurs at a relatively high concentration (>1.5%) of added amylopectin, with the formation of protein aggregates. The experimental results, from confocal scanning light microscopy and image analysis, show a remarkable agreement with the Vrij depletion theory[133] for polymer and colloid mixtures. Amylopectin gives different microstructures in contrast with the gums usually studied. Using the experimental procedures of Sperry[134] and the theoretical approach developed by Lekkerkerker *et al.*[135] this difference is ascribed to a low value (<0.5) of the polymer:colloid size ratio, $\xi$, ($= \sigma_p/\sigma_c$) for milk protein–amylopectin systems, related to the attraction range of polymers (diameter $\sigma_p$) and colloids (diameter $\sigma_c$).

It must be noted that these interactions have no effect on the gelatinization endotherm and temperatures of native granules when gelatinized in blends with other biopolymers as described above.

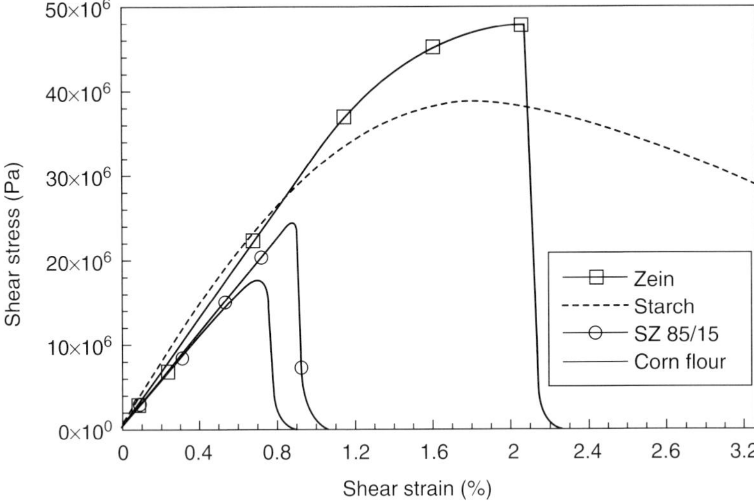


**Fig. 2.17**   Typical shear stress–strain curve (obtained by three point bending) of moulded glassy zein (8% moisture content, wet basis), amorphous starch, starch-zein blend (85:15) and corn flour (12% moisture content, wet basis).[136]

Mixing starch with proteins at low moisture contents such as in extrusion also leads to phase separation, which can be easily detected by confocal light microscopy. At low zein levels (5–15%), starch gives a continuous phase with a homogeneous distribution of zein particles (5–50 mm). Blends with larger zein contents present continuous zein phase with delimited amorphous starch domains.[136] An interesting feature is the behavioural difference between storage proteins and starch in the glassy state. At ambient temperature glassy starch with 12% moisture is ductile whereas storage proteins present an elastic limit, with a fragile rupture observed for zein (Fig. 2.17) and gluten.[137] The fragility of the blends can be attributed to the phase separation between starch and zein, with a threshold close to 20–30% zein content. Above 50%, zein behaviour predominates because of the continuity of the protein.

## *2.4.2 Sugars*

Oligosaccharides are known to modify starch behaviour during gelatinization by shifting gelatinization towards higher temperatures, increasing or decreasing the enthalpy change, decreasing the viscosity and reducing swelling and solubility.[69,119,138,139]

The general trend is that sugars delay gelatinization to higher temperatures (>90°C) and increase gelatinization enthalpy in the following order: water alone < ribose < fructose < glucose << maltose < sucrose. The swelling factors in the presence of sugar were higher compared with water alone for sugar concentrations below 25% and reduced at sugar concentrations >25%.[139] Starch systems containing sugars presented pronounced shear-thinning behaviour. Mixing glucose and fructose gives results near to those gained with equivalent amounts of one single sugar, although minor differences could suggest non-additive effects.[140]

The mechanistic interpretation is based upon coupling DSC and X-ray diffractometry at large and small angles. The shifting of the endotherm upwards reflects the shifting of the mobility threshold of amorphous growth ring regions of the starch granules.[141] At this threshold, the mobility and the swelling of the amorphous growth regions reach a maximum level, beyond which native granular structure is irreversibly altered. This effect of sugars can be explained by the concept of antiplasticization, demonstrated with several plasticizers.[94,142,143] Sugar, in the presence of starch and water, behaves as a plasticizing cosolvent with water, such as the competition between sugar and water leads to a plasticizing complex of higher molecular weight (and thereby lower free volume) than water alone. The plasticization of the amorphous regions that immediately precedes the gelatinization is less pronounced than with water alone. Consequently the gelatinization temperature is shifted to higher values to bring thermally induced mobility. An interesting observation is also that when the DSC scan is performed in conditions giving rise to a double endotherm for starch gelatinization, the double endotherm is shifted into one single endotherm. This observation favours nonuniform moisture distribution between starch granules during gelatinization.

This behaviour of antiplasticization is also present for amorphous starchy materials,[142,143] where a strain hardening is observed at low plasticizer contents.

For gelatinized starches and granular remnants (ghosts), maltose and maltotriose induce a reduction of recrystallization.[144] Threitol and xylitol give the most significant reduction of recrystallization of potato starch.

## 2.4.3 Amylose complexation with small molecules

Amylose has the unique feature of forming complexes with a variety of molecules. Monoacyl lipids and emulsifiers as well as smaller ligands such as alcohols or flavour compounds are able to induce the formation of left-handed amylose single helices, also known as V amylose. V amylose is a generic term for crystalline amyloses obtained as single helices cocrystallized with compounds such as iodine, dimethylsulfoxide (DMSO), alcohols or fatty acids. Although such compounds are required for the formation of the V-type structure, they are not systematically included in the amylose helix. The resulting helical conformation and crystalline packing depend on the nature of the ligands and conditions for the formation of the complex. The specific interactions of amylose with lipids or flavour compounds have a strong impact on food quality. Such complexation could also have important potential in the field of health foods when applied to protection and vectoring of vitamins and micronutrients.

### 2.4.3.1 Lipids

Starch and lipids are present in both native lipid-containing starch granules and foods, where the dominant interaction between them gives the well-known amylose–lipid complex.

The biosynthesis *in situ* of amylose–lipid complexes in starch, with naturally occurring fatty acids and phospholipids, has been demonstrated,[48,145] but their crystalline nature is still debated; such structures would be of limited extent and insufficiently perfect to be detected by X-ray diffraction in native starch. Gernat *et al.*[146] determined amounts ranging from 15 to 25% of the total crystalline phase for amylose-rich starches from maize and wrinkled pea. High-amylose starches are known to contain greater amounts of lipids. It must be remembered that the extraction, purification and drying of starch may readily induce amylose–lipid complex formation upon heating. Nevertheless, experimental evidence was presented by Morrison *et al.*[145] that such complexes might be present in native starches in the amorphous state. In that case amylose chains are involved in a separate (isolated single helices) inclusion compound or the arrangements are too limited in extent to show the Vh-type diffraction diagram. Such structures can reorganize in a detectable crystalline packing (propagation) in the presence of heat and moisture.[147] Morrison[47] determined amounts of lipid-complexed amylose ranging from 15–20% of total amylose for maize and wheat to 37–59% for waxy barley.

Polar lipids greatly affect the behaviour of starch for the development of viscous and gelling properties.[47] The technological importance of such interactions is clearly demonstrated when comparing cereal starches (excluding waxy genotypes) with lipid-free starches such as tuber and legume starches. Differences in functional properties have been extensively studied and most often analysed on a quantitative basis. Ultrastructural aspects, however, have not been reported, due to the difficulty of analysing such complex molecules. The intensity and nature of phase transitions (annealing, melting, polymorphic transitions, recrystallization, etc.) induced by hydrothermal treatments in crystalline structures are related to temperature and water content.[148] Despite its small concentration, the lipid phase mainly present in cereal starches has a large influence on starch properties, particularly in complexing amylose. Many authors have shown that complex formation occurs during heat/moisture treatments, especially during gelatinization of starches naturally containing lipids[149,150] or when lipids are added to defatted starches[64,151] or pure amylose-free natural lipids.[152] Both naturally occurring and heat-formed complexes have specific properties such as the decrease in amylose solubility or the increase in gelatinization temperatures.[46,47,128] Polar lipids, for example fatty acids and their monoglyceride esters, are used in food technology for their ability to reduce stickiness, improve freeze–thaw stability[153] and retard retrogradation. The foremost example is probably the use of fatty acids and monoglycerides as anti-staling agents in bread and biscuits: incorporation of such additives in the dough induces a slower crystallization (retrogradation) of the amylopectin fraction and therefore retards the staling of bread.[154,155]

Interaction of amylose with fatty acids and monoglycerides yields more or less crystalline structures with the Vh crystalline type, also obtained with linear alcohols.[156–158] In such structures, the chain conformation consists of a left-handed helix containing six residues per turn and with a pitch height of 0.792–0.805 nm (i.e. a rise per monomer of between 0.132 and 0.136 nm). In the case of amylose–lipid complexes it is assumed that the aliphatic part of the lipid is included inside the amylose helix (Fig. 2.18) whereas the polar group lies outside, being too big to be included.[145,159] In the Vh-type, the most common form obtained by complexation of amylose with lipids, the single helices are packed in an orthorhombic unit cell ($a = 1.37$ nm, $b = 2.37$ nm, $c = 0.805$ nm) with the space group $P2_12_12_1$ and 16 water molecules within the unit cell. The amylose lipid complexes can be crystalline or amorphous depending on the temperature at which they form.[147,160] The two forms cannot be distinguished by DSC because they yield very similar melting/decomplexing enthalpy and temperature. Such behaviour is apparently due to a major contribution of the intramolecular energy in the formation of a single helix relative to the total energy of complex formation

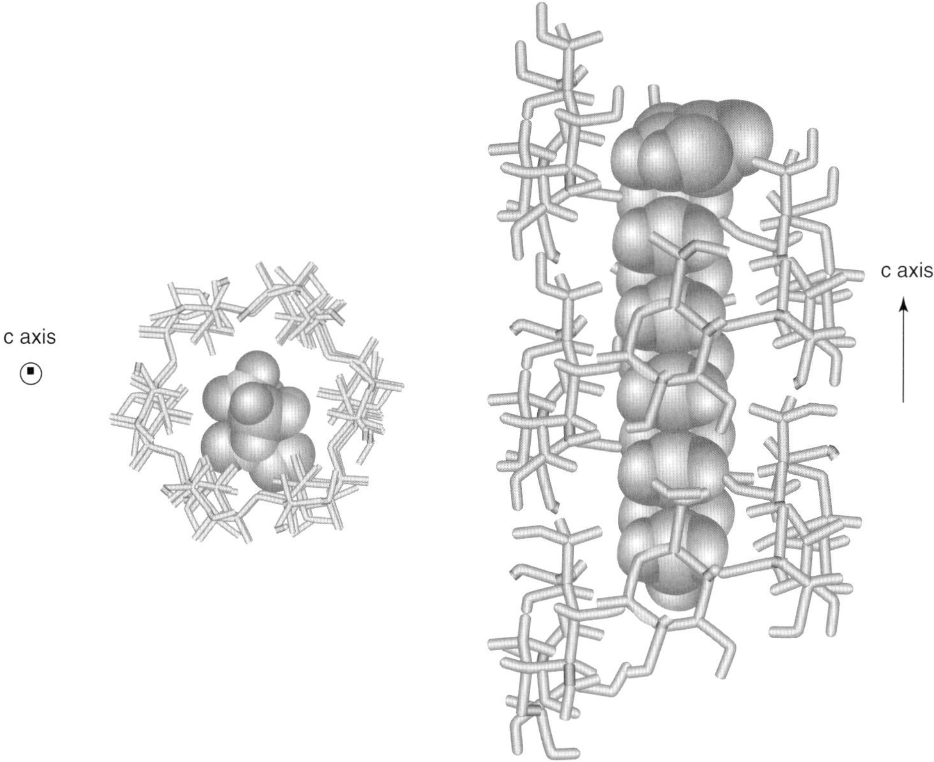



**Fig. 2.18**  Molecular modelling representation of amylose–fatty acid complexes showing the inclusion of the aliphatic part ($C_{12}$) of a fatty acid inside the hydrophobic cavity of the amylose single helix.[148]

(packing energy), that is, interhelical bonding seems to contribute very little. The amorphous form consists of individual complexed single helices, which are not involved in a crystalline packing. The formation of Vh crystalline structures[62] was observed by synchrotron X-ray diffraction in native maize starch heated at intermediate and high moisture contents (between 19 and 80%). With a heating sample holder, including a calorimetric response, it was possible to record X-ray diffraction and DSC curves simultaneously at 2–3 degrees per minute. The crystallization of amylose–lipid complexes was demonstrated upon heating after gelatinization at 110–115°C in excess water (Fig. 2.19). Therefore it was confirmed that the second endotherm in the thermogram of lipid-containing starches can be assigned to the melting of crystalline amylose–lipid complexes formed upon heating. For intermediate water contents, mixed A+Vh (or B+Vh for high-amylose starch) diffraction diagrams were recorded. Two mechanisms can be involved in amylose complexing: the first mechanism relates to crystallization of the amylose and lipid released during starch gelatinization, and the second to crystalline packing of separate complexed amylose chains (amorphous complexes) present in native cereal starches.

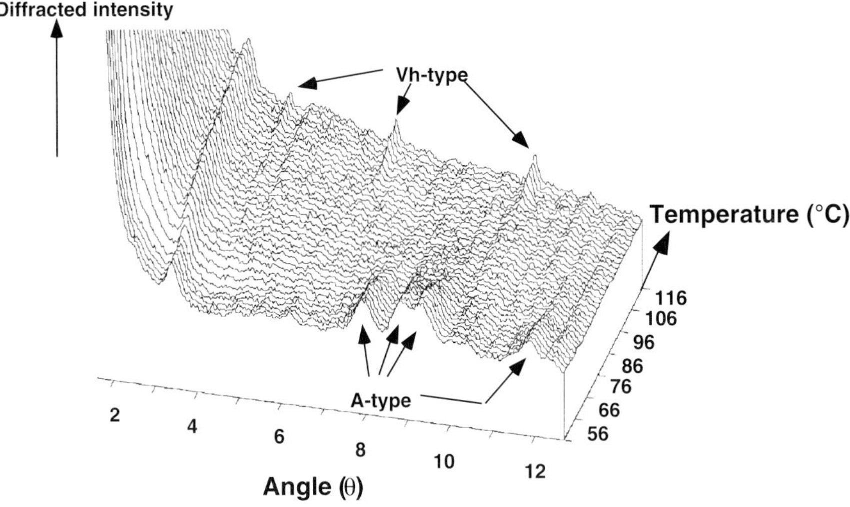


**Fig. 2.19**　Synchrotron monitoring of the gelatinization of maize starch at 65% $H_2O$ (wet basis) and crystallization of amylose–lipid complexes (X-ray diffraction diagrams recorded every 2°C).

### *2.4.3.2 Alcohols, aroma and flavours*

Besides monoacyl lipids, amylose, the linear fraction of starch, forms crystalline complexes (known under the generic name of V amylose) with a variety of small ligands. Apart from very specific complexes obtained with iodine, the best documented complexes were obtained with alcohols (Fig. 2.20). Linear alcohols yield the well-known Vh structure already mentioned for amylose–lipid complexes. Complexes induced by *n*-butanol yield a unit cell that is larger than that of the Vh type.[161] Similar observations were made with V-isopropanol amylose[162] and, to a lesser extent, V-glycerol complexes.[163] For these three types of structures, it was possible to transform crystals into the Vh type by solvent exchange. It was therefore hypothesized that amylose helices were also made of six D-glycosyl units per turn. However, the cavity of the helix may be too small to host the ligand, which conceivably lies between helices. Amylose with a larger helical diameter is formed in the presence of α-naphthol. In this case, the helix consists of eight D-glycosyl units per turn and the bulky molecules are included in the helical cavity and between helices.[164,165]

Complexing of amylose by aroma is frequently induced by food processing and is thought to be one of the possible mechanisms for aroma retention in starch-containing food systems.[166–168] Therefore, knowledge of the interactions between aroma and the matrix is essential to optimize formulations in food processing. This could be obtained through the determination by X-ray scattering of the crystalline type of the complex formed, which influences the nature of inclusion (inter-/intra-helical), and through the thermostability using DSC. In several cases, resulting complexes exhibit a structure different from that of the classical Vh amylose. Powder X-ray diffraction patterns of these complexes were first attributed to helices with seven D-glycosyl units per turn. Thus, the largest cavity of the sevenfold helix would allow the inclusion of larger molecules in cross-section, as is the case for several flavour compounds. A correlation was also found between the helix diameter and

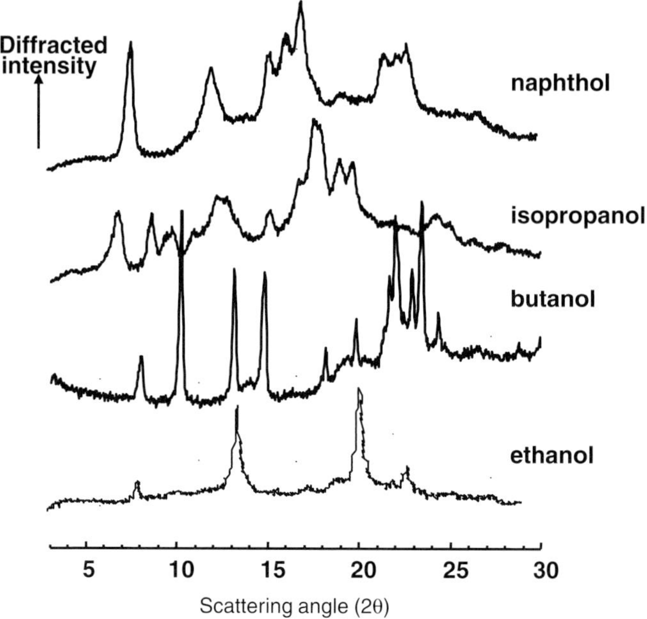


**Fig. 2.20**   X-ray diffraction diagrams of amylose complexes with ethanol, butanol, isopropanol and naphthol.

the amount of flavour compound necessary for saturation of the amylose helices with six, seven and eight glucose units per turn, respectively.[166] Recently, by using both electron and X-ray diffractions on lamellar crystals of pure amylose complexed with some aromas like fenchone, menthone or geraniol, Nuessli *et al.*[169] showed that the X-ray diffraction diagram initially attributed to a sevenfold helix corresponded to a V-isopropanol type with the same sixfold helix but a larger unit cell than the Vh type. Therefore, depending on the size of the ligand, it could be included within and/or between helices.

As a complement to X-ray diffraction, solid state $^{13}$C-cross-polarization magic angle spinning/nuclear magnetic resonance (CPMAS/NMR) is also a very efficient noninvasive technique to assess the helical conformation or the crystalline structure of native and complexed starch. Gidley and Bociek[170] demonstrated that C-1 and C-4 glycosidic sites were more sensitive to conformational changes than the C-2, C-3 and C-5 carbons because C-1 and C-4 showed higher chemical shift dispersions under various conformations of the glycosidic linkage in α-(1→4) glucans.[170,171] These findings echoed previous proposals by Horii and co-workers in 1983; they showed correlations between the C-1 and C-4 chemical shifts of α-(1→4) glucans and their torsion angles $\phi$ and $\psi$. Moreover, it has been observed that hydration of samples induces a signal resolution enhancement. Such an approach was recently applied to the different crystalline types of amylose–alcohol complexes[172,173] and to aromas like linalol and menthone.[174] Sequential washing of the powdered complexes with ethanol before and after desorption permitted the probing of intra- and inter-helical inclusions. High-resolution magic angle spinning (HRMAS) recordings were also used to compare the chemical shifts of free and bound aroma, and suggested the involvement of some hydrogen bonding in the amylose complexing. Moreover, it showed that free aroma was completely removed by ethanol washings. Using CPMAS and X-ray scattering experiments, it was demonstrated that the V-isopropanol type was retained for linalool whatever the treatment used. On the contrary, the V-isopropanol type shifts towards the Vh type for menthone after ethanol washing before the desorption step, reflecting the disappearance of interhelical associations between menthone and amylose. The stability of the complex prepared with linalool shows that this ligand is more strongly linked to amylose helices. The discrepancies observed in the chemical shifts attributed to carbons C-1 and C-4 in CPMAS spectra of V-isopropanol and Vh forms could be attributed either to a deformation of the single helix (with possible inclusion of the ligand inside) or to the presence of the ligand between helices (only water molecules are present in the Vh form).

## 2.5 Starch as a nutrient

The degradation of native granules is beyond the scope of this chapter as all foods are cooked to some degree leading to gelatinized or even retrograded starches.

### 2.5.1 Classification

The qualitative aspect of starch modification is very complex and still a matter of debate. The glycaemic index (GI) concept is now accepted for ranking foods with respect to their potential for raising blood glucose. The starch fraction that is available for absorption in the small intestine is measured as the sum of sugars and starch, excluding resistant starch (RS),[175] this latter being 'the sum of the starch and the products of hydrolysis of the starch not absorbed in the small intestine of the healthy man'. Ranking of foods by the glycaemic responses elicited when equicarbohydrate portions are consumed has provided a unique and sometimes controversial perspective on the issue of carbohydrate quality.

Food-processing technologies used for cereal products lead to the complete or partial gelatinization of starch granules, thereby making them highly or partially digestible. Noncrystalline starch can be broken down in the stomach and the small intestine and then converted into blood glucose. Consequently, most cereal food-products generally have a high GI value and low amount of RS. Relationships between the rate or extent of starch digestion and the GI have been established by investigations of *in vitro* amylolytic hydrolysis.[176–179] Consequently, Englyst *et al.*[175] have proposed a classification of starchy foods into rapidly digestible starch (RDS), slowly digestible starch (SDS) and resistant starch (RS). More recently, another fraction corresponding to the amount of rapidly available glucose (RAG) has been described.[180] This classification was based on data from *in vitro* amylolysis of starchy foods.

In foods the extent of hydrolysis of gelatinized starches ranges up to 100%. The second trend is that the extents of hydrolysis of retrograded starch gels are lower than their freshly gelatinized counterparts. It is difficult to rank foods due to differences in enzyme concentration, time of hydrolysis and source of enzyme.

The degree of starch conversion has become a popular way to quantify starch disorganization in foods. It corresponds to the decrease in the gelatinization endotherm before and after processing. Nutritionists should keep in mind that gelatinization enthalpy is a net thermodynamic quantity of granule swelling, crystallite melting, hydration and even recrystallization (exothermic) when complexable lipids are present.

## 2.5.2 Resistant starch

The 'resistant starch' (RS) fraction is undegraded in the upper gut[181] and fermented in the colon,[182] and is extensively studied because of its potential impact on health. It induces significant production of butyrate, and leads to a decrease in faecal pH. Both phenomena have beneficial effects on health, notably on colon pathology.[183,184]

Although resistant starch is defined on the basis of a physiological phenomenon during digestion, this nutritional component in fact encompasses different types of molecular organization (Fig. 2.21). The different types of products identified as resistant starches were classified by Englyst *et al.*[175] in three classes (RS1 to RS3).

RS1 consists of physically inaccessible starch composed of starch granules or macromolecules entrapped in cell walls (legumes) or protein networks (pasta products). The diffusion step is a major limiting factor with regard to the macromolecular nature of amylases and the partial accessibility of starchy substrates. The importance of the diffusion event is difficult to evaluate in view of the scant knowledge about the porosity of starchy foods. The presence of other components like lipids, soluble fibres or proteins can decrease the diffusion kinetics by increasing the viscosity. The accessibility is controlled by the specific area and embedding of the substrate by cell walls[185] or protein network.[186]

RS2 contains native starches, intrinsically resistant to α-amylolysis, essentially B-type starches (potato, high-amylose maize, wrinkled pea). RS2 is unexpected in foods because most foods are cooked.

RS3 consists of retrograded starches formed during cooling and storage of gelatinized starches, in bread and baked potato, for example,[187] or by gelation of *de novo* molecules.[82]

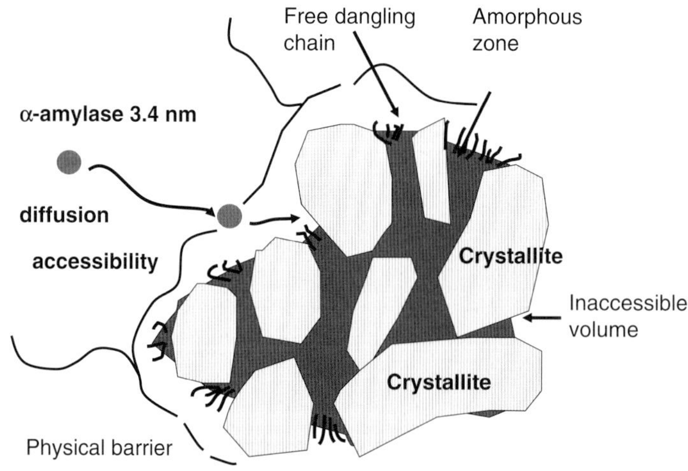


**Fig. 2.21**   Mechanisms of α-amylolysis of insoluble starchy substrates.

In such products, amylose molecules, one of the two constitutive macromolecules of starch, reassociate upon cooling in networks strongly resistant to $\alpha$-amylases. These differences in susceptibility to amylolysis are observed between lintnerized A and B starches or between A- and B-type crystals recrystallized from amylose short chains. There is no step of rapid hydrolysis for these highly crystalline samples. Nevertheless, A-type crystals are completely degraded, while B-type spherulites or crystals are never degraded more than 10% .[188]

A fourth class, RS4, corresponding to chemically modified starches, has been added,[189] broadening the initial classification of Englyst *et al.*[175]

A new type of resistant starch, made by debranching and precipitation of concentrated solutions of maltodextrins, was patented recently.[190] The digestion of this substrate was examined by using a mix of the ileal content of eight humans. More than 50% of $\alpha$-glucans ingested reached the terminal ileum. Results were compared with *in vitro* digestion by porcine pancreatic $\alpha$-amylase. Both *in vivo* and *in vitro* resistant starch contents and structural features were similar, showing that the *in vitro* methods employed were adequate to estimate the fraction reaching the terminal ileum of such a particular substrate. This product presents A-type crystallinity, in contrast to all other known starches classified as RS. Moreover, this product consists of low DP chains and yields similar electron diffraction patterns to those of A-type low DP crystals.[191] A study of the partial hydrolysis by $\alpha$-amylase of A-type model substrates consisting of short chains around DP 15 (crystalline lamellae issued from the lintnerization of waxy maize starch and lamellar crystals grown from solution) was carried out using transmission electron microscopy. It was shown that in such highly crystalline substrates, double helices were hydrolysed by their sides and not their ends (Fig. 2.22). It was concluded that the resistance of this new type of RS was due to its particularly dense and compact morphology, resulting from the epitaxial growth of elementary crystalline A-type platelets. In the resulting morphology, the accessibility of double helices to $\alpha$-amylase is strongly reduced by aggregation.[191]


## 2.6 Conclusions

Most starch transformations can now be described in terms of physical and chemical factors, based on a fundamental understanding. The final state of a food is the result of a myriad of various stages in processing and storage (Fig. 2.23), each with characteristic kinetic and thermodynamic factors.

With this range of available parameters, it is now possible for each technology to develop a state diagram. A state diagram represents a plot of composition vs temperature, showing the boundaries of solid and liquid phases (states) that are in thermodynamic equilibrium.

This concept of a phase diagram[192] has recently been extended to encompass thermodynamic states that are away from equilibrium, kinetically controlled metastable states. These supplemented phase diagrams, or dynamic phase diagrams, allow a qualitative analysis of food systems, when mapping composition–time–temperature.

Within the frame of a general review on starch, all key factors necessary to build a dynamic phase diagram have been presented and should be adapted to each case. Data mining of published studies should enable the building of complex dynamic phase diagrams. Glass

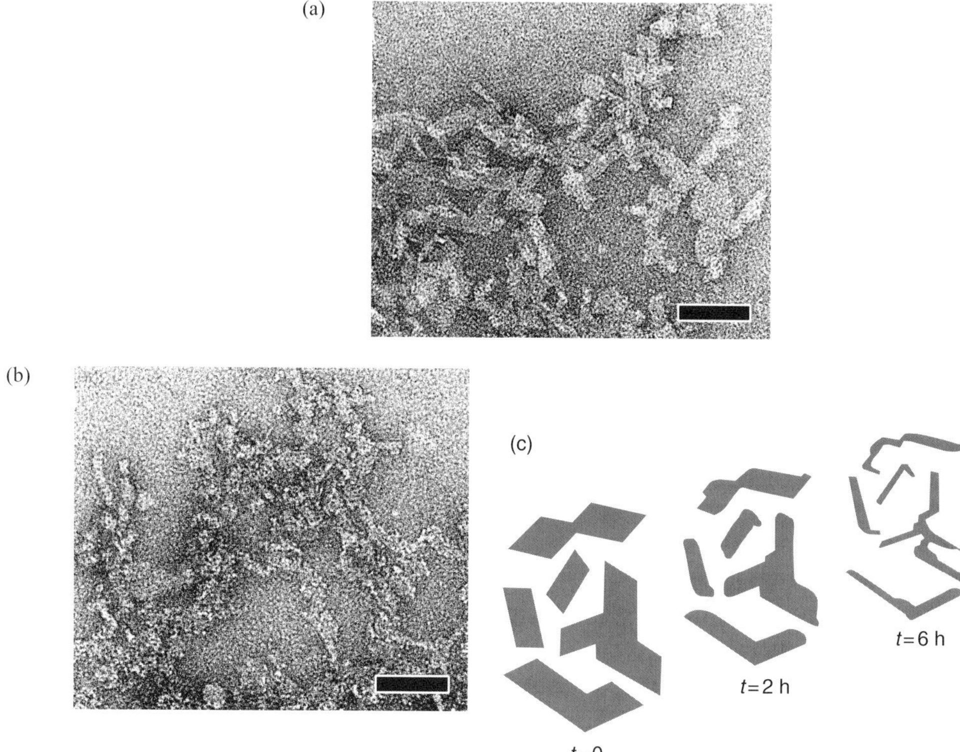


**Fig. 2.22**   (a,b) TEM images showing the morphological changes of waxy maize starch nanocrystals after 2 and 6 hours of α-amylolysis respectively (negative staining; bars: 50 nm). (c) Schematic drawing describing the morphological evolution of lintners during the enzymatic hydrolysis. Double helices lie perpendicular to the main plane of lamellae, therefore their length (crystal thickness) is retained during hydrolysis.

transition has been considered in the last ten years as the central factor. A sucrose–water state diagram has revealed the relative locations of glass, solidus, liquidus and vaporus curves, useful in explaining the cookie and cracker technologies.[193] Time is a critical parameter when modelling kinetically controlled phenomena in the rubbery state. With starch the question is very complex due to the large number of interactions in complex systems and the potential modifications of component thermal properties that might result from phase separation.[194] The fact that a specific structural level is responsible for a property (e.g. texture, shelf life) enables the use of state diagrams as a rational basis for the selection of the best technological paths (heat and mass transfers) or formulation (structure-function) for a desired quality goal.

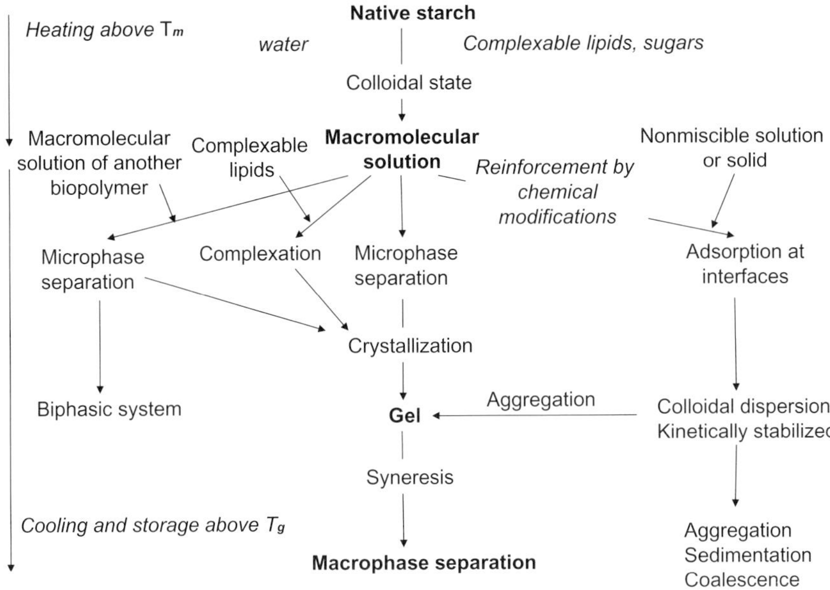



**Fig. 2.23**   Diagrammatic representation showing the various ways in which native starch may evolve.

## 2.7 References


1 Galliard, T. & Bowler, P. (1987) Morphology and composition of starch. In: Galliard, T. (ed.) *Starch: Properties and Potential*, pp. 55–78. John Wiley and Sons, Toronto.

2 Tester, R.F., Karkalas, J. & Qi, X. (2004) Starch – composition, fine structure and architecture. *J. Cereal Sci.* **39**, 151–165.

3 French, D. (1972) Fine structure of starch and its relationship to the organization of the granule. *J. Jap. Soc. Starch Sci.* **19**, 8–25.

4 Buleon, A., Colonna, P., Planchot, P. & Ball, S. (1998) Starch granules: structure and biosynthesis. *Int. J. Biol. Macromol.* **23**, 85–112.

5 Donald, A., Kato, L., Perry, P. & Waigh, T. (2001) Scattering studies of the internal structure of starch granules. *Starch/Staerke* **53**, 504–512.

6 Cameron, R.E. & Donald, A.M. (1993) A small-angle X-ray scattering study of the absorption of water into the starch granule. *Carbohydr. Res.* **244**, 225–236.

7 Oostergetel, G.T. & van Bruggen, E.F. (1993) The crystalline domains in potato starch granules are arranged in a helical fashion. *Carbohydr. Polym.* **21**, 7–12.

8 Gallant, D., Bouchet, B. & Baldwin, P. (1997) Microscopy of starch: evidence of a new level of granule organization. *Carbohydr. Polym.* **32**, 177–191.

9 Waigh, T., Perry, P., Riekel, C. *et al.* (1998) Chiral side-chain liquid-crystalline polymeric properties of starch. *Macromolecules* **31**, 7980–7984.

10 Waigh, T., Kato, L., Donald, A. *et al.* (2000) Side chain liquid crystalline model for starch. *Starch/Staerke* **52**, 450–460.

11 Gerard, C., Planchot, V., Buléon, A. & Colonna, P. (2001) Amylolysis of maize double mutants. *J. Sci. Food Agric.* **81**, 1281–1287.

12 Imberty, A., Chanzy, H., Perez, S. *et al.* (1988) The double-helical nature of the crystalline part of A-starch. *J. Mol. Biol.* **201**, 365–378.

13  Imberty, A., Buleon, A., Tran, V. & Perez, S. (1991) Recent advances in knowledge of starch structure. *Starch/Staerke* **43**, 375–384.

14  Horii, F., Yamamoto, H., Hirai, A. & Kitamaru, R. (1987) Structural study of amylose polymorphs by cross-polarization-magic-angle spinning, 13C-N.M.R. spectroscopy. *Carbohydr. Res.* **160**, 29–40.

15  Tomlinson, P. & Denyer, K. (2003) Starch synthesis in cereal grains. *Adv. Bot. Res.* **40**, 1–61.

16  Banks, W. & Greenwood, C.T. (1975) *Starch and its Components.* Edinburgh University Press, Edinburgh.

17  Takeda, Y., Hizukuri, S., Takeda, C. & Suzuki, A. (1987) Structures of branched molecules of amyloses of various origins and molar fractions of branched and unbranched molecules. *Carbohydr. Res.* **165**, 139–145.

18  Hizukuri, S., Takeda, Y., Abe, J. *et al.* (1997) Analytical development: molecular and macromolecular characterization. In: Frazier, P.J., Richmond, P. & Donald, A.M. (eds) *Starch: Structure and Functionality*, pp. 121–128. Royal Society of Chemistry, Cambridge UK.

19  Yashushi, Y.B., Takenouchi, T. & Takeda, Y. (2002) Molecular structure and some physical properties of waxy and low-amylose barley starches. *Carbohyd. Polym.* **47**, 159–167.

20  Roger, P. & Colonna, P. (1996) Molecular weight distribution of amylose fractions obtained by aqueous leaching of corn starch. *Int. J. Biol. Macromol.* **19**, 51–61.

21  Fishman, M.L. & Hoagland, P.D. (1994) Characterization of starches dissolved in water by microwave heating in a high pressure vessel. *Carbohydr. Polym.* **23**, 175–183.

22  Yalpani, M. & Sandford, P.A. (1987) Commercial polysaccharides – recent trends and developments. In: Yalpani, M. (ed.) *Industrial Polysaccharides, Genetic Engineering, Structure/Property Relations and Applications*, pp. 311–335, Elsevier Scientific Publishing, Amsterdam.

23  Yu, X., Houtman, C. & Atalla, R.H. (1996) The complex of amylose and iodine. *Carbohydr. Res.* **292**, 129–141.

24  Peat, S., Whelan, W.J. & Thomas, G.J. (1952) Evidence of multiple branching in waxy-maize starch. *J. Chem. Soc., Chem. Commun.* 4546–4548.

25  Hizukuri, S. (1985) Relationship between the distribution of the chain length of amylopectin and the crystallite structure of starch granules. *Carbohydr. Res.* **141**, 295–305.

26  Hizukuri, S. (1986) Polymodal distribution of the chain length of amylopectin and its significance. *Carbohydr. Res.* **147**, 342–347.

27  Hanashiro, J. Abe, J. & Hizukuri, S. (1996) A periodic distribution of the chain length of amylopectin as revealed by high-performance anion-exchange chromatography. *Carbohydr. Res.* **283**, 151–159.

28  Manners, D.J. (1989) Recent developments in our understanding of amylopectin structure. *Carbohydr. Polym.* **11**, 87.

29  French, D. (1984) Organizations of starch granules. In: Whistler, R.L., Bemiller, J.N. & Parschall, E.F. (eds) *Starch, Chemistry and Technology,* pp. 183–247. Academic Press, New York.

30  Robin, J-P., Mercier, C., Charbonniere, R. & Guilbot, A. (1974) Lintnerized starches. Gel filtration and enzymatic studies of insoluble residues from prolonged acid treatment of potato starch. *Cereal Chem.* **51**, 389–406.

31  Gerard, C., Planchot, V., Colonna, P. & Bertoft, E. (2000) Relationship between branching density in amylopectin and crystalline structure of A- and B-type in maize mutant starches. *Carbohydr. Res.* **326**, 130–144.

32  Bertoft, E. (2004) On the nature of categories of chains in amylopectin and their connection to the superhelix model. *Carbohydr. Polym.* **57**, 211–224.

33  Davis, H., Skrzypek, W. & Khan, A. (1994) Iodine binding by amylopectin and stability of the amylopectin–iodine complex. *J. Polym. Sci. Part B* **32**, 2267–2274.

34  Caldwell, R.A. & Matheson, N.K. (2003) $\alpha$-(1-4) chain distributions of three-dimensional, randomly generated models, amylopectin and mammalian glycogen: comparisons of chromatograms of debranched chains of these polysaccharides and models with random dendromeric models with the same chain lengths (CL, ICL, ECL) and fractions of A chains. *Carbohydr. Polym.* **54**, 201–213.

35 Takeda, Y., Shibahara, S. & Hanashiro, I. (2003) Examination of the structure of amylopectin molecules by fluorescent labeling. *Carbohydr. Res.* **338**, 471–475.

36 Burchard, W. (1983) Static and dynamic light scattering from branched polymers and biopolymers. *Adv. Polymer* Sci. **48**, 1–124.

37 Baba, T. & Arai, Y. (1984) Structural characterization of amylopectin and intermediate material in amylomaize starch granules. *Agric. Biol. Sci.* **48**, 1763–1775.

38 Colonna, P. & Mercier, C. (1984) Macromolecular structure of wrinkled and smooth pea starch components. *Carbohydr. Res.* **126**, 233–247.

39 Wang, L.Z. & White, P.J. (1994) Structure and properties of amylose, amylopectin and intermediate material of oat starches. *Cereal Chem.* **71**, 263–268.

40 Yoon, J-W. & Lim, S-T. (2003) Molecular fractionation of starch density-gradient ultracentrifugation. *Carbohydr. Res.* **338**, 611–617.

41 Gerard, C., Barron, C. Planchot, V. & Colonna, P. (2001) Amylose determination in genetically modified starches. *Carbohydr. Polym.* **44**, 19–27.

42 Schofield, J.D. (1994) Wheat proteins: structure and functionality in milling and breadmaking. In: Bushuck, W. & Rasper, V. (eds) *Wheat Production, Composition and Utilization,* pp. 73–78. Blackie, Glasgow.

43 Greenwell, P. & Schofield, J.D. (1986) A starch granule associated with endosperm softness in wheat. *Cereal Chem.* **63**, 379–380.

44 Morris, C.F. (2002) Puroindolines: the molecular genetic basis of wheat grain hardness. *Plant Mol. Biol.* **48**, 633–647.

45 Turnbull, K.M., Marion, D., Gaborit, T. *et al.* (2003) Early expression of grain hardness in the developing wheat endosperm. *Planta* **216**, 699–706.

46 Morrison, W.R. (1988) Lipids in cereal starches: a review. *J. Cereal Sci.* **8**, 1–15.

47 Morrison, W.R. (1995) Starch lipids and how they relate to starch granule structure and functionality. *Cereal Foods World* **40**, 437–446.

48 Morrison, W.R., Tester, R.F., Snape, C.E. *et al.* (1993) Swelling and gelatinisation of cereal starches. IV. Some effects of lipid-complexed amylose and free amylose in waxy and normal barley starches. *Cereal Chem.* **70**, 385–391.

49 Morrison, W.R., Milligan, T.P. & Azudin, M.N. (1984) A relationship between the amylose content and lipids content from diploid cereals. *J. Cereal Sci.* **2**, 257–271.

50 Tester, F. & Morrison, W.R. (1990) Swelling and gelatinisation of cereal starches. I. Effects of amylopectin, amylose and lipids. *Cereal Chem.* **67**, 551–557.

51 Lim, S-G., Kasemsuwan, T. & Jane, J-L. (1994) Characterization of phosphorus in starches by $^{31}$P-nuclear magnetic resonance spectroscopy. *Cereal Chem.* **71**, 488–493.

52 Blanshard, J.M.V. (1987) Starch granule structure and function: a physicochemical approach. In: Gaillard, T. (ed) *Starch Properties and Potential,* Critical Reports on Applied Chemistry, vol. 13, pp. 16–54. John Wiley and Sons, Toronto.

53 Hermansson, A.M. & Svegmark, K. (1996) Developments in the understanding of starch functionality. *Trends Food Sci. Technol.* **7**, 345–353.

54 Colonna, P., Buleon, A., Mercier, C. & Lemaguer, M. (1982) *Pisum sativum* and *Vicia faba* carbohydrates. IV – Granular structure of wrinkled pea starch. *Carbohydr. Polym.* **2**, 43–51.

55 Inouchi, N., Glover, D.V., Sugimoto, Y. & Fuwa, H. (1991) DSC characteristics of gelatinization of starches of single-, double- and triple-mutants and their normal counterpart in the inbred Oh43 Maize (*Zea mays* L.) background. *Starch/Staerke* **43**, 468–472.

56 Marchant, J.L. & Blanshard, J.M.V. (1978) Studies of the dynamics of the gelatinization of starch granules employing a small angle light. *Starch/Staerke* **30**, 257–264.

57 Vergnes, B., Della Valle, G. & Colonna, P. (2003) Rheological properties of biopolymers and applications to cereal processing. In: Kaletunc, G. & Breslauer, K.J. (eds) *Characterization of Cereals and Flours,* pp. 209–265. Marcel Dekker, New York.

58 Vergnes, B., Villemaire, J-P., Colonna, P. & Tayeb, J. (1987) Interrelationships between thermomechanical treatment and macromolecular degradation of maize starch in a novel rheometer with preshearing. *J. Cereal Sci.* **5**, 189–202.

59  Lourdin, D., Della Valle, G. & Colonna, P. (1995) Influence of amylose content on starch films and foams. *Carbohydr. Polym.* **27**, 261–270.

60  Jenkins, P.J. & Donald, A.M. (1998) Gelatinisation of starch. A combined SAXS/WAXS/DSC and SANS study. *Carbohydr. Res.* **308**, 133–147.

61  Garcia, V., Colonna, P., Lourdin, D. *et al.* (1996) Thermal transitions of cassava starch at intermediate moisture contents. *J. Thermal Analysis* **47**, 1213–1228.

62  Le Bail, P., Bizot, H., Ollivon, M. *et al.* (1999) Monitoring the crystallization of amylose-lipid complexes during maize starch melting by synchrotron x-ray diffraction. *Biopolymers* **5**, 99–110.

63  Donovan, J.W. (1979) Phase transitions of starch-water system. *Biopolymers* **18**, 263–275.

64  Biliaderis, C.G., Page, C.M., Maurice, T.J. & Juliano, B.O. (1986) Thermal characterization of rice starches: a polymeric approach to phase transitions of granular starch. *J. Agric. Food Chem.* **34**, 6–14.

65  Russell, P.L. (1987) Gelatinisation of starches of different amylose/amylopectin content. A study by differential scanning calorimetry. *J. Cereal Sci.* **6**, 133–145.

66  Flory, P.J. (1953) *Principles of Polymer Chemistry.* Cornell University Press, Ithaca, NY.

67  Whittam, M.A., Noel, T.R. & Ring, S.G. (1990) Melting behaviour of A- and B-type starches. *Int. J. Biol. Macromol.* **12**, 359–362.

68  Prasad, A. & Mandelkern, L. (1989) Equilibrium dissolution temperature of low molecular weight polyethylene fractions in dilute solution. *Macromolecules* **22**, 914–920.

69  Crochet, P., Beauxis-Lagrave, T., Nole, T.R. *et al.* (2005) Starch crystal solubility and starch granule gelatinization. *Carbohydr. Res.* **340**, 107–113.

70  Slade, L. & Levine, H. (1988) Non equilibrium of native granular starch. Part I: Temperature location of the glass transition temperature associated with gelatinization of a-type cereal starches. *Carbohydr. Polym.* **8**, 183–191.

71  Tan, I., Wee, C.C., Sopade, P.A. & Halley, P.J. (2004) Investigation of the starch gelatinisation phenomena in water-glycerol systems: application of modulated temperature differential scanning calorimetry. *Carbohydr. Polym.* **58**, 191–204.

72  Couchman, P. & Karasz, F.E. (1987) A classical thermodynamic discussion of the effect of composition on glass transition temperatures. *Macromolecules* **11**, 117–119.

73  Lourdin, D., Coignard, L., Bizot, H. & Colonna, P. (1997) Influence of equilibrium relative humidity and plasticizer concentration on the water content and glass transition temperature. *Polymer* **38**, 5401–5406.

74  Zeleznak, K.J. & Hoseney, R.C. (1987) The glass transition of starch. *Cereal Chem.* **64**, 121–124.

75  Whittam, M.A., Noel, T.R. & Ring, S. (1991) Melting and glass/rubber transition of starch polysaccharides. In: *Food Polymers, Gels and Colloids*, pp. 277–288. Royal Society of Chemistry, Cambridge, UK.

76  Waigh, T., Gidley, M., Komanshek, B. & Donald, A. (2000) The phase transformations in starch during gelatinisation: a liquid crystalline approach. *Carbohydr. Res.* **328**, 165–176.

77  Slade, L. & Levine, H. (1987) Structural stability of intermediate moisture foods – a new understanding. In: Blanshard, J.M.V. & Mitchell, J.R. (eds) *Food Structure: its Creation and Evaluation*, pp. 115–147. Butterworth, London.

78  Doublier, J.L. (1987) A rheological comparison of wheat, maize, faba bean and smooth pea starches. *J. Cereal Sci.* **5**, 247–262.

79  Della Valle, G., Vergnes, B., Colonna, P. & Patria, A. (1997) Relations between rheological properties of molten starches and their expansion by extrusion. *J. Food Eng.* **31**, 277–296.

80  Katelunc, G. & Breslauer, K.J. (1993) Glass transition of extrudates: relationship with processing induced fragmentation and end-product attributes. *Cereal Chem.* **70**, 548–552.

81  Leloup, V., Colonna, P. & Ring, S.G. (1990) Studies on probe diffusion and accessibility in amylose gels. *Macromolecules* **23**, 862–866.

82  Rolland-Sabaté, A., Colonna, P., Potocki-Véronèse G. *et al.* (2004) Elongation and insolubilization of α-glucans by action of *Neisseria polysaccharea* Amylosucrase. *J. Cereal Sci.* **40**, 17–30.

83  Leloup, V.M., Colonna, P. & Ring, S.G. (1992) Physicochemical aspects of resistant starch. *J. Cereal Sci.* **16**, 253–266.

84  Ring, S.G., I'Anson, K.J. & Morris, V.J. (1985) Static and dynamic light scattering studies of amylose solutions. *Macromolecules* **18**, 182–187.

85  Ring, S., Colonna, P., I'Anson, K.J. *et al.* (1987) Gelation and crystallisation of amylopectin. *Carbohydr. Res.* **162**, 277–293.

86  Miles, M., Morris, V.J. & Ring, S.G. (1985) Gelation of amylose. *Carbohydr. Res.* **135**, 257–269.

87  Miles, M.J., Morris, V.J., Orford, P.D. & Ring, S.G. (1985) The roles of amylose and amylopectin in the gelation and retrogradation of starch. *Carbohydr. Res.* **135**, 271–281.

88  Doublier, J.L. & Choplin, L. (1989) A rheological description of amylose gelation. *Carbohydr. Res.* **193**, 215–226.

89  Gidley, M. & Bulpin, P. (1989) Aggregation of amylose in aqueous systems: the effect of chain length on phase behaviour and aggregation kinetics. *Macromolecules* **22**, 341–346.

90  Clark, A.H., Gidley, M.J., Richardson, P.K. & Ross, Murphy S.B. (1989) Rheological studies of aqueous amylose gels: the effects of chain length and concentration on gel modulus. *Macromolecules* **22**, 346–351.

91  Ortega-Ojedo, F.E., Larsson, H. & Eliasson, A.C. (2004) Gel formation in mixtures of amylose and high amylopectin potato starch. *Carbohydr. Polym.* **57**, 55–66.

92  Kalichevsky, M. & Ring, S. (1987) Incompatibility of amylose and amylopectin in aqueous solution. *Carbohydr. Res.* **162**, 323–328.

93  Leloup, V.M., Colonna, P. & Buleon, A. (1991) Influence of amylose–amylopectin ratio on gel properties. *J. Cereal Sci.* **13**, 1.

94  Slade, L. & Levine, H. (1991) Beyond water activity: recent advances based on an alternative approach to the assessment of food quality and safety. *Crit. Rev. Food Sci. Nutr.* **30**, 115–360.

95  Silverio, J.H., Fredriksson, R., Andersson, R. *et al.* (2000) The effect of temperature cycling on the amylopectin retrogradation of starches with different amylopectin unit-chain length distribution. *Carbohydr. Polym.* **42**, 175–184.

96  Paris, M., Bizot, H., Emery, J. *et al.* 2001) NMR local range investigations in amorphous starchy substrates. 1 – Structural heterogeneity probed by 13C CP-MAS NMR. *Int. J. Biol. Macromol.* **29**, 127–136.

97  Farhat, I.A. & Blanshard, J.M. (2001) Modeling the kinetics of starch retrogradation. In: Chinachoti, P. & Vodovotz, J. (eds) *Bread Staling*, pp. 163–172. CRC Press, Boca Raton, FL.

98  Del Nobile, M.A., Martoriella, T., Mocci, G. & La Notte, E. (2003) Modeling the starch retrogradation kinetic of durum wheat bread. *J. Food Eng.* **59**, 123–128.

99  Mestres, C., Colonna, P. & Buleon, A. (1988) Gelation and crystallization of maize starch after pasting, drum-drying or extrusion cooking. *J. Cereal Sci.* **7**, 123–134.

100  Franks, F. (1991) Water sorption and glass transition behaviors of freeze-dried sucrose-dextran mixtures. In: Levine, H. & Slade, L. (eds) *Water Relationships in Foods*, pp. 1–19. Plenum Press, New York.

101  Slade, L. & Levine, H. (1991) Beyond water activity: Recent advances based on an alternative approach to the assessment of food quality and safety. *Crit. Rev. Food Sci. Nutr.* **30**, 115–360.

102  Slade, L. & Levine, H (1993) The glassy state phenomenon in food molecules. In: Blanshard, J.M.V. & Lillford, P.J. (eds) *The Glassy State in Foods,* pp. 35–101. Nottingham University Press, Nottingham, UK.

103  Slade, L. & Levine, H. (1993) Water relationships in starch transitions. *Carbohydr. Polym.* **21**, 105–113.

104  Roos, Y.H. (1995) Food components and polymers. In: Taylor, S.L. (ed) *Phase Transitions in Foods*, pp. 133–135. Academic Press, San Diego, CA.

105  Rahman, M.S. (2006) State diagrams of foods: its potential use in food processing and product stability. *Trends Food Sci. Technol.* **17**, 129–141.

106  Bizot, H., Le Bail, P., Leroux, B. *et al.* (1997) Calorimetric evaluation of the glass transition in hydrated, linear and branched polyanhydroglucose compounds. *Carbohydr. Polym.* **32**, 33–50.

107  Kalichevsky, M.T., Jaroszkiewicz, E.M., Ablett, S. *et al.* (1992) The glass transition of amylopectin measured by DSC, DMTA and NMR. *Carbohydr. Polym.* **18**, 77–88.

108  Noel, T.R. & Ring, S.G. (1992) A study of the heat capacity of starch/water mixtures. *Carbohydr. Res.* **227**, 203–213.

109  Levine, H. & Slade, L. (1988) Thermomechanical properties of small-carbohydrate water glasses and rubbers – kinetically metastable systems at sub-zero temperatures. *J. Chem. Soc. Faraday Trans.* **84**, 2619–2635.

110  Richards, G.N. & Vandenburg, J.Y. (1995) Fructose-grafted amylose and amylopectin. *Carbohydr. Res.* **268**, 201–207.

111  Struik, L.C.E. (1978) *Physical Aging in Amorphous Polymers and Other Materials.* Elsevier, Amsterdam.

112  Simon, S.L., Plazek, D.J., Sobieski, J.W. & McGregor, E.T. (1997) Physical aging of a polyetherimide: Volume recovery and its comparison to creep and enthalpy measurements. *J. Polym. Sci. Part B: Polym. Phys. Ed.* **35** (6), 929–936.

113  Shogren, R.L. (1992) Effect of moisture content on the melting and subsequent physical ageing of corn starch. *Carbohydr. Polym.* **19**, 83–90.

114  Hoseney, R.C., Zeleznak, K. & Lai, C.S. (1986) Wheat gluten: a glassy polymer. *Cereal Chem.* **63**, 285–286.

115  Sauvageot, F. & Blond, G. (1991) Effect of water activity on crispness of breakfast cereals. *J. Text. Studies* **22**, 423–442.

116  Le Bail, P., Bizot, H. & Buléon, A. (1993) 'B' to 'A' type phase transition in short amylose chains. *Carbohydr. Polym.* **21**, 99–104.

117  Seow, C.C. & Teo, C.H. (1993) Annealing of granular rice starches. Interpretation of the effect of phase transitions associated with gelatinization. *Starch/Stärke* **45**, 345–351.

118  Borde, B., Bizot, H., Vigier, B. & Buleon, A. (2002) Calorimetric analysis of the structural relaxation in partially hydrated amorphous polysaccharides. II Phenomenological study of physical ageing. *Carbohydr. Polym.* **48**, 111–123.

119  Evans, I.D. & Haiman, D.R. (1982) The effect of solutes on the gelatinization of potato starch. *Starch/Staerke* **34**, 224–231.

120  Tolstoguzov, V. (2003) Thermodynamic considerations of starch functionality in foods. *Carbohydr. Polym.* **51**, 99–111.

121  Abeysekera, R.M. & Robards, A.W. (1995) Microscopy as an analytical tool in the study of phase separation of starch –gelatin binary mixtures. In: Harding, S.E., Hill, S.E. & Mitchell, J.R. (eds) *Biopolymer Mixtures,* pp. 143–160. Nottingham University Press, Nottingham, UK.

122  Christianson, D.D., Hodge, J.E., Osborne, D. & Detroy, R.W. (1981) Gelatinization of wheat starch as modified by xanthan gum, guar gum and cellulose gum. *Cereal Chem.* **58**, 513–517.

123  Funami, T., Kataoka, Y., Omoto, T. *et al.* (2005) Food hydrocolloids control the gelatinization and retrogradation behavior of starch. 2a. Functions of guar gums with different molecular weights on the gelatinization behavior of corn starch. *Food Hydrocoll.* **19**, 15–24.

124  Closs, C.B., Conde-Petit, B., Roberts, I.D. *et al.* (1999) Phase separation and rheology of aqueous starch–galactomannan systems. *Carbohydr. Polym.* **39**, 67–77.

125  Breaden, P.W. & Wikkhoft, E.M.A. (1971) Bread staling. III. Measurement of the redistribution of moisture in bread by gravimetry. *J. Sci. Food Agric.* **22**, 647–649.

126  Ottenhof, M.A. & Farhat, I.A. (2004) The effect of gluten on the retrogradation of wheat starch. *J. Cereal Sci.* **40**, 269–274.

127  Eliasson, A.C. (1983) Differential scanning calorimetry studies on wheat starch-gluten mixtures. II Effect of gluten and sodium stearoyl lactylate on starch crystallization during ageing of wheat starch gels. *J. Cereal Sci.* **1**, 207–213.

128  Eliasson, A.C., Carlson, T.L., Larsson, K. & Miezis, Y. (1981) Some effects of starch lipids on the thermal and rheological properties of wheat starch. *Starch/Stärke* **33**, 130–134.

129  Sevenou, O. (2002) Starch: its relevance to dough expansion during baking. Doctoral dissertation, University of Nottingham, UK.

130 Master, A.M. & Steeneken, P.M. (1997) Rheological properties of highly cross-linked waxy maize starch in aqueous suspensions of skim milk components. *Carbohydr. Polym.* **32**, 297–305.

131 Abu-Jdayil, B., Mohamed, H.A. & Eassa, A. (2004) Rheology of wheat starch-milk-sugar systems: effect of starch concentration, sugar type and concentration and milk fat content. *J. Food Eng.* **64**, 207–212.

132 De Bont, P.W., van Kempen, G.M.P. & Vrecker, R. (2002) Phase separation in milk protein and amylopectin mixtures. *Food Hydrocolloids* **16**, 127–138.

133 Vrij, A. (1976) Polymers at interfaces and the interactions in colloidal dispersions. *Pure Appl. Chem.* **44**, 471–483.

134 Sperry, P.R. (1984) Morphology and mechanism in latex flocculated by volume restriction. *J. Colloids Interfac. Sci.* **99**, 97–108.

135 Lekkerkerker, H.N.W., Poon, W.C.K., Pusey, Pn. *et al.* (1992) Phase behaviour of colloid + polymer mixtures. *Europhysics Lett.* **20**, 559–564.

136 Chanvrier, H., Colonna, P., Della Valle, G. & Lourdin, D. (2005) Structure and mechanical behaviour of corn flour and starch-zein based materials in the glassy state. *Carbohydr. Polym.* **59**, 109–119.

137 Attenburrow, G.E., Davies, A.P., Goodband, R.M. & Ingman, S.J. (1992) The fracture behavior of starch and gluten in the glassy state. *J. Cereal Sci.* **16**, 1–12.

138 Eliasson, A.C. (1992) A calorimetric investigation of the influence of sucrose on the gelatinisation of starch. *Carbohydr. Polym.* **18**, 131–138.

139 Ahmad, F.B. & Williams, P.A. (1999) Effect of sugars on the thermal and rheological properties of sago starch. *Biopolymers* **50**, 401–412.

140 Sopade, P.A., Halley, P.J. & Junning, L.L. (2004) Gelatinisation of starch in mixtures of sugars. II. Application of differential scanning calorimetry. *Carbohydr. Polym.* **58**, 311–321.

141 Perry, P.A. & Donald, A.M. (2002) The effect of sugars on the gelatinisation of starch. *Carbohydr. Polym.* **49**, 155–165.

142 Gaudin, S., Lourdin, D., Le Botlan, D. *et al.* (1999) Plasticization and mobility in starch-sorbitol films. *J. Cereal Sci.* **29**, 273–284.

143 Gaudin, S., Lourdin, D., Forsell, P. & Colonna, P. (2000) Antiplasticisation and oxygen permeability of starch-sorbitol films. *Carbohydr. Polym.* **43**, 33–37.

144 Smits, A.L.M., Kruiskamp, P.H., van Soest, J.J.G. & Vliegenthart, J.F.G. (2003) The influence of various small plasticizers and malto-oligosaccharides on the retrogradation of partly gelatinized starch. *Carbohydr. Polym.* **51**, 417–424.

145 Morrison, W., Law, R. & Snape, C. (1993) Evidence for inclusion complexes of lipids with v-amylose in maize, rice and oat starches. *J. Cereal Sci.* **18**, 107–109.

146 Gernat, C., Radosta, S., Anger, H. & Damashun, G. (1993) Crystalline parts of three different conformations detected in native and enzymatically degraded starches. *Starch/Stärke* **45**, 309–314.

147 Biliaderis, C.G. (1992) Structures and phase transitions of starch in food systems. *Food Technol.* **6**, 98–109. [9 pp. not consecutive.]

148 Buleon, A., Le Bail, P., Colonna, P. & Bizot, H. (1998) In: Reid, D.S. (ed) *The Properties of Water in Foods* (ISOPOW 6), pp. 160–178. Blackie Academic and Professional, London and New York.

149 Kugimiya, M., Donovan, J.W. & Wong R.Y. (1980) Phase transitions of amylose–lipid complexes in starches: a calorimetric study. *Starch/Stärke* **32**, 265–270.

150 Kugimiya, M. & Donovan, J.W. (1981) Calorimetric determination of the amylose content of starches based on formation and melting of the amylose–lysolecithin complex. *J. Food Sci.* **46**, 765–777.

151 Biliaderis, C.G., Page, C.M. & Maurice, T.J. (1986) On the multiple melting transitions of starch/monoglyceride systems. *Food Chem.* **22**, 279–295.

152 Biliaderis, C.G., Page, C.M., Slade, L. & Sirett, R.R. (1985) Thermal behavior of amylose–lipid complexes. *Carbohydr. Polym.* **5**, 367–389.

153 Mercier, C., Charbonniere, N., Grebaut, J. & de la Gueriviere, J.F. (1980) Formation of amylose–lipid complexes by twin-screw extrusion cooking of manioc starch. *Cereal Chem.* **57**, 4–9.

154 Krog, N. (1971) Amylose complexing effect of food grade. *Starch/Stärke* **22**, 206–210.

155 Champenois, Y., Colonna, P., Buleon, A. *et al.* (1995) Gelatinisation et rétrogradation de l'amidon dans le pain de mie. *Scie. Aliments* **15**, 593–614.

156 Sarko, A. & Zugenmaier, P. (1980) Crystal structure of amylose and its derivatives. In: French, A.D. & Gardner, K.C. (eds) *Fiber Diffraction Methods.* ACS Symposium Series, Vol. 141, pp. 459–482. American Chemical Society, Washington DC.

157 Rappenecker, G. & Zugenmayer, P. (1981) Detailed refinement of the crystal structure of Vh-amylose. *Carbohydr. Res.* **89**, 11–19.

158 Whittam, M.A., Orford, P.D., Ring, S.G. *et al.* (1989) Aqueous dissolution of crystalline and amorphous amylose–alcohol complexes. *Int. J. Biol. Macromol.* **11**, 339–344.

159 Godet, M.C., Buléon, A., Tran, V. & Colonna, P. (1993) Structural features of fatty acids-amylose complexes. *Carbohydr. Polym.* **21**, 91–95.

160 Godet, M.C., Bizot, H. & Buleon, A. (1995) Crystallization of amylose-fatty acid complexes prepared with different amylose chain lengths. *Carbohydr. Polym.* **27**, 47.

161 Helbert, W. & Chanzy, H. (1994) Single crystals of V amylose complexed with n-butanol or n-pentanol: structural features and properties. *Int. J. Biol. Macromol.* **16**, 207–213.

162 Buléon, A., Delage, M.M., Brisson, J. & Chanzy, H. (1990) Single crystals of V amylose complexed with isopropanol and acetone. *Int. J. Biol. Macromol.* **12**, 25–33.

163 Hullemann, S.H.D., Helbert, W. & Chanzy, H. (1996) Single crystals of V amylose complexed with glycerol. *Int. J. Biol. Macromol.* **18**, 115–122.

164 Yamashita, Y. & Monobe, K. (1971) Single crystal of amylose V complexes. III Crystals with $8_1$ helical configuration. *J. Polym. Sci.* (A-2) **9**, 1471–1481.

165 Winter, W.T., Chanzy, H., Putaux, J.L. & Helbert, W. (1998) Inclusion compounds of amylose. *Polym. Prep.* **39**, 703.

166 Rutschmann, M.A. & Solms, J. (1990) Formation of inclusion complexes of starch with different organic compounds. IV. Ligand binding and variability in helical conformations of V amylose complexes. *Lebensm. Wiss. Technol.* **23**, 84–87.

167 Nuessli, J., Conde-Petit, B., Trommsdorff, U.R. & Escher, F. (1996) Influence of starch flavour interactions on rheological properties of low concentration starch systems. *Carbohydr. Polym.* **28**, 167–170.

168 Nuessli, J., Sigg, B., Conde-Petit, B. & Escher, F. (1997) Characterization of amylose-flavour complexes by DSC and X-ray diffraction. *Food Hydrocoll.* **11**, 27–34.

169 Nuessli, J., Putaux, J.L., Le Bail, P. & Buleon, A. (2003) Crystal structure of amylose complexes with small ligands. *Int. J. Biol. Macromol.* **33**, 227–234.

170 Gidley, M.J. & Bociek, S.M. (1988) Carbon-13 CP/MAS NMR studies of amylose inclusion complexes, cyclodextrins and the amorphous phase of starch granules: relationships between glycosidic linkages conformation and solid-state carbon-13 chemical shifts. *J. Am. Chem. Soc.* **110**, 3820–3829.

171 Marchessault, R.H., Taylor, M.G., Fyfe, C.A. & Veregin, R.P. (1985) Solid-state 13C-CP-MAS NMR of starches. *Carbohydr. Res.* **144**, C1–C5.

172 Kawada, J. & Marchessault, R. (2004) Solid state NMR and X-ray studies on amylose complexes with small organic molecules. *Starch/Staerke* **56**, 13–19.

173 Le Bail, P., Rondeau-Mouro, C. & Buléon, A. (2005) Structural investigation of amylose complexes with small ligands: helical conformation, crystalline structure and thermostability. *Int. J. Biol. Macromol* **35**, 1–7.

174 Rondeau-Mouro, C., Le Bail, P. & Buleon, A. (2004) Structural investigation of amylose complexes with small ligands: inter- or intra-helices associations. *Int. J. Biol. Macromol.* **34**, 309–319.

175 Englyst, H.N., Kingman, S.M. & Cummings, J.H. (1992) Classification and measurement of nutritionally important starch fractions. *Eur. J. Clinical Nutr.* **46**, 33–50.

176  Holm, J., Björck, I., Ostrowska, S. *et al.* (1983) Digestibility of amylose-lipid complexes in-vitro and in-vivo. *Starch/Stärke* **35**, 294–297.

177  Holm, J. & Björck, I. (1992) Bioavailability of starch in various wheat-based bread products: evaluation of metabolic responses in healthy subjects and rate and extent of in vitro starch digestion. *Am. J. Clin. Nutr.* **55**, 420–429.

178  Bornet, F., Fontvieille, A-M., Rizkalla, S. *et al.* (1989) Insulin and glycemic responses from different starches according to food processing. Correlation to in vitro starch α-amylase susceptibility. *Amer. J. Clin. Nutr.* **50**, 315–323.

179  Englyst, K.N., Vinoy, S., Englyst, H.N. & Lang, V. (2003) Glycaemic index of cereal products explained by their content of rapidly and slowly available glucose. *Br. J. Nutr.* **89**, 329–340.

180  Englyst, K.N., Englyst, H.N., Hudson, G.J. *et al.* (1999) Rapidly available glucose in foods: an in vitro measurement that reflects the glycemic response. *Am. J. Clin. Nutr.* **69**, 448–454.

181  Colonna, P., Buléon, A. & Leloup, V. (1992) Limiting factors of starch hydrolysis. *Eur. J. Clin. Nutr.* **46**, S17–S32.

182  Faisant, N., Buléon, A., Colonna, P. *et al.* (1995) Resistant starch: a modified method adapted to high RS products. *Br. J. Nutr.* **73**, 111–123.

183  McIntyre, A., Gibson, P.R. & Young, G.P. (1993) Butyrate production from dietary fiber and protection against large-bowel cancer in a rat model. *Gut* **34**, 386–391.

184  Scheppach, W., Bartran, H.P. & Richter, F. (1995) Role of short chain fatty acids in the prevention of colorectal cancer. *Eur. J. Cancer* **31A**, 1077–1080.

185  Liljeberg, H., Grandfeldt, Y. & Bjorck, I. (1992) Metabolic responses to starch in bread containing intact kernels versus milled flour. *Eur. J. Clin. Nutr.* **46**, 561–575.

186  Colonna, P., Barry, J.L., Cloarec, D. *et al.* (1990) Enzymatic susceptiblity of starch from pasta. *J. Cereal Sci.* **11**, 59–70.

187  Eerlingen, R.C. & Delcour, J.A. (1995) Formation, analysis, structure and properties of type III enzyme resistant starch. *J. Cereal Sci.* **22**, 129–138.

188  Planchot, V., Colonna, P. & Buléon, A. (1997) Enzymatic hydrolysis of α-glucan crystallites. *Carbohydr. Res.* **298**, 319–326.

189  Seib, P. & Woo, K.S. (1998) Food grade starch resistant to alpha-amylase. Patent no. WO98/54973.

190  Kettlitz, B.W., Coppin, J.V.J.M., Roper, H.W.W. & Bornet, F. (1996) Highly fermentable resistant starch. Patent no. EP–0846704A2.

191  Pohu, A., Putaux, J.L., Planchot, V. *et al.* (2004) Origin of the limited α-amylolysis of debranched maltodextrins crystallized in the A form: a TEM study on model substrate. *Biomacromolecules* **5**, 119–125.

192  McKenzie, A.P. (1977) Non-equilibrium freezing behavior of aqueous systems. *Phil. Trans. R. Soc.* **278**, 167–189.

193  Slade, L. & Levine, H. (1995). Water and the glass transition – dependence of the glass transition on composition and chemical structure: Special implications for flour functionality in cookie baking. *J. Food Eng.* **24**, 431–509.

194  Katelunc, G. (2003) Construction of state diagrams. In: Katelunc, G. & Breslauer, K.J. (eds) *Characterization of Cereals and Flours,* pp. 151–171. Marcel Dekker, New York.

# Chapter 3
# Water Transport and Dynamics in Food

*Brian Hills*

## 3.1 Introduction

Research on structure–function relations in food naturally focuses on the biopolymers comprising the food matrix and their interactions with sugars, lipids, preservatives and flavours. Yet water also has an essential role in almost every aspect of food science, including the processing response of raw materials, food stability, organoleptic properties and microbial safety. It is therefore surprising that, despite thousands of research publications, our fundamental understanding of the role of water in food remains largely empirical. Part of the problem lies in the extreme complexity of most foods, which, being microheterogeneous, multicomponent and multiphase systems makes it hard to predict how water partitions between the various components and microphases. There are even greater problems predicting water transport during complex processing operations like extrusion cooking, baking, drying, rehydration and freezing. Nor do the difficulties occur only during processing. Operations such as baking and air-drying invariably create nonequilibrium distributions of water that slowly readjust during storage. This slow water redistribution is often associated with undesired textural changes, such as staling, and reduced food shelf life. Of course, similar problems occur during the storage of nonprocessed foods like fruit and vegetables. Here it is the (sub)cellular redistribution of water between membrane-bound compartments that affects quality through altered turgor pressure, accelerated spoilage reactions and tissue breakdown.

Other difficulties in understanding water transport and dynamics in food result from the wide range of distance and time scales involved. On the molecular scale water correlation times range from picoseconds to milliseconds depending on the nature of the biopolymer–water and/or solute–water interactions. On the microscopic scale transport takes place on the millisecond timescale or longer, depending on the pore structure and connectivity of the food matrix. On the macroscopic scale moisture transport during processing and storage has timescales extending from seconds to many months, depending on which, of many factors, control the effective water diffusivity and/or matrix dynamics. The lack of suitable experimental tools for monitoring water dynamics over this enormous range of distance and time scales exacerbates the problem. As we shall see, different techniques give limited information on particular time and distance scales so great care is needed when comparing and interpreting the data.

This chapter will focus primarily on water transport and dynamics on the microscopic and molecular distance scales. Water transport on the macroscopic distance scale during processes such as drying can be measured noninvasively with magnetic resonance imaging[1] or, destructively, by mechanical slicing and weighing. The experimental macroscopic water concentration profiles can usually be modelled with conventional mass-transfer formalisms based on diffusion theory, two-film theory and penetration theory described in standard chemical engineering texts.[2,3] Invariably, these macroscopic mass-transfer formalisms introduce empirical transport coefficients such as effective water diffusion coefficients and interfacial mass transfer coefficients that have a complicated, and usually poorly characterized, functional dependence on the time- and space-dependent local water content, temperature, pressure and composition. While it is possible, in simple systems, to measure this functional dependence the real scientific challenge lies in understanding how these macroscopic water transport coefficients are determined by the changing matrix microstructure, the changing microscopic water distribution and, ultimately, by water–biopolymer–solute interactions at the molecular level. This is an increasingly urgent task because biopolymer structure and microstructure can now be engineered, or at least modified, by an impressive array of genetic, enzymatic, chemical and mechanical techniques, but if we do not understand how the modifications alter water–biopolymer interactions and water dynamics then we have no way of predicting how the modified biopolymer or matrix microstructure will affect functional behaviour such as texture and shelf life. The converse is also true, in that without a detailed understanding of water–biopolymer interactions we cannot predict how a biopolymer or microstructure should be modified to create a desired functional behaviour. Nor can we predict how the rate and intensity of processing operations, such as drying or baking, should be altered to create desired end-product functionality.

Although we will focus on water relations in processed foods, processing and storage are only the middle parts of the whole field-to-fork food chain and it should not be forgotten that water transport in plants growing in the field also affects the quality of the raw food materials entering the processing operation. Varying rainfall, temperature and soil composition as well as harvesting time can all affect water transport in the growing plant tissue, and this, in turn, affects ripening, processing response and product functionality. Cassava starch is an interesting example of this type of problem.[4] Cassava starch is used extensively in industry because of its unique thickening properties, its high purity, low cost and its ability to form clear viscous pastes. Unfortunately, the size, composition and gelatinization behaviour of cassava starch granules extracted from the tapioca plant display unpredictable variation, depending on the environmental conditions at the time of harvest. This variability causes difficulties in optimizing the commercial viability of cassava starch, especially in countries like Thailand, where more than a million hectares are planted annually for the cassava industry. Developing predictive models for optimum harvest time from measured rainfall and seasonal temperature profiles is therefore another aspect requiring a deeper understanding of the role of water migration and transport in the developing plant tissue.

Water transport is also important at the other end of the field-to-fork chain, namely in the consumption stage. It cannot be overemphasized that, besides a food's appearance, most consumers' attitudes to novel processed food, especially low-fat and sugar-free foods, are based on the organoleptic properties of food as it is masticated in the mouth. Indeed, there is little point in optimizing the harvesting, processing and storage steps if the resulting product has the texture and taste of, say, cardboard. Mastication not only breaks down a food's structure, but by increasing its surface area it permits aqueous saliva to ingress rapidly into the food. This hydrates and softens the food matrix, dissolves flavours and helps release aromas into the headspace, all of which affect the texture, taste and aroma sensations that determine food acceptability. Understanding the role of water (saliva) as the food matrix is hydrated and destroyed in the mouth is therefore vitally important for a food's commercial viability. These mastication aspects lie outside the scope of the present chapter, but the interested reader will find further discussion in refs 5–7.

In this chapter we therefore take a fundamental approach and begin by considering the physicochemical factors controlling water distribution and transport on the microscopic distance scale and then, in later sections, consider water–matrix interactions on the molecular distance scale.

## 3.2 Statistical thermodynamics and the microscopic water distribution

Many of the factors affecting water transport on the microscopic distance scale are illustrated by the thermal processing of a model system comprising pea starch granules embedded in an egg albumin matrix. This system has at least two microphases and three biopolymer components (amylose, amylopectin and albumin) so it is representative of many microheterogeneous foods. The two microphases are clearly seen in the optical micrograph in Fig. 3.1, which shows the system after it has been heated and allowed to cool. Although the micrographs do not show the water distribution we expect that as the temperature is first raised the native starch granules swell and draw water out of the albumin phase until the starch gelatinization temperature of about 60°C is reached. Thereafter, we expect the release of amylose from the granules to carry water back into the albumin phase where there is microscopic phase separation between the amylose and albumin mixed polymer phases (see Fig. 3.1, inset). If the temperature is further raised the proteins in the egg albumin phase begin to denature and cross-link and bind water. There is therefore a backward and forward movement of water between microphases on the microscopic distance scale. The amount of water migration determines the gelatinization temperature and the rate of phase separation, so it is a prime factor in controlling the processing response in this system. Until the factors controlling this redistribution are understood we cannot predict the effects of modifying the structure of the starch granules or the effect of altering the composition of the albumin phase. Moisture migration between components and microphases during storage also affects shelf life by altering rates of starch retrogradation and phase separation and, therefore ultimately, texture and microbial stability.

Let us begin the theoretical analysis of this system by imagining we keep the temperature constant and wait until an equilibrium state is reached, defined here as the situation where change in moisture concentration in all microphases and components ceases because the water chemical potential is everywhere uniform. Of course, it may be the case that moisture transport has ceased, not because it has reached thermodynamic equilibrium but because kinetic factors have so arrested or slowed down transport that they can no longer be sensibly measured. This might be the case if one or more microphases enters the glassy state or becomes so concentrated that moisture diffusion is rate limiting. Interfacial barriers such as surface structures at the starch granule–albumin interface could also act as moisture permeability barriers. Slow macromolecule motion and entanglement could also be the rate-limiting factor in moisture redistribution. All these factors mean that, in general, it may not be safe to assume thermodynamic equilibrium in complex foods. However, to progress the analysis we assume thermodynamic control and that the water chemical potential, $\mu$, in all phases and components is equal. For the case of starch granules in egg albumin, we can therefore write, at equilibrium,

$$\mu_{starch} = \mu_{albumin} = \mu_{water\ vapour} \tag{3.1}$$

where we assume that the system is in contact with water vapour in the air-spaces and $\mu_i$ is the water chemical potential in microphase i. Defining the water activity as $\mu_i = \mu_0 + RT\ln a_i$ we can further write:

$$\mu_0 + RT\ln a_{starch} = \mu_0 + RT\ln a_{albumin} = \mu_0 + RT\ln(p/p_0) \tag{3.2}$$

where we have used the fact that the water activity in the vapour phase is, at least for a perfect gas, given by the ratio of the vapour pressure, $p$, to that for bulk water at the same temperature, $p_0$. Equation 3.2 shows that, at equilibrium, $a_{starch} = a_{albumin} = (p/p_0)$. In other words, that the water activity must everywhere be the same and given by the equilibrium ratio of vapour pressures. Measurement of the sorption isotherms, $a_i(W_i)$ for each separate component, i, such as egg albumin or starch, should therefore be sufficient to predict the equilibrium water content, $W_i$, in each component and microphase. However, we can take the analysis deeper by asking how the component sorption isotherms depend on water–biopolymer and water–solute interactions and matrix microstructure. One approach to this question is to use the formalism of equilibrium statistical mechanics to calculate how the water chemical potential depends on microstructure.

Consider, first, the free energy change, $\Delta F$, caused by the ingress of water into the starch granule. This can be written as the sum of the mixing and elastic contributions:

$$\Delta F = \Delta F_{mixing} + \Delta F_{elastic} \tag{3.3}$$

Here $\Delta F_{mixing}$ is the free energy change resulting from the mixing of the water with the starch granule matrix, which, as usual, has enthalpy ($H$) and entropy ($S$) contributions:

$$\Delta F_{mixing} = \Delta H_{mixing} - T\Delta S_{mixing} \tag{3.4}$$

The enthalpy change can be estimated by counting the change in the number of water–biopolymer interactions using the lattice model of polymer physics.[8] In the lattice model, each voxel in a three-dimensional cubic lattice is occupied either by a water molecule or by a segment of a polymer chain of equal volume to a water molecule. In the simplest model the assignment is random, subject only to the constraint that polymer segments must connect into a chain of defined length. For a single starch granule we assume, for simplicity, that there is only a single large chain of segments and we make no distinction between amylose and amylopectin. According to this lattice model, the enthalpy of mixing is given as

$$\Delta H_{mixing} = kT\chi\, n_w v_w \tag{3.5}$$

where $\chi$ is the interaction energy per water molecule in units of $kT$, in other words it is the difference in energy of a water molecule immersed in pure polymer (starch) compared with one surrounded by water (i.e. in pure water); $n_w$ is the number of water molecules and $v_w$ is the volume fraction of the water in the granule. By the usual statistical method of counting configurations, the lattice model, together with Stirling's approximation, gives an entropy of mixing of the form:[8]

$$\Delta S_{mixing} = -k n_w \ln v_w \tag{3.6}$$

In writing Equation 3.6 we have once again assumed that we are dealing with only a single cross-linked amylose and amylopectin network. Substituting Equations 3.5 and 3.6 into Equation 3.4 gives:

$$\Delta F_{mixing} = kT\{\chi\, n_w v_w + n_w \ln v_w\} \tag{3.7}$$

It is important to note that this derivation assumes random arrangement of water in the starch granule. As we shall see, this assumption is not strictly correct because there are different water–starch interaction sites having different interaction potentials that are not randomly occupied as water content increases. Nevertheless, it is instructive to proceed with the derivation because it shows how molecular interactions play a fundamental role in controlling microscopic water transport, in this case, water ingress into the starch granule microphase. The free energy of mixing (Equation 3.7) predicts that water will continue to ingress into the granule network until the whole granule is diluted and forms a uniform solution. Of course this does not happen because the amylopectin and amylose cross-links eventually prevent further swelling and dilution. This is expressed by the elastic free energy term in Equation 3.3, which assumes that as the starch granule network swells it behaves as an elastic material. As before we therefore write:

$$\Delta F_{el} = \Delta H_{el} - T\Delta S_{el} \tag{3.8}$$

and proceed to calculate the elastic enthalpy and entropy contributions. The enthalpy change associated with stretching the network (in the absence of the mixing term) can usually be neglected compared with the entropy change, which can be calculated from the statistical theory of rubber elasticity[8] as:

$$\Delta S_{el} = -(k v_c/2)[3\alpha^2 - 3 - \ln\alpha^3] \tag{3.9}$$

Here $v_c$ is the effective number of chains in the starch granule and $\alpha$ is the fractional degree of swelling, assumed isotropic. In other words the granule radius, $R$, increases to $\alpha R$, on swelling. The derivation assumes a random, Gaussian distribution of end-to-end starch chain vectors, which is, of course, only an approximation to reality.

The chemical potential, $\mu_w$, of the water inside the granule is obtained from its definition as the free energy per mole, as:

$$\mu_w - \mu_0 = N(\partial \Delta F_{mix}/\partial n_w)_{T,P} + (\partial \Delta F_{el}/\partial \alpha)_{T,P}(\partial \alpha/\partial n_w)_{T,P} \tag{3.10}$$

where $N$ is the Avogadro constant. The derivative $(\partial \alpha/\partial n_w)_{T,P}$ can be evaluated by noting that $\alpha^3$ is the ratio of the swollen granule volume to the volume of the unswollen polymer, $V_0$, so that,

$$\alpha^3 = [V_0 + (n_w v_w/N)]/V_0 = 1/v_s \tag{3.11}$$

where $v_s$ is the volume fraction of the starch biopolymers. Differentiation and substitution finally gives an expression for the water chemical potential inside the granule:

$$\mu_w - \mu_0 = RT\{\ln(1 - v_s) + v_s + \chi v_s^2 + V_w(v_c/V_0)(v_s^{1/3} - v_s/2)\} \tag{3.12}$$

where $V_w$ is the molar volume of the water. The water activity, $a_w$, of the water inside the granule is obtained from Equation 3.12 and its definition is: $\ln a_w = (\mu_w - \mu_0)/RT$.

This somewhat lengthy derivation of the water activity inside the starch granule highlights several important features. First, if we assume the water in the egg albumin phase surrounding the granule has a chemical potential $\mu_{out}$ then water will redistribute between the granule and albumin microphases until the term in brackets in Equation 3.12 equals $(\mu_{out} - \mu_0)$ and the polymer concentration inside the granule assumes an equilibrium value, $v_{s\,(equil)}$. If the albumin phase is replaced by pure water then the term in brackets will be zero and the water activity inside the granule is unity. This equilibrium is possible because the entropy change in the elastic free energy term has the opposite sign to the mixing term. In fact the last term in Equation 3.12 represents the increase in water free energy from the elastic reaction of the network and this can be expressed as an osmotic pressure term, $\pi_{net} v_s$, where $\pi_{net}$ is the effective osmotic pressure created by the network that raises the chemical potential of the water. To be explicit, we can write:

$$\pi_{net} v_s = RT V_w(v_c/V_0)(v_s^{1/3} - v_s/2) \tag{3.13}$$

Equations 3.12 and 3.13 allow us, at least in principle, to predict the effect of varying the activity of the outer water phase, for example, by replacing it with a concentrated biopolymer solution or gel, on the granule swelling and hence microscopic water redistribution.

While the statistical mechanical approach based on the lattice model gives insight into the factors controlling water migration, it only reveals part of the story. We still require a deeper understanding of water–biopolymer interactions on the molecular level. This can be seen in the appearance of unknown parameters such as water–biopolymer interaction energy, $\chi$, in Equation 3.12. We will discuss this in Section 3.8 below, but at this point it is clear that the lattice formalism is not yet developed to the point where it can predict the effect of modifying the starch granule structure, such as the ratio of amylose and amylopectin, on the degree of equilibrium swelling. We will see in Section 3.4 that a similar situation holds for the nonequilibrium situation and the microscopic transport equations for the water ingress into the granule.

## 3.3 Experimental probes of the microscopic water distribution

Finding suitable experimental tools for monitoring water (re)distribution between different phases and components on the microscopic distance scale continues to be a challenge. While microscopic techniques such as optical and electron microscopy and atomic force microscopy (AFM) give detailed information about the microstructure of the food matrix, as exemplified by Fig. 3.1, they fail to show how water is distributed between the microphases. Nuclear magnetic resonance (NMR) is undoubtedly the most promising experimental tool for doing this, but even the NMR methods are problematic and require further development. In some simple cases the one-dimensional distribution of water proton transverse relaxation times, the so-called 'relaxation spectrum' obtained by constrained inverse Laplace transformation of the Carr–Purcell–Meiboom–Gill (CPMG) echo decay envelope,[1] gives useful information about microscopic water distribution. Figure 3.2 shows the relaxation spectrum for a suspension of potato starch granules in a 30% bovine serum albumin (BSA) solution as a function of thermal processing temperature. In this experiment, which was designed to mimic the system in Fig. 3.1, the starch suspension was heated to the indicated temperature for 20 minutes, then cooled to 23°C, where it was measured with the CPMG pulse sequence with a 90–180 pulse spacing of 240 µs. The resulting decay curve was analysed as a continuous distribution of exponentials using the standard Resonance Instruments Molecular Biotools (Oxford, UK) WINDXP software. Each peak in Fig. 3.2 corresponds to water in a different microphase, though it is not always straightforward to assign them to specific structures.

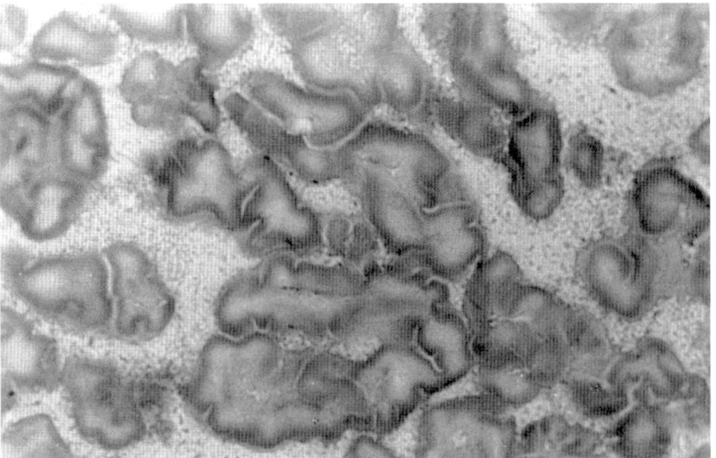
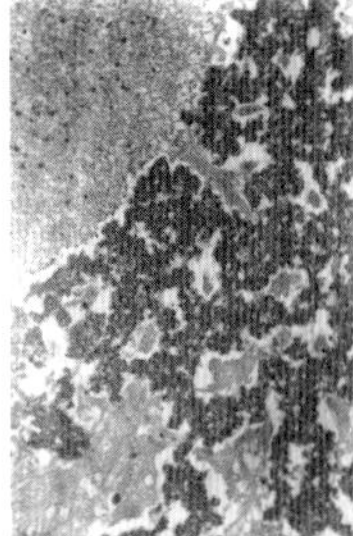

**Fig. 3.1**   An optical micrograph of partially gelatinized pea starch granules embedded in an egg albumin matrix. The inset shows phase separation of extragranular amylose and albumin proteins. Courtesy of Dr Mary Parker.

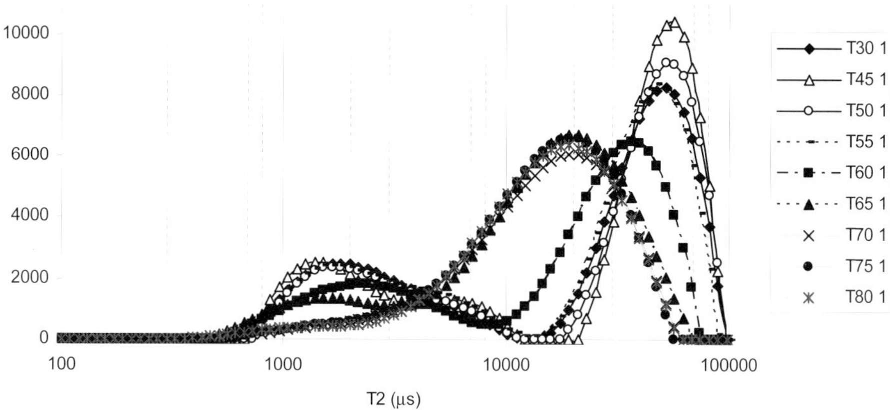



**Fig. 3.2**   The distribution of water proton transverse relaxation times for potato starch granules embedded in a 30% BSA solution after heating to the indicated temperatures. Amplitudes are in arbitrary units.

The peak at c. 50 ms in the unheated sample is believed to correspond to water in the BSA solution outside the granules, and the broad peaks between 1 and 10 ms to water inside the starch granules.[9,10] Heating the sample at or above the gelatinization temperature of c. 60°C causes the release of amylose and water from the granule into the surrounding BSA phase and this can be quantified as an increase in the area of the peaks above c. 6 ms.

Difficulties in peak assignment and peak overlap are two obvious shortcomings of this approach, as is the unknown extent to which water diffusion between different microphases averages the signal. The latter can be minimized by working at temperatures just above freezing, and peak separation can be altered to some extent by modifying the pulse spacing and by varying the extent of diffusive attenuation by imposition of constant field gradients during the CPMG acquisition. Alternatively, deuterium relaxometry in a $D_2O$ exchanged system could give insight by removing the signal from nonexchanging biopolymer protons.

The recent development of two-dimensional $T_1$-$T_2$ correlation spectroscopy[11] and other multidimensional NMR relaxation-diffusion techniques[12] may offer a way to circumvent some of the difficulties in the one-dimensional relaxation spectra. Plate 1 shows the $T_1$-$T_2$ correlation spectrum of a 10% suspension of native potato starch granules in a 25% BSA solution with well-resolved water peaks. Resolving the relaxation time peaks in two dimensions in this way not only can remove peak overlap but also can resolve peaks arising from the nonexchanging biopolymer protons. Whether water peaks are clearly resolved in more complex systems such as that in Fig. 3.1 remains to be investigated, though preliminary results[13,14] suggest this is possible.

Unfortunately water diffusion sets a limit to the ability of the NMR relaxation method to resolve signal from water in different microphases and compartments. This is because fast diffusion of water between microphases and compartments on the NMR measurement timescale of a few milliseconds averages the signal so that the relaxation time peaks merge together. In a time, $t$, water molecules in pure water diffuse a distance $(6Dt)^{1/2}$, which is about 10 µm in 10 ms, which is a typical NMR measurement timescale. This means that, unless there is a barrier to water diffusion, the NMR relaxation method has a spatial resolution limit of about 10 microns.

In principle the NMR relaxation data, including the effects of interphase diffusion, can be interpreted for any arbitrary microstructure by numerically solving the Bloch–Torrey equations for the time evolution of the water proton transverse magnetization densities. The Bloch–Torrey equations are simply the relaxation-diffusion equations for the magnetization and can be written thus:

$$\partial M^{+}_{i}/\partial t = -i\omega_{i}M^{+}_{i} - \gamma_{i}M^{+}_{i} + D_{i}\nabla^{2}M^{+}_{i} \quad (i = 1,2) \tag{3.14}$$

Here $M^{+}(\mathbf{r},t)$ is the complex sum $(M_{x}(\mathbf{r},t) + iM_{y}(\mathbf{r},t))$ in a voxel located at $\mathbf{r}$, and there is a corresponding complex conjugate equation for $M^{-}_{i}$. These equations are best envisaged with a vectorial representation of the water magnetization, whereby each volume element is associated with a vector representing its complex water proton magnetization density. The term in $\omega_{i}$ describes precession of the vector at the resonance frequency, $\omega_{i}$. The second term describes the decrease in the length of the magnetization density vector at a rate, $\gamma_{i}$; while the third term describes the diffusion of the vectors between voxels. These coupled equations must be solved subject to initial conditions appropriate to the NMR pulse sequence. For example, a hard 90° pulse excites $x$-magnetization in all compartments, but selective excitation experiments can be designed so that transverse (or longitudinal) magnetization is excited in only one type of microphase. For example, the Goldman–Shen pulse sequence can be used to excite magnetization in the microphase with the longest transverse relaxation time.[1] The Bloch–Torrey equations are solved with spatial boundary conditions appropriate to the microphase size and geometry, and it is in this way that the water relaxation behaviour can be used to probe water distribution on a microscopic distance scale. Numerical methods, such as the Monte Carlo technique, are most suited for this computation.[15] The Monte Carlo method is worthy of further comment because it has hardly begun to be exploited in relaxation and diffusion simulations and yet has enormous computational flexibility. In the Monte Carlo method, the microstructure is digitized inside a cubic lattice that is coarse-grained into $N^{3}$ volume elements that are small compared with the distance scale of the microstructure, but large compared with molecular distances. Magnetization density vectors are then randomly distributed over the voxels and assigned a resonance frequency, $\omega_{i}$, relaxation rate, $\gamma_{i}$, and diffusion coefficient, $D_{i}$, corresponding to their position. Initially the effect of a single, hard, 90° pulse is simulated by assigning to each magnetization vector an initial $x$-magnetization equal to the inverse of the number of vectors while assuming the initial $y$-magnetization is zero. The subsequent time evolution is determined by giving every magnetization vector a probability, $p$, of jumping randomly to a neighbouring volume element in each of the three Cartesian directions in a time step, $\Delta t$, with zero probability of jumping outside the cube. The jump probability, $p$, is calculated from the diffusion coefficient as $p = D_{i}\Delta t/(\Delta x)^{2}$, where $\Delta t$ is the incremental time step and $\Delta x$ is the volume element length. The length of each vector decays exponentially with a relaxation rate and precession frequency corresponding to its position. The new total magnetization after time $\Delta t$ is obtained by summing over all vectors, taking into account their precession and relaxation. The effects of various pulse sequences can be introduced directly as phase changes on the individual magnetization vectors during the simulation. External applied gradient pulses, proton exchange and surface relaxation of water magnetization can all be simulated by appropriate changes in the basic model.[1] The Monte Carlo method has been applied to water diffusion into Sephadex microspheres and for studying combined diffusion and proton exchange between two model lamellar phases.[1] However, the potential of this approach for modelling NMR water relaxation and diffusion in reconstructed microheterogeneous media has yet to be fully explored.

Up to this point we have considered only water-saturated systems lacking any gas phase. However, a gas phase exists in sponge-like foods such as breads, biscuits and cakes, so we need to consider the effect of water diffusion through the gas microphase. This can be done by assuming fast exchange of water molecules between the liquid and vapour microphases so that local equilibrium exists. If so, the local concentrations of water in the liquid phase, $w_{liq}$ and in the vapour phase, $w_{vap}$, are related by the equilibrium sorption isotherm:

$$w_{vap} = f(w_{liq}) \tag{3.15}$$

The transport can then be described by a coupled diffusion equation of the form

$$\partial[w_{vap} + w_{liq}]/\partial t = \nabla[D_{vap}\nabla w_{vap} + D_{liq}\nabla w_{liq}] \tag{3.16}$$

which, combined with the sorption isotherm, gives an effective diffusion equation of the form $\partial w_{liq}/\partial t = \nabla D_{eff}\nabla w_{liq}$ where $D_{eff} = D_{liq} + D_{vap}(df/dw_{liq})$ which assumes $w_{liq} \gg w_{vap}$. The problem in unsaturated porous matrices therefore reduces to measurements of the sorption isotherm (Equation 3.15), and the effective vapour- and liquid-phase diffusion coefficients. Once again we see that the microscopic formalisms drive us to yet smaller distance scales to understand the functional form of the sorption isotherm and the effective water and vapour diffusivities, all of which will depend in a complicated way on microstructure and the detailed nature of the water–matrix interaction.

## 3.4 The water self-diffusion propagator

So far the analysis of water transport on the microscopic distance scale has involved the assignment of water self-diffusion coefficients to each microphase, including the vapour phase. An alternative approach focuses not on the diffusion coefficient itself but on the water self-diffusion propagator. The water self-diffusion propagator, or more precisely, the conditional displacement probability, $P_s(\mathbf{r}/\mathbf{r}',t)$, is the conditional probability for finding a water molecule at location $\mathbf{r}'$ at a time $t$ when it was originally at $\mathbf{r}$ at time zero. The conditional probability flux, $\mathbf{J}$, is obtained from the propagator as,

$$\mathbf{J} = -D\nabla P_s \tag{3.17}$$

where $D$ is the self-diffusion coefficient. Because the total conditional probability is conserved, the continuity equation applies,

$$\nabla\mathbf{J} = -\partial P_s/\partial t \tag{3.18}$$

so that, combining Equations 3.17 and 3.18 gives:

$$\partial P_s / \partial t = D\nabla^2 P_s \qquad\qquad (3.19)$$

which shows the relationship between the diffusion propagator and the self-diffusion coefficient. There are many advantages to working with the self-diffusion propagator, $P_s$, rather than with the self-diffusion coefficient, $D$. In complex spatially microheterogeneous systems there are no analytic solutions to the diffusion equation so it is far easier to use numerical simulations based, for example, on the Monte Carlo or cellular automata methods to directly calculate $P_s$ for ensembles of water molecules randomly located in the matrix than it is to try to solve diffusion equations.

Analytic expressions for the average propagator have been derived for water diffusion in a variety of simple morphologies appropriate to microstructured foods. These include matrices of connected pores,[16] fractal geometries[16] and multicompartment systems.[17] In this context the example of starch granules in a protein matrix can be modelled as a two-compartment system where the water transport is related to the intrinsic water self-diffusion coefficients in the separate starch and albumin microphases. The interested reader will find more details about these models in refs 1, 16 and 17. However, in most cases, such analytical models are only crude representations of the actual sample microheterogeneity. Current research is therefore aimed at calculating the propagator numerically for realistic morphologies. A particularly promising area of current research involves numerically reconstructing the morphology of microheterogeneous matrices.[18] The reconstruction process involves numerically generating random matrices having the same morphological features, such as porosity and spatial correlation functions, as the real system. These reconstructed morphologies can then be used as templates for numerical calculation of the diffusion propagator using, for example, the finite element methods of FEMLAB[19] or some other numerical approach such as the Monte Carlo or cellular automaton methods. Alternatively the actual digitized confocal images and cross-sections could be used directly in the propagator simulations.

Another motive for preferentially calculating the self-diffusion propagator rather than the diffusion coefficient is that the self-diffusion propagator can be measured directly with NMR pulsed field gradient methods even in complex heterogeneous systems. To this we now turn.

## 3.5 Experimental probes of the water self-diffusion propagator

Figure 3.3 shows the simplest NMR pulsed-gradient spin-echo pulse sequence for measuring the self-diffusion propagator.[1,16] After excitation with a $90°$ radiofrequency pulse the water proton magnetization throughout the sample is phase-labelled with a short pulsed field gradient of amplitude $G$ and duration $\delta$. The phase is refocused into a spin echo at a time $2\tau$ by a $180°$ radiofrequency pulse at a time $\tau$. If the water molecules do not translate in the time, $\Delta$, between the two gradient pulses then the phase change introduced by the first gradient pulse is exactly reversed by the second gradient pulse and there is no echo attenuation apart from the usual transverse relaxation processes. However, in general, diffusion of a water molecule from $\mathbf{r}$ to $\mathbf{r}'$ during the time $\Delta$ introduces a phase change $\mathbf{q}.(\mathbf{r}' - \mathbf{r})$ where $\mathbf{q}$ is the pulsed gradient area, $\gamma G\delta$, and $\gamma$ is the proton gyromagnetic ratio. This phase change means that the transverse magnetization is no longer completely refocused so the echo amplitude

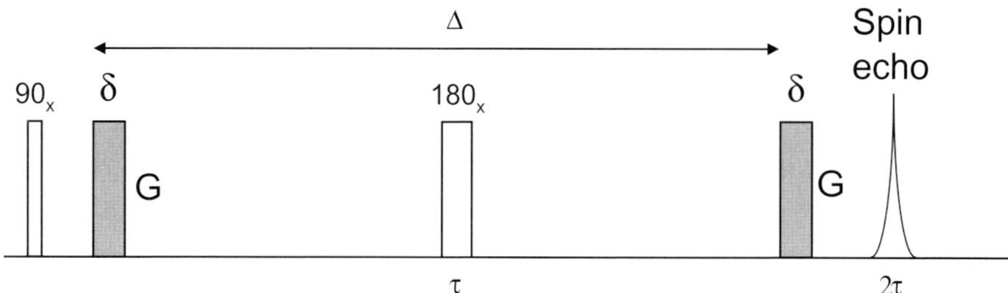


**Fig. 3.3**   The simplest pulsed-gradient spin-echo pulse sequence used to measure the water self-diffusion propagator. The shaded pulses denote gradients of amplitude $G$ and duration $\delta$ with a separation, $\Delta$, which defines the diffusion time.

is attenuated. The echo attenuation caused by water diffusion throughout the whole sample is obtained by integration:

$$S(q,\Delta) = \iint drdr'\rho(r)P_s(\mathbf{r}|\mathbf{r}',\Delta)\exp[i2\pi\mathbf{q}.(\mathbf{r}' - \mathbf{r})] \qquad (3.20)$$

Here $\rho(r)$ is the water proton density, giving the initial distribution, and the exponential factor gives the phase change, but, most significantly, we see that this phase change is weighted by the self-diffusion propagator, $P_s(\mathbf{r}|\mathbf{r}',\Delta)$.

To take the analysis further it is usual to assume equilibrium and translational invariance (i.e. that the results are independent of the initial position of the water molecules) so that Equation 3.20 can be transformed to the displacement variable, $\mathbf{R} = \mathbf{r}' - \mathbf{r}$, by introducing the average propagator,

$$P_{av}(\mathbf{R},\Delta) = \int dr\rho(r)P(\mathbf{r}|\mathbf{r}',\Delta) \qquad (3.21)$$

With this transformation, Equation 3.20 becomes the Fourier relationship:

$$S(q,\Delta) = \int dR\, P_{av}(\mathbf{R},\Delta)\exp(i\mathbf{q}.\mathbf{R}) \qquad (3.22)$$

which means that the average propagator can be obtained by inverse Fourier transformation of the NMR echo amplitude measured as a function of increasing wavevector, $\mathbf{q}$:

$$P_{av}(R,\Delta) = \int d\mathbf{q}\, S(\mathbf{q},\Delta)\exp(-i\mathbf{q}.\mathbf{R}) \qquad (3.23)$$

The resulting diffusion propagator can be compared with the theoretical calculations discussed in the previous section. Alternatively, the calculated propagator can be inserted into Equation 3.22 and the resulting echo amplitude, $S(q,\Delta)$, compared with experiment.

## 3.6 Water transport in nonequilibrium microheterogeneous systems

So far we have assumed thermodynamic equilibrium or, at least, that water concentrations throughout the food matrix do not change on the measurement timescale, though, of course, water self-diffusion still occurs even in the equilibrium state. We will now remove the equilibrium constraint and try formulating transport equations describing the microscopic water transport from one microphase to another, using, as before, the example of water migration from the albumin protein phase into a swelling starch granule. One approach assumes the microscopic validity of Darcy's law[20] describing macroscopic fluid flow through porous media under an external pressure gradient. For the swelling kinetics of the starch granules Darcy's law assumes the form:

$$v = -(\kappa/\upsilon\rho)\nabla\pi_{sp} \qquad (3.24)$$

which relates the water transport velocity, $v$, into the starch granule, to the swelling pressure, $\pi_{sp}$, driving water into the granule. Here $\rho$ is the water density, $\kappa$ is the permeability tensor of the starch granule amylopectin and amylose matrix, and $\upsilon$ is the kinematic viscosity of the water. The swelling pressure, $\pi_{sp}$, is the difference between the osmotic pressure, $\pi$, and the pressure created by elastically expanding the cross-linked starch biopolymer network, $\pi_{net,}$ considered in Section 3.2:

$$\pi_{sp} = \pi - \pi_{net} \qquad (3.25)$$

The momentum balance in Equation 3.24 needs to be complemented with the mass transfer balance,

$$\partial w/\partial t + \nabla.\rho v = 0 \qquad (3.26)$$

where $w$ is the moisture content in the granule. The problem with this approach is that $\kappa$, the permeability tensor of the starch granule, is unknown and will vary with the degree of swelling and cross-linking density, all of which makes quantitative prediction of the rate of microscopic water transport very difficult. Equations 3.24 and 3.26 can be written in the form of a diffusion equation if we write the swelling pressure in the form $\pi_{sp} = C(w)w$, then

$$\partial w/\partial t = \nabla.D_{eff}(w)\nabla w \qquad (3.27)$$

where $D_{eff} = C(\kappa/\upsilon)$ is the effective water diffusivity, which, in general, depends on the local water content both through the permeability coefficient and the coefficient $C$. Once again we see that the microscopic formalism forces us to consider water–matrix interactions on the molecular distance scale to understand the functional form of the phenomenological coefficients $D_{eff}$, $C$ and $\kappa$.

The Darcy's law approach obscures the important role played by the kinetics of the swelling biopolymer network in controlling water transport. In many cases it is the slow rate of polymer disentanglement associated with swelling that is rate limiting and that determines the rate of water transport. An alternative approach developed by Thomas and Windle[21] emphasizes this aspect by introducing a viscous flow rate for water transport, $1/\eta$, that decreases exponentially with increasing water concentration,

$$\eta = \eta_0 \exp(-MW) \tag{3.28}$$

Here $M$ is a constant and $W$ is the ratio of water volume fraction to the equilibrium volume fraction after complete swelling. $W$ is assumed to obey first-order kinetics that depends on the ratio of swelling pressure to local viscosity:

$$\partial W / \partial t = k\pi(r,t)/\eta(r) \tag{3.29}$$

where $\pi$ is the local swelling pressure at position r and time t, and $k$ is a constant. When the biopolymer relaxation is slow compared with water diffusion these equations predict a sharp water 'front' penetrating into the biopolymer matrix at a constant velocity determined by the rate of polymer relaxation.[21]

It is important to note that, so far, we have assumed that the only driving force for water transport is the gradient in water chemical potential, which is tantamount to assuming that frictional forces between the water molecules and the food matrix are so large that water transport is an incoherent process characterized by a diffusion propagator. This is certainly the case in the starch–egg albumin system because we have tacitly assumed it is water-saturated and contains little or no air. However, many materials, including wet granular foods, not only exhibit matrix heterogeneity but also contain both microscopic aqueous and gas phases dispersed throughout the food matrix. In such water-unsaturated, coarse-grained material water-lattice frictional forces are not necessarily capable of preventing rapid localized coherent percolative flow of the aqueous phase through the porous matrix, so that other forces, such as interfacial capillary forces, need to be included in the transport equation. One way of doing this in the Darcy formalism is to include a capillary pressure term in the momentum balance:

$$v(r,t) = -L\nabla\mu - (K_r K^0/\eta)\nabla p \tag{3.30}$$

The first term is the diffusive flux resulting from the gradient in water chemical potential already considered, while the second describes coherent flow resulting from a capillary pressure gradient. $K^0$ is the Darcy permeability coefficient for the pure fluid, while $K_r$ is the relative Darcy permeability coefficient for water and air. $\eta$ is the shear viscosity and $L$ is the Onsager coefficient for diffusion.[22] As before, we can combine this momentum balance with the mass balance equation:

$$\phi\partial(\rho S)/\partial t - \nabla\cdot(\rho v) \tag{3.31}$$

to derive a generalized diffusion-type equation for water transport in an unsaturated micro-porous matrix. Here $\phi$ is the porosity and $S(r,t)$ is the spatially varying degree of saturation ($0 \leq S \leq 1$). We then arrive at a generalized diffusion equation of the form:

$$\partial S/\partial t = \nabla.[D_{\mu}(S) + D_{c}(S)]\nabla S \tag{3.32}$$

where

$$D_{\mu}(S) = L\phi^{-1}(\partial\mu/\partial S)_{\mathrm{T}} \tag{3.33}$$

$$D_{c}(S) = K^{0}K_{r}(S)\phi^{-1}\eta^{-1}(\partial P/\partial S)_{\mathrm{T}} \tag{3.34}$$

$D_{\mu}(S)$ is the usual diffusion coefficient arising from the gradient in chemical potential, while $D_{c}(S)$ is an effective diffusion coefficient arising from the capillary pressure term. We only need to think of the rapid ingress of water into a dry biscuit when it is dunked in tea or coffee or the less trivial example of the ingress of saliva into low-water-content foods in the mouth during consumption to appreciate that, because of percolative coherent flow, $D_{c}(S)$ is usually much larger than $D_{\mu}(S)$.

## 3.7 The state of water in nanopores

The discussion so far has concerned water distribution and transport in food matrices on the microscopic distance scale of more than about a micron. For example, the starch granules referred to in the previous sections have diameters ranging typically between 1 and 20 microns depending on the type of starch and its degree of swelling. We now reduce the distance scale by several orders of magnitude to the nanoscale and consider the state of water in nanopores having diameters between about 0.5 and 50 nm inside the starch granule and in the aggregated egg albumin protein gel. More generally we need to consider water dynamics inside the nanopores of biopolymer membranes, aggregates and gels.

The state of water in the nanopores of synthetic membranes, such as the cellulose esters, has been widely studied both because of its relevance to understanding reverse osmosis, nanofiltration and ultrafiltration, and also because such synthetic materials help us understand more complex water–biopolymer interactions. Molecular dynamics simulations give the most detailed information about nanopore water, and numerous simulations have been performed, and reviewed.[23] A noteworthy study by Spohr and coworkers[24] modelled the effect of systematically increasing the amount of water from 19 to 95% hydration in a 2-nm radius cylindrical pore having a hydrophilic silica surface. The simulation ran for a respectable time of 1 nanosecond at room temperature. In this example the water was strongly adsorbed to the pore surface, so that at the lowest hydration level studied (19%) the radial density profile showed only a monolayer of adsorbed water with only a hint of a second layer. Surprisingly, the monolayer only became saturated at hydration levels exceeding about 58%. The second layer saturated at around 77% hydration but was already present at 38%. Only at 95% hydration was the pore filled with a continuous density profile and even then, the density was less than that of bulk water, though the number of hydrogen bonds was seen to be constant over the central region of the pore at the highest hydration level. A notable feature of the simulation was the appearance of mobile 'patches' of water molecules on the silica surface when the monolayer was unsaturated (19% hydration). These clusters had a continuous

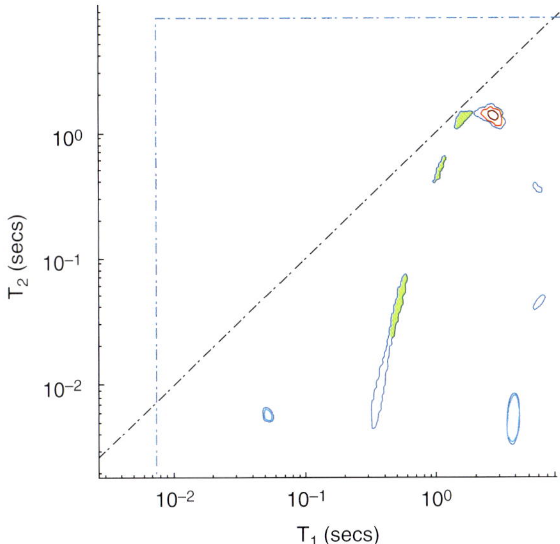



**Plate 1** The two-dimensional $T_1$–$T_2$ spectrum of 10% native potato starch granules suspended in a 25% BSA solution at 298 K. The dashed diagonal line shows the condition $T_1 = T_2$. The elongated peak arises from water inside the starch granules, whereas the peak with the longest $T_2$ arises from water in the BSA solution. Other minor peaks are believed to arise from mobile nonexchanging BSA and starch protons.

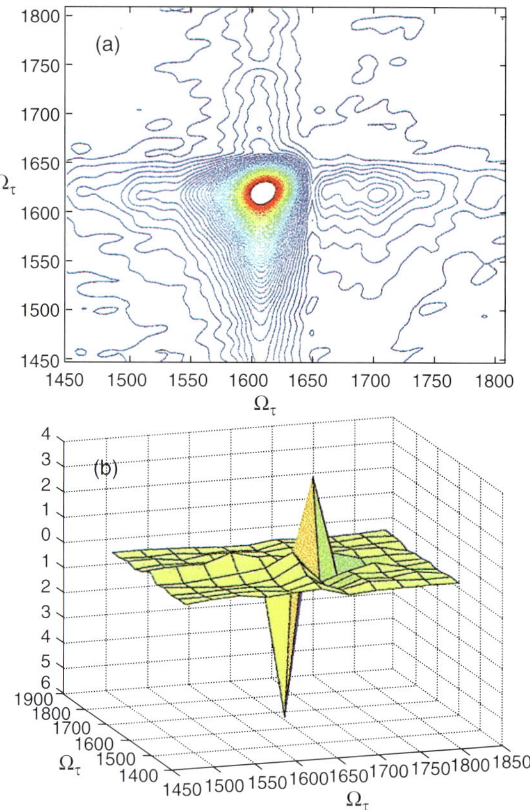



**Plate 2**  Two-dimensional-IR spectra for *n*-methylacetamide-d in $D_2O$. (a) The absolute value of the 2D-IR spectrum, showing a single peak in both frequency dimensions. (b) The real part of the 2D-IR spectrum, showing the fundamental and anharmonically shifted peaks, which have opposite signs. From Asplund, M.C., Zanni, M.T. & Hochstrasser, R.M. (2000) Two-dimensional infrared spectroscopy of peptides by phase controlled femtosecond vibrational photon echoes. *Proc. Natl. Acad. Sci. U. S. A.* **97**, 8219–8224, with permission. Copyright 2000 National Academy of Sciences, USA.

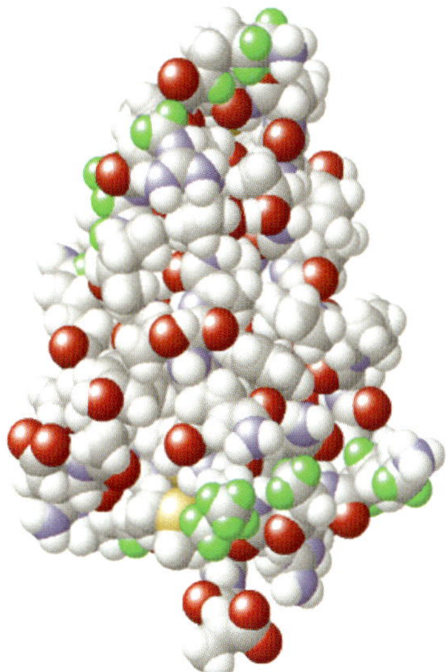

**Plate 3**   Model of a BPTI protein molecule showing surface BPTI protons that give NOEs with water. Those coloured red have substantial contributions for proton chemical exchange; those in green are free from proton exchange artefacts. From B. Halle, sited at www.fkem2.1th.se/research/areas/projects/noe_cross_relaxation.

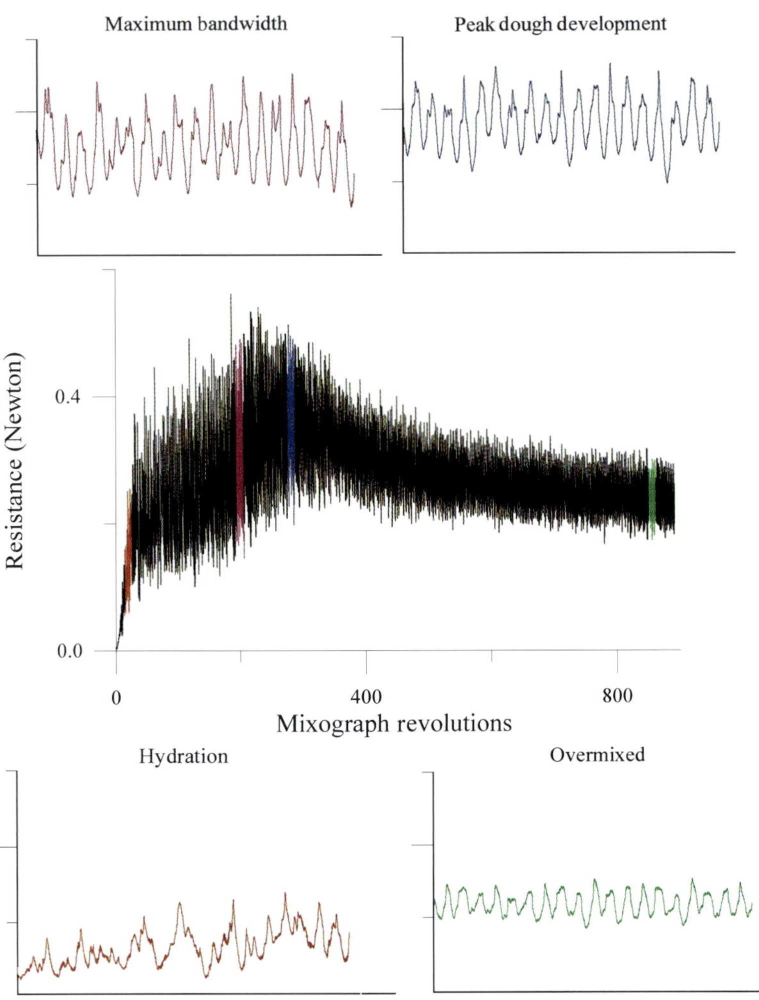



**Plate 4** Snapshots of the structure of the corresponding high-resolution mixogram for four different stages in the development and breakdown of a dough. From Anderssen, R.S., Bekes, F., Gras, P.W. et al (2004) with permission.

**Plate 5**   A dough being mixed on a prototype 2-gram Mixograph. The pin to the right is the moving pin. The thick rubber-like viscoelastic formation of the dough about the fixed pins is clearly illustrated. Reproduced by kind permission of Professor Chuck Walker, University of Kansas.

distribution of sizes ranging between 1 and about 100 molecules. Such clusters had been previously postulated to explain neutron diffraction data and appear to form preferentially where the water can hydrogen bond silicon oxygen atoms on the surface. The short-time water self-diffusion coefficient (mean square displacement over a nanosecond timescale) along the direction of the cylindrical pore axis was calculated for various cylindrical shells outward from the pore centre as a function of hydration level (see Table 1 in ref. 24). This showed that each shell was associated with a characteristic diffusion coefficient that was lowest in the monolayer and increased towards the pore centre. Water molecules at the centre in the near-saturated pores actually had a greater mobility than in bulk water because they were less restrained by hydrogen bonding at the 'surface' of the distribution.

The spectral density of the translational motion of the hydrogen atoms obtained by integrating the velocity autocorrelation function is even more revealing. The results, which agree well with inelastic neutron scattering measurements, show that in the central part of the water-saturated pore, the spectral density is very close to that of bulk water. In particular, the spectral density of hydrogen atoms shows peaks at 60 and 500 cm$^{-1}$ corresponding to the translational and reorientational dynamics of the water molecules. However, near the pore surface the 60 cm$^{-1}$ peak almost vanishes as the translational motion is suppressed by strong surface adsorption, but more surprisingly, the maximum of the rotational contribution shifts to higher frequencies, from 500 to 600 cm$^{-1}$. This presumably shows more rapid reorientation of adsorbed water on the surface, though this reorientation is probably not isotropic but characterized by anisotropic rotation about a molecular symmetry axis or hydrogen bond.

The reorientational spectral density of water confined in nanopores can, in principle, be determined with water-oxygen-17 magnetic relaxation dispersion (MRD) techniques. This measures the dependence of the water-oxygen-17 longitudinal relaxation rate, $R(\omega_0)$, on the oxygen-17 resonance frequency, $\omega_0$. In the simplest model, where bulk water exchanges with nanopore water characterized with a single reorientational correlation time, $\tau$, it can be shown that:

$$R(\omega_0) = R_{\text{bulk}} + 0.2J(\omega_0) + 0.8J(2\omega_0) \tag{3.35}$$

where $R_{\text{bulk}}$ is the frequency-independent relaxation rate of bulk water outside the nanopore, and the spectral density, $J(\omega)$, is:

$$J(\omega) = \alpha + \beta\{\tau/[1 + (\omega_0\tau)^2]\} \tag{3.36}$$

Here $\alpha$ gives the frequency-independent contribution from any bulk-like water inside the nanopore, and $\beta$ the contribution of any nanopore water with a reduced correlation time, $\tau$. Before proceeding it is very important to note the frequency window available with water oxygen-17 dispersions, which is determined by $\omega_0$ and is typically between 1 and 100 MHz. This means that nanopore water with correlation times less than about a nanosecond will give no measurable frequency dispersion. Bearing in mind that the molecular dynamics simulation of Spohr and coworkers[24] discussed previously had a total run-time of 1 nanosecond, we would not expect to see any frequency dependence in the oxygen-17 dispersion for nanopore water. Nevertheless, the surprising fact is that a strong dispersion is observed, at least for strongly adsorbed molecules like water in the nanopores of a silica glass.[25] These motions cannot refer to the reorientation of the water on the surface, which, the simulations suggest, occur on a subnanosecond timescale, but rather to the much slower motions required to average out residual quadrupolar interactions caused by surface-induced anisotropy in the orientational probability distribution of water molecules. This much slower averaging occurs by repeated desorption and readsorption between the surface and bulk phases in a slow stochastic process described by Lévy walks in a mechanism called RMTD (reorientation mediated by translational displacement).[26] While such slow motions are academically interesting they do not throw much light on the dynamic state of nanopore water. We shall refer to this point again in a later section dealing with the molecular distance scale.

Infrared spectroscopy is a much more powerful tool for studying the state of water in nanopores because its short measurement timescale (subpicosecond) essentially takes 'snapshots' of the water hydrogen bond configurations in the absence of reorientational averaging (whose correlation time in pure water is a few picoseconds). This technique therefore operates on a timescale that is actually too short directly to observe water dynamics; nevertheless it gives great insight into the state of the nanopore water because the water hydroxyl (OH) stretching frequency decreases with increasing strength of hydrogen bonding. For example, isolated water molecules in the vapour phase have antisymmetric and symmetric OH stretching frequencies centred at 3755 and 3657 cm$^{-1}$ respectively. In the bulk liquid state the OH stretching linewidth broadens and shifts to lower frequencies between 3400 and 3300 cm$^{-1}$. In ice-like water (sometimes called 'stretched water'), where the hydrogen bonding is much stronger, the stretching frequency is further reduced to 3300–3200 cm$^{-1}$. In 1995 Murphy and de Pinho[27] published a careful ATR-FTIR (attenuated total reflection-Fourier transform infrared) study of the state of water in a series of six cellulose acetate membranes with decreasing pore sizes. Three cellulose acetate membranes used for the ultrafiltration of macromolecules had mean pore sizes of 7.65, 4.37 and 3.0 nm and three cellulose acetate membranes used for nanofiltration had mean pore sizes of 0.61, 0.49 and 0.34 nm. A comparison of the ATR spectra of these six water-saturated membranes concluded that the hydrogen bonding strength of the water clusters, and possibly the average cluster size, decreases from a value close to that in bulk water in the largest 7.65-nm pore to weakly hydrogen bonded water and small clusters in the smallest 0.34-nm pore. The analysis was placed on a quantitative basis in a later study[28] in 1999. The ATR $\nu$(OH) stretching band was deconvoluted into four Gaussians centred at 3560, 3450, 3370 and 3270 cm$^{-1}$, the first corresponding to monomers and dimers, the last to large ice-like clusters with strong H-bonding. In this later study membranes were prepared with different water permeability coefficients and the effect of water structure-making and -breaking ions was investigated. It was found that cellulose acetate (CA) membranes associated with a decrease in the relative area of the large ice-like clusters at 3270 cm$^{-1}$ and an increased relative area of the monomer 3560 cm$^{-1}$ peak were also associated with a reduction in the membrane permeability coefficients both for pure water and for NaCl and an increase in the salt rejection coefficient. Moreover, equilibration of the CA membranes with a structure-breaking ion, such as 100 mM KCl, increased the relative area of the monomer peak; conversely a structure-forming salt such as 100 mM MgCl$_2$ increased the relative area of the ice-like strongly H-bonded water cluster. These ATR data therefore suggest that the network of water hydrogen bonding is perturbed only on distance scales below about 7 nm and that the hydrophobic surface causes a weakening of the hydrogen bonding strength and a reduction in mean cluster size.

In case these conclusions appear straightforward, even unsurprising, it is worth mentioning the controversial viewpoint championed by Philippa Wiggins to explain a phenomenon called 'microosmosis'.[29] Wiggins proposed that because water at a hydrophobic surface cannot make as many hydrogen bonds its enthalpy and free energy are increased relative to bulk water. To compensate, one of two events occurs. If possible, the high-free-energy water is removed by bringing hydrophobic surfaces together, which gives rise to a 'hydrophobic force'. Alternatively, if this is not possible, the water lowers its chemical potential back to the bulk value by reducing its density. It is proposed that to lower its density water assumes more of the 'stretched' clusters configuration of ice-1-like water and is therefore characterized by more linear, stronger H-bonding and longer rotational and translational correlation times. The simulations and ATR measurements on nanopore water would suggest that such low-density water would not extend more than a few nanometres from the hydrophobic surface and therefore be localized inside nanopores in (bio)polymer matrices. It is further hypothesized that the solvent properties of the low-density nanopore water differ from those of bulk water. Structure-making ions and small hydrophobic solutes are excluded from the low-density water in hydrophobic nanopores whereas structure-breaking ions and small hydrophilic solutes are preferentially partitioned into the pore. Such partitioning would normally occur over the very short timescale determined by solute diffusion and rapidly establish an equilibrium state where the water chemical potential is the same inside and outside the pore. However, if the repartitioning kinetics is accompanied by volume changes in the porous polymer network the process can be rate limited by slow biopolymer chain rearrangements. In such cases it may take many days to establish equilibrium and involve alternate swelling and contraction of the whole polymer gel matrix containing the nanopores and redistribution of the water and ionic solutes. This complex sequence of events has been termed 'microosmosis'.

Somewhat surprisingly, a number of simple macroscopic experiments on the Biogel-P series of gel beads appear to demonstrate the reality of microosmosis.[30] The nanopores in the polyamide Biogel-P series of gels have diameters between 1 and 3 nm and have hydrophobic surfaces comprising stretches of hydrocarbon chain with weakly H-bonding carbonyl and amide groups. Consistent with the microosmosis hypothesis it was observed that the density of water in the Biogel-P gel beads ranged between 0.98 (P-30) and 0.973 g cm$^{-3}$ (P-2), which falls between the bulk water value (1.0) and that of ice (0.92). No change in bead volume was observed when the beads were placed in a solution of a structure-making ion (100 mM NaCl) but the gel beads showed monotonic swelling over a period of 15 days when placed in a solution of a structure-breaking ion (10 mM KCl). Most remarkably, oscillations in internal gel bead volume over the observation period of 5 days were observed in a solution containing both structure-making and structure-breaking ions (160 mM NaCl, 30 mM KCl) (Fig. 3.4). The oscillations are believed to be caused by rapid partitioning of ions between the inside and outside of the nanopore, which causes a nonequilibrium osmotic pressure gradient driving oscillatory transport of water between the nanopores and bulk phases at a rate limited by the kinetics of swelling or contraction of the polymer matrix.

Despite the macroscopic data on the water density and swelling in the Biogel-P gels, it has to be said that the ATR or molecular dynamics evidence for the central hypothesis that low-density water exists in nanopores of hydrophobic polymers such as CA and Biogel-P and is characterized by stronger H-bonding and increased rotational and translational correlation

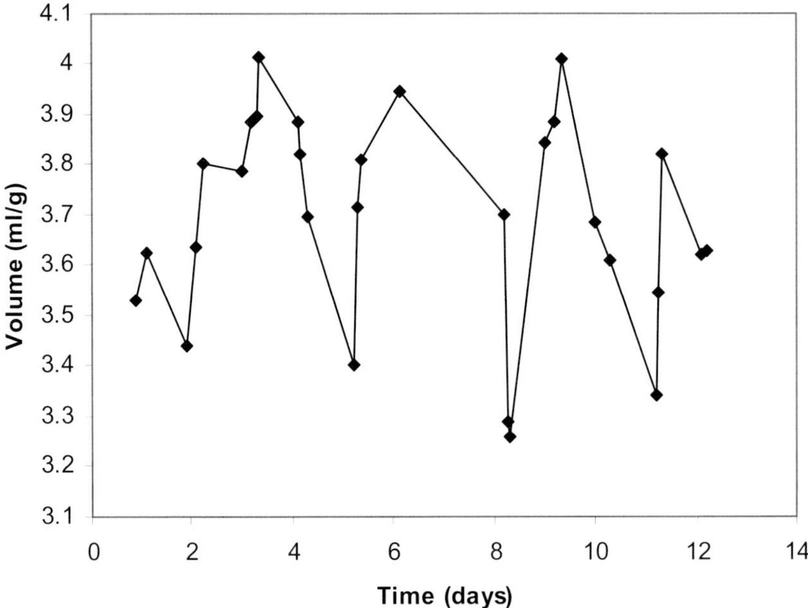

**Fig. 3.4**   The time course of the internal volume of Biogel P-6 (150–300 μm) in the presence of Ringer solution, pH 7.4. Replotted from the data of Wiggins,[29] with permission.

times appears somewhat lacking. In fact a deconvolution analysis of the ATR spectra in the active layer of a water-saturated cellulose acetate membrane with a pore diameter of 0.34 nm (CA-316–86) shows that the water–water hydrogen bonds have a frequency of 3378 cm$^{-1}$, which is higher than that for bulk water (3340 cm$^{-1}$), showing, in fact, that the water in the nanopore has weaker H-bonding than in bulk water.[28] Nevertheless, Wiggins has speculated that the phenomenon of microosmosis might have an important biological role in controlling ion fluxes across membranes, such as the opening and closing of membrane channels, the operation of proton pumps and even in the contraction of muscle fibres.[31] Clearly, further research is needed to test the central tenets of the microosmosis hypothesis.

## 3.8 Experimental probes of water–biopolymer interactions

Section 3.2 showed how the water–biopolymer interaction energy, $\chi$, played an essential role in controlling the equilibrium microscopic distribution of water in multicomponent, multiphase systems. Likewise, in Section 3.6 we saw how water transport on the microscopic distance scale depended on the phenomenological coefficients $D_{\text{eff}}$, $C$ and $\kappa$, and therefore on water–matrix interactions and, despite the 'stretched water' controversy, all the spectroscopic and molecular dynamics studies of water in nanopores agree that the dynamic state of the water over a few nanometres from the pore surface depends sensitively on the chemical nature of the polymer surface and on the amount of water in the pore. In this section we therefore consider how water–polymer interactions affect the water dynamics, beginning with experimental aspects.

Infrared spectroscopy is not only a powerful tool for monitoring nanopore water, it also gives detailed site-specific information about water–polymer interactions.[32] To get this information we must work in the low-water-content regime and progressively rehydrate a thin (2–3 μm) dry film of the (bio)polymer by equilibration with water vapour in a hydration cell while the infrared absorption spectrum is recorded at each level of hydration. Because water vapour gives readily identifiable sharp absorption peaks, the relative intensity of the vapour peaks to those of a saturated atmosphere gives a direct measure of the equilibrium vapour pressure ratio, $p/p_0$, in the cell, and therefore of the water activity in the film at each level of hydration. A set of 'hydration spectra' as a function of water activity is then obtained by subtracting both the water vapour spectrum and the spectrum for the dry film. These hydration spectra, $S(p/p_0,\omega)$, are deconvoluted into weighted combinations of three or four 'elementary hydration spectra', such that:

$$S(p/p_0,\omega) = \Sigma_i c_i(p/p_0)s_i(\omega) \tag{3.37}$$

Here $s_i(\omega)$ are the elementary hydration spectra and $c_i(p/p_0)$ are the weighting coefficients. Figure 3.5 shows the three elementary hydration spectra, labelled A to C, used to fit the hydration spectra of a BSA film. The choice of the elementary hydration spectra is, admittedly, somewhat arbitrary, but is constrained by the requirement that the coefficients $c_i$ are positive monotonically increasing functions of $(p/p_0)$. In practice they represent the spectra

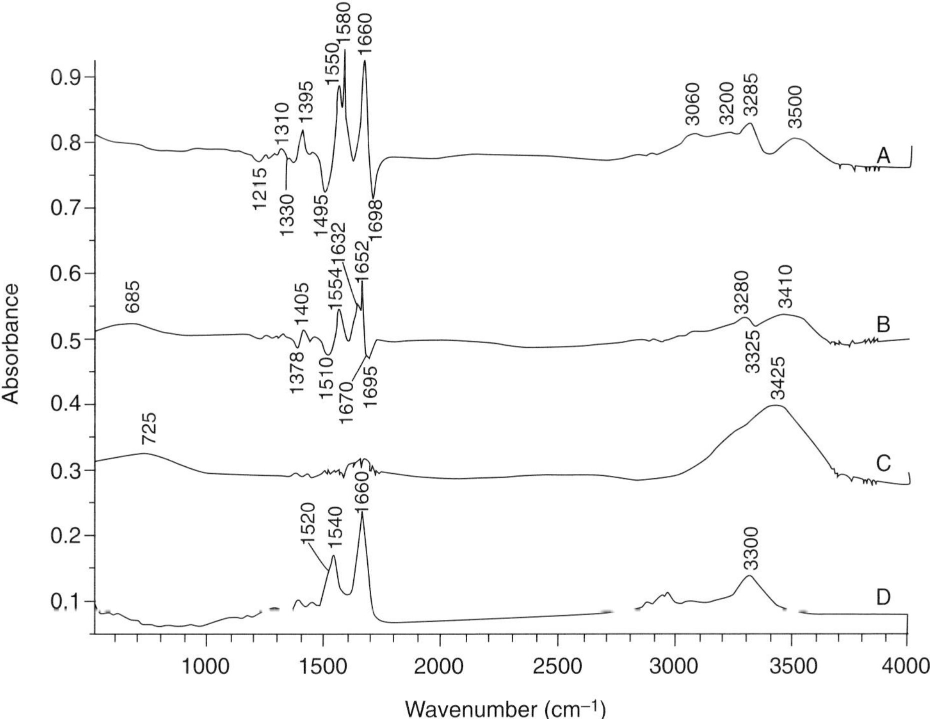

Fig. 3.5   The three elementary hydration spectra of a BSA film. Spectrum D is that of the dried BSA film. From Grdadolnik and Marechal,[32] with permission.

of various states of water as they build up by adsorption into the film. For example, the elementary spectrum, C, first appearing on the dried film is that of strongly adsorbed water hydrogen-bonding exposed hydrophilic groups on the polymer, whereas elementary spectrum A, which is the last elementary hydration spectrum to appear in the expansion at the highest values of $(p/p_0)$, differs very little from that of bulk liquid water and therefore arises from water hydrogen-bonding to the monolayer of adsorbed water on the polymer surface. Site-specific information about the hydration mechanism giving rise to each elementary hydration spectrum can be obtained by comparison of the elementary hydration spectra with the temperature dependence of the dried film spectra. This shows frequency shifts characteristic of the functional groups in the polymer. The numbers of water molecules contributing to each such site can be obtained from the intensity of the libration band at 600–700 cm$^{-1}$ in each elementary spectrum. As an example, Fig. 3.6 shows two possible hydration mechanisms that could contribute to the elementary hydration spectrum C in Fig. 3.5, and Fig. 3.7 shows plots of the coefficients, $c_i(p/p_0)$, for BSA. This very powerful approach, which gives a detailed picture of polymer hydration, has so far been applied to only a limited number of biopolymers including BSA,[32] Hyaloronan-Na$^{+}$[33] and polymethacrylate-based hydrogels.[34] When more polymers and hydration mechanisms have been studied it may be possible to begin formulating general rules for hydration mechanisms. These would, in principle, permit prediction of the effect of modifying the polymer structure on the sorption isotherm.

It is worth noting that two-dimensional vibrational spectra can be acquired using phase-controlled femtosecond infrared (IR) pulses.[35] Such spectra are entirely analogous to the multidimensional high-resolution NMR spectra acquired using multipulse radiofrequency techniques. The phase-locked femtosecond pulses are generated by splitting an incident beam

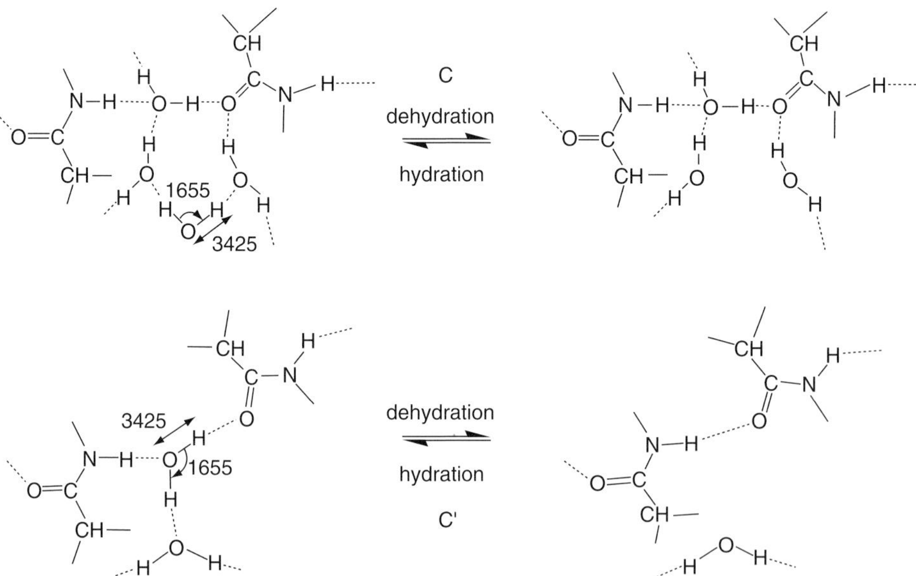

**Fig. 3.6**   The two possible hydration/dehydration mechanisms corresponding to elementary spectrum C in Fig. 3.5, which is the spectrum of the configuration on the left-hand side minus the spectrum of the configuration on the right-hand side. From Grdadolnik and Marechal,[32] with permission.

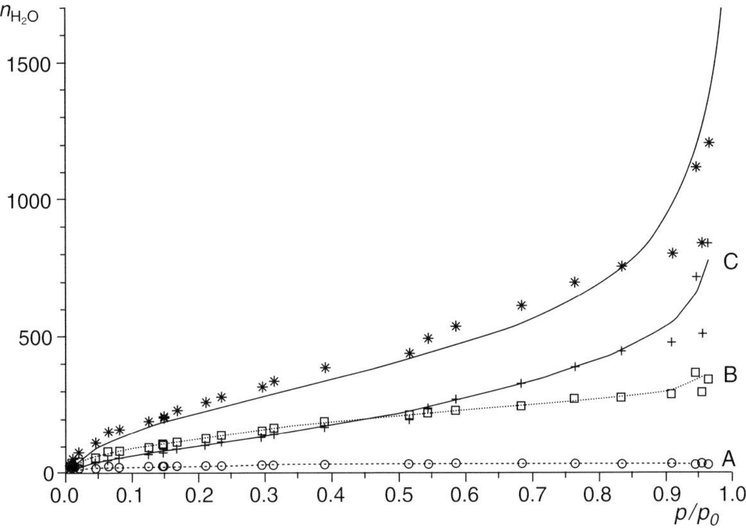



**Fig. 3.7**   The coefficients, $c_i$, expressed as the number of embedded water molecules in one BSA molecule as a function of water activity at 27°C. The letters refer to the elementary hydration spectra in Fig. 3.5. Asterisk points show the sum that compares well with the literature sorption isotherm denoted by a continuous line. From Grdadolnik and Marechal,[32] with permission.

and introducing variable time delays by a mechanical translation stage. In an analogous way to NMR spin echoes, vibrational photon echoes can be detected. By independently varying the time separation (in femtoseconds, $10^{-15}$ s) of the incident pulses a two-dimensional IR spectrum can be obtained by Fourier transformation. Cross-peaks in the spectrum appear whenever two or more vibrational modes are coupled in some way. An example for a dipeptide, acyl-proline-$NH_2$, is shown in Plate 2. As with NMR, the linewidths are determined by both longitudinal (population relaxation) and transverse (dephasing of the spatial vibrational coherences) relaxation functions. Such Fourier transformed 2D-IR spectra provide information on couplings between vibrational modes in different spatial parts of a molecule, or molecules, on the experimental timescale of less than a few picoseconds. At present the technique is still being tested on simple molecules, such as the dipeptide in Plate 2, but it is anticipated that, in the not too distant future, it will be applied to the hydration of small polymers where it has the potential of revealing hydration mechanisms in exquisite detail.

Of course, supplementary information on the hydration mechanisms deduced by vibrational spectroscopy can be obtained from X-ray diffraction, provided the biopolymer is available in crystalline form. Positions of the internal water molecules are readily detected in the X-ray structure, as well as many of the surface water molecules. In the latter case, this does not mean that the surface residence time is long, merely that, on average, water molecules in the protein crystals occupy preferential time-averaged positions and orientations even on the surface. Moreover, water that is classified as 'ordered' in the X-ray patterns from protein crystals is not necessarily 'ordered' when the protein is in the solution state. For example, X-ray studies of lysozyme[36] identified approximately 140 ordered water molecules per lysozyme but, as we shall see, MRD (magnetic relaxation dispersion) and molecular dynamics simulations both give subnanosecond lifetimes for the surface water and only three internal water molecules with lifetimes longer than 10 ns. Care must therefore be taken when using X-ray data from crystals.

Although X-ray diffraction as well as one- and, potentially, two-dimensional infrared techniques are powerful tools for identifying site-specific hydration mechanisms, they give no information about the dynamic state of the water as it interacts with the (bio)polymer. Fortunately this information can, in favourable circumstances, be obtained by working on the slower timescale of NMR. For site-specific information about water–protein interactions it is, of course, necessary to resolve as many individual protein proton peaks in the NMR spectrum as possible. This limits the NMR technique to dilute solutions of relatively small polymers, such as small globular proteins, because the short transverse relaxation times in concentrated biopolymer solutions and gels cause broadening and overlap of the spectral lines. The NMR technique therefore operates at the opposite end of the water-content regime to the transmission infrared studies described earlier. To help resolve the numerous protein proton peaks it is also necessary to work at high magnetic field strengths corresponding to proton resonance frequencies of 600 MHz or higher, and to separate further the peaks by working with two- or three-dimensional pulse sequences.

The high-resolution NMR method is based on the nuclear Overhauser effect (NOE) and is implemented using two- or three-dimensional pulse sequences, such as NOESY (NOE exchange spectroscopy) and ROESY (rotating frame exchange spectroscopy), which operate in the laboratory and rotating frames respectively.[37] Unfortunately the information on hydration water dynamics extracted from the NOESY and ROESY data is very sensitive to the details of the theoretical model used to interpret the data, so there remains some uncertainty about what the methods actually imply about water dynamics. To understand this difficulty it is necessary to examine the theory underlying the NOE, which lies at the heart of the high-resolution NMR method.

The NOE involves 'cross-relaxation', in other words, the transfer of longitudinal magnetization between water and protein protons mediated by dipole–dipole interactions. The correlation times for the water are extracted by noting that at short mixing (i.e. magnetization transfer) times the intensities of the NOESY cross-relaxation peaks in the two-dimensional spectra are proportional to the cross-relaxation rates, $\sigma^{NOE}$ respectively, which is, in turn, related to the spectral density of the water translational motion as:

$$\sigma^{NOE} = 6J(2\omega_0) - J(0) \tag{3.38}$$

Because the spectral density function, $J(\omega)$, is a positive, monotonically decreasing function of increasing frequency, $\sigma^{NOE}$ can be positive or negative, depending on the correlation time. In the limit of fast molecular motion, when $J(2\omega_0) = J(\omega_0) = J(0)$, $\sigma^{NOE}$ is positive. Conversely, in the slow-motion regime when $J(0) \gg J(\omega_0) > J(2\omega_0)$, $\sigma^{NOE}$ is negative. Here 'fast' and 'slow' refer to the timescale of $1/\omega_0$, which, at a typical proton frequency of 600 MHz, is about 300 ps. The sign of the NOESY cross-peaks therefore sets limits on the motional timescale. The ROESY data are useful because at short mixing times the ROESY peaks for magnetization transfer by dipolar interaction (NOE) are always of opposite sign to those caused by proton (chemical) exchange. NOESY peaks arising from proton exchange rather than dipolar interaction are therefore readily distinguished in the ROESY spectrum.

Pioneering work by Otting *et al.*[38] on the NOESY spectra of a small globular protein, BPTI (bovine pancreatic trypsin inhibitor), showed two qualitatively different types of hydration site. After the proton exchange peaks had been eliminated using the ROESY spectrum (these exchanging protons are indicated by red in Plate 3) a well-defined small number of water molecules in the interior of the protein had negative NOE cross-relaxation rates and long residence times between $10^{-2}$ and $10^{-8}$ seconds. By contrast, the surface hydration water had positive NOE cross-relaxation rates and apparent residence times ($\tau$) in the subnanosecond range with the exposed protons shown in green in Plate 3.

The theoretical difficulties arise in relating the NOESY data to the dynamics of the surface hydration water. Because the interaction between two dipoles separated by a distance $R$ decreases as $R^{-6}$ it was assumed in the early work that the NOE method only probes water molecules in close proximity to protein protons, in other words only water either buried inside the protein or surface water in the monolayer within about 0.5 nm of the protein surface. However, more recent work[39] has questioned this assumption. Although the pairwise dipolar interaction falls off sharply as $R^{-6}$, the number of water molecules at a given distance from an exposed surface protein proton increases as $R^2$. Moreover, the characteristic time for orientational modulation of the internuclear vector, **R**, by water translational diffusion through a given solid angle also increases as $R^2$. This means that the effective NOE surface interaction for an exposed protein proton is not short-range but actually decreases at the much slower rate of $R^{-2}$ and involves not just water molecules in the surface monolayer but numerous water molecules in multilayers extending from the protein surface and having motional characteristics differing little from those of bulk water. The zero-frequency spectral density, $J(0)$, in Equation 3.38 is then most sensitive to the distance of closest approach, $d$, of water to a protein proton, which follows because

$$J(0) \sim \int_d^\infty dR\, R^{-2} = d^{-1} \tag{3.39}$$

If this interpretation is correct then, contrary to earlier conclusions, $J(0)$ for surface water (as apart from water buried in the protein) reflects slow modulation of dipolar couplings to numerous bulk water molecules and is almost independent of the lifetime of the water in the monolayer (or first hydration sheath). In other words, the NOE data are insensitive to the dynamics of surface hydration water and report mainly on the distance of closest approach, which depends on the exposure of each protein proton to water. Clearly more theoretical and experimental work is required to clarify this situation.

The dynamics of hydration water can, in principle, be studied independently with water oxygen-17 MRD.[40] This particular water isotope is singled out because the interpretation of proton (and deuteron) MRD data in biopolymer systems is complicated by fast exchange between water and exchangeable protons (or deuterons) on the biopolymer surface.[41,42] Under proton decoupling conditions, or at pH values away from neutrality, water oxygen-17 nuclei relax by intramolecular rotational modulation of the electric quadrupole interaction. The technique therefore probes the reorientational motion of the hydration water, and the earlier discussion on nanopore water pointed out that because $\omega_0$ for oxygen-17 is typically between 1 and 100 MHz only water with correlation times longer than about a nanosecond will give a measurable frequency dispersion. The fact that such dispersions are observed in many aqueous biopolymer systems therefore points to reorientational motions slow on the nanosecond timescale. However, the origin of these motions has also been a source of widely differing interpretation over the years. Some of the earliest models involved specific long-lived (c. $10^{-6}$ s) hydration sites on the protein surface[43] or long-ranged hydrodynamic coupling between protein and water rotation.[44] Subsequently, theories were presented for a relatively long-lived ($>10^{-8}$ s) locally anisotropic surface hydration layer with water characterized by a fast local, near-isotropic reorientational correlation time of c. 10 ps and a slower correlation time of the order of several nanoseconds required for averaging of residual quadrupolar interactions.[45] The most recent model assigns the observed dispersions not to surface hydration water but to the few water molecules 'buried' inside the clefts and cavities of the globular protein structure. In this model the surface hydration water has sub-nanosecond residence times, so no relaxation dispersion arises from surface water and these water molecules contribute only to the frequency-independent $\alpha$ term in Equation 3.36. For globular protein solutions $\alpha$ can be related to the number of these surface water molecules, $N_\alpha$, and their average correlation time, $<\tau_\alpha>$, as

$$\alpha = R_{bulk} \cdot (N_\alpha/N_T)[(<\tau_\alpha>/\tau_{bulk}) - 1] \tag{3.40}$$

where $N_T$ is the known water:biopolymer mole ratio and the bulk water correlation time, $\tau_{bulk}$, is about 17 ps. In this model the observed frequency dispersion (the $\beta$ term in Equation 3.36) arises from slow modulation of the internal hydration water on a timescale longer than a nanosecond. If this interpretation is correct, the correlation time, $\tau$, in the $\beta$ term in Equation 3.36 can be written:

$$1/\tau = 1/\tau_w + 1/\tau_{bio} \tag{3.41}$$

where $\tau_w$ is the residence time of the internal water and $\tau_{bio}$ is the rotational correlation time of the biopolymer and we have assumed these motions are uncorrelated.

An application of the revised 'internal water' model to the MRD data for BPTI concluded that more than 95% of the water molecules at the protein surface had very short correlation times, with mean rotational correlation times (c. 20 ps) only about twice that of bulk water (c. 10 ps).[37,46] Because rotational and translational motions of water molecules are both governed by hydrogen bond dynamics, this implies that the translational motion and therefore the mean residence time at the surface is also retarded by no more than a factor of two compared with bulk water. The $\beta$-dispersion gave long residence times of $170 \pm 20$ µs for a singly buried water molecule (known as W122) at 27°C and provides bounds for the residence times for at least two other buried water molecules of 10 ns $\ll \tau < 1$ µs.

If the frequency dispersion arises from a few long-lived water molecules inside the globular protein structure, then unfolding the protein by thermal denaturation should remove these internal waters and result in a loss of the frequency dispersion. This is indeed observed[40] for ribonuclease A, and this lends strong support for the model. Nevertheless, this 'internal water' model may still be an oversimplification because it takes no account of the long time tail in the orientational correlation function of the surface hydration water caused by the need to average the residual electric quadrupole interaction. As mentioned in the context of nanopore water, this averaging requires repeated desorption and readsorption between the surface and bulk phases in a slow stochastic process described by Levy walks and the

RMTD mechanism mentioned previously. Future work should therefore aim to determine the relative importance of the dispersive contributions arising from 'internal' water and that from residual quadrupolar interactions in the surface water.

The need to vary the field strength in the MRD technique can be avoided using triple-quantum filtered water oxygen-17 longitudinal and transverse relaxation.[47] The oxygen-17 nucleus is spin 5/2 so that its transverse and longitudinal relaxation is, in general, multiple exponential except in the limit of fast water motion when it averages into a single exponential. For the particular case of water in globular protein solutions it is assumed that the relaxation can be described by the fast exchange of a fraction, $p_s$, of 'internal' water molecules characterized by an average correlation time, $\tau_s$, with a fraction $(1 - p_s)$ of water with a relaxation rate, $R_{bulk}$, that includes all the bulk water phase as well as the surface hydration water. The global transverse relaxation matrix of water, assuming fast exchange, is then given as:

$$\mathbf{R}_2 = p_s\mathbf{R}_{2s} + (1 - p_s)\mathbf{U}R_{bulk} \tag{3.42}$$

where $\mathbf{U}$ is the unit matrix and there is a similar expression for the longitudinal relaxation with the subscript 2 replaced by 1. A detailed analysis of the relaxation matrices[48] shows that by measuring both the longitudinal and transverse relaxation with triple-quantum filtered pulse sequences it is possible to determine uniquely $p_s$ and $\tau_s$ and thereby characterize the 'internal' water. In this way it was unambiguously shown that there are four slow internal water molecules in lysozyme characterized by an average correlation time of 4 ns in a 15% lysozyme solution, increasing to 6 ns in a 40% solution at 28°C. These results agree well with those of MRD. A similar study on the larger BSA molecule gave correlation times of 5.7 ns for 40 water molecules in a 5% solution, changing to 9.0 ns for 21 water molecules in a more concentrated 15% solution.[49]

The dynamics of surface hydration water has been probed on the femtosecond ($10^{-15}$ s) timescale with an optical technique based on the time-resolved fluorescence spectrum of tryptophan side chains on a protein.[50] The ground state of tryptophan has only a small dipole moment, but UV excitation into the fluorescing electronic excited $^1L_a$ state creates a large static dipole at an energy level that is very sensitive to the extent of hydrogen bonding with solvent water molecules. Using femtosecond time-resolved UV excitation a dipole is suddenly created on the tryptophan and the time courses of the frequency shifts in the time-resolved fluorescence spectrum caused as the hydrating water molecules reorient and relax around the newly created dipole are recorded. In a simple solution of tryptophan in bulk water the correlation function for the frequency shifts can be fitted with two exponential decays with time constants of 180 fs (20%) and 1.1 ps (80%). These decays are believed to correspond with inertial rotational motion of the water molecules about a solute-water axis (180 fs) and more collective motions of the water causing rotational diffusion (1.1 ps). When the experiment was repeated with the single tryptophan side chain located on the surface of the protease Subtilisin *Carlsberg* these biexponential decay times lengthened to 800 fs (61%) and 38 ps (39%) respectively. The 800-fs decay was assigned to inertial (mainly librational) motions of water molecules in the surface hydration layer, and the 38-ps correlation time was assigned to the time required for collective 'diffusive' motions of the surface water molecules to relax. These collective motions reflect both the rotational and translational motions needed to make and break hydrogen bonds, so it is not surprising that it also corresponds, in order of magnitude, to the residence times determined by MRD and NOESY experiments and by molecular dynamics simulations, to which we now turn.

## 3.9 Molecular dynamics simulations of water–biopolymer interactions

The uncertainties in the interpretation of the spectroscopic data, especially of the state of surface hydration water, have motivated a number of molecular dynamics simulations. A molecular dynamics simulation of the hydration dynamics of BPTI was undertaken over a 1.4-ns time period.[51] Five interior water molecules were found not to have changed their position during the simulation run-time, which is consistent with the long correlation times (>10 ns) determined by the NOE and MRD methods. In agreement with the MRD data, the average residence time of surface water molecules at charged protein atoms was 19 ps, which is longer by a factor of two than the residence time of one water molecule in the hydration shell of another in bulk water (c. 10 ps) but shorter than the average residence time of water at the backbone carbonyl groups (43 ps). This last result is opposite to the intuitive feeling that hydrogen bonding to a charged atom should lengthen residence times. Clearly, in such cases entropic factors need to be taken into account. The maximum residence time of any surface water molecule was several hundred picoseconds and never more than 1 ns. These molecular dynamics results agree well with the NOE results and support the assignment of MRD dispersions to the interior water molecules.

Similar conclusions were found in a more recent molecular dynamics simulation of the hydration dynamics of lysozyme over a 9-ns time period.[52] Three water molecules inside the protein did not move during the 9-ns simulation run whereas water in hydrophilic pores and superficial clefts had a mean residence time of c. 0.7 ns. In contrast the surface water had a subnanosecond mean residence time.

An interesting molecular dynamics simulation of a model peptide (*N*-acetyl-leucine-methylamide), which has a hydrophilic backbone and a hydrophobic side chain, showed spatially heterogeneous water dynamics, with water residence times near the hydrophilic backbone close to those of pure water (c. 10–11 ps) but correlation times at the hydrophobic backbone about three times faster (3.5–4 ps).[53]

Despite their relatively small numbers, the interior water molecules have a vital role to play in stabilizing protein tertiary conformations, and in protein folding. They also serve to fill and stabilize cavities and modulate enzyme–ligand specificity. It is not therefore surprising that considerable effort has been devoted to trying to predict the hydration sites of these internal water molecules from knowledge of the protein sequence. These predictive models include energy-based calculations[54] and rule-based approaches such as those using the directionality of the hydrogen bond. Most recently artificial neural networks have been used.[55] These were trained with 40 protein sequences using crystal structures in the protein databank[56] and tested on 77 proteins, all of which had more than 32 crystallographic water sites. A predictive success rate of 77% was thereby achieved.

The functional role of surface water in controlling protein conformations is less clear, though it obviously affects the dependence of biopolymer systems on state variables such as concentration, temperature and pressure, to which we now turn.

# 3.10 The dependence of water dynamics on state variables

## 3.10.1 Low-water-content systems

As a food is dried the free bulk water is removed first; then surface multilayer water and finally, at water contents below 15–20% w/w, the few remaining internal water molecules that are integral to biopolymer structure are removed. This progressive loss of water affects not only the correlation times and residence lifetimes of these various water fractions but also the dynamic state of the biopolymers and solutes comprising the food matrix. Removal of surface hydration water results in slowed rates of enzymic reactions, including spoilage reactions, which is one reason why many foods are dried to extend their shelf life. Nevertheless our understanding of the effect of reduced water content on water and biopolymer dynamics is still at a primitive level. There are few systematic measurements of water correlation times at low water contents partly because most of the experimental techniques discussed in Section 3.8 do not work in this regime. High-resolution methods such as NOESY and ROESY fail due to line broadening, and MRD or triple quantum water oxygen-17 methods become problematic because the oxygen-17 relaxation times become prohibitively short once the bulk water phase has been removed. Femtosecond fluorescence spectroscopy is, in principle, possible but there do not appear to be any reports of systematic low-water studies with this technique.

Fortunately, some progress can be made with simple NMR water proton transverse relaxation measurements because, in low-water-content systems proton chemical exchange between water and biopolymer is usually so slow that it can be neglected and the complications of dipolar cross-relaxation do not arise in the transverse relaxation mode. Moreover, the transverse relaxation time of the biopolymer protons is usually much less than that of the more mobile water molecules so it is straightforward to separate the water and polymer transverse relaxation. When bulk and surface water exist in the system the transverse relaxation time is sufficiently long to allow the transverse relaxation to be probed with the conventional CPMG sequence with a short echo spacing. But when only surface and internal water remain the FID must be used to measure the transverse relaxation. This simple approach has been applied to gelatine gels for water contents down to about 5% w/w and some of the data are reproduced in Fig. 3.8.[57] Figure 3.8 shows three distinct regimes corresponding to the removal of bulk water, multilayer surface water and progressively longer-lived internal water. Average water correlation times of the internal and surface water could be extracted with a simple single-correlation time model of the water spectral density. This showed interior water residence lifetimes lengthening to 0.1 µs at the very lowest water contents as progressively more strongly hydrogen-bonded water is removed.[57] Such long water correlation times bring the water dynamics into the time-window of proton field cycling relaxometry (MRD), which, in principle, permits direct measurement of the water spectral density. This has been done for a 90.1% w/w sucrose-water solution in the glassy state at 230.9 K,[58] but not, it appears, for biopolymer systems. In the case of the sugar glass, cross-relaxation of longitudinal magnetization between the water and sucrose protons is so fast that the observed longitudinal relaxation is single exponential and can be written as a weight-averaged double dispersion:

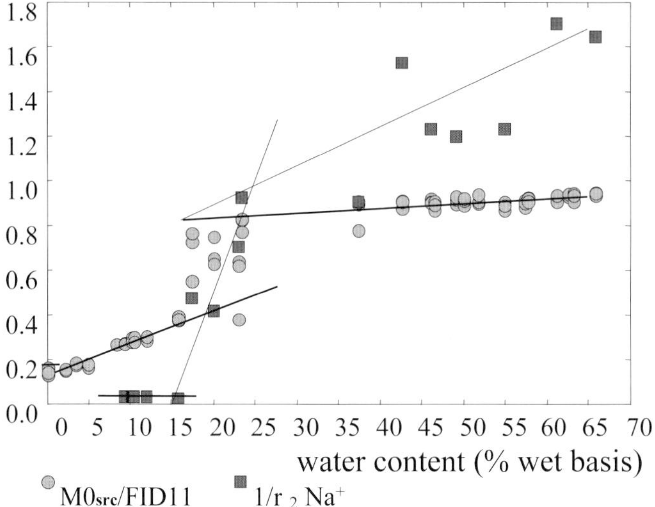



**Fig. 3.8**  The ratio of the amplitudes of the slow relaxing component (src) corresponding to water ($M0_{src}$) to the solid protein proton component after a short 11-microsecond time delay (FID11) in the FID of gelatine gels plotted against water content. The squares show the transverse relaxation times of sodium ions in the same gels. Note the three regimes corresponding to removal of bulk, surface and inner water states. From Vackier *et al.*,[57] with permission.

$$R(\omega) = P_w R_w(\omega) + (1 - P_w)R_s(\omega) \tag{3.43}$$

Here $P_w$ is the proton fraction of the water, $R_w(\omega)$ is the water longitudinal relaxation rate and $R_s(\omega)$ is the sugar longitudinal relaxation rate, both of which can be modelled with single correlation time spectral densities resembling Equation 3.34. Figure 3.9 shows the double dispersion together with the fit of Equation 3.43. The result shows that, even in the glassy state at 230.9 K, the water retains a high degree of mobility with a correlation time of about 138 ns. Whether this approach can be applied to concentrated biopolymer systems in the glassy or 'jammed' state remains to be investigated.

In principle, molecular dynamics simulations can be used to model water dynamics in concentrated biopolymer systems, though only for evolution times of a few nanoseconds, but it is clearly impractical to run such simulations for all the plethora of diverse food matrices of interest. What is needed is a unifying theoretical approach to water relations in each microphase of a complex food matrix. Returning, once again, to the example of starch granules embedded in egg albumin, we saw in Sections 3.5 and 3.6 how sorption isotherms describe the equilibrium water distribution and how effective diffusion coefficients describe the nonequilibrium transport of water between the starch and protein microphases. However, to be useful as predictive tools, these approaches still require knowledge of the concentration dependence of the water activity and intrinsic diffusion coefficients of water inside the starch granule and albumin protein microphases. One approach to this problem, called 'multistate' theory, explicitly acknowledges the existence of the various hydration states of water discussed in Sections 3.8 and 3.9, namely, that there is fast exchange of water molecules between the free 'bulk' state, the state classified as multilayer surface water whose

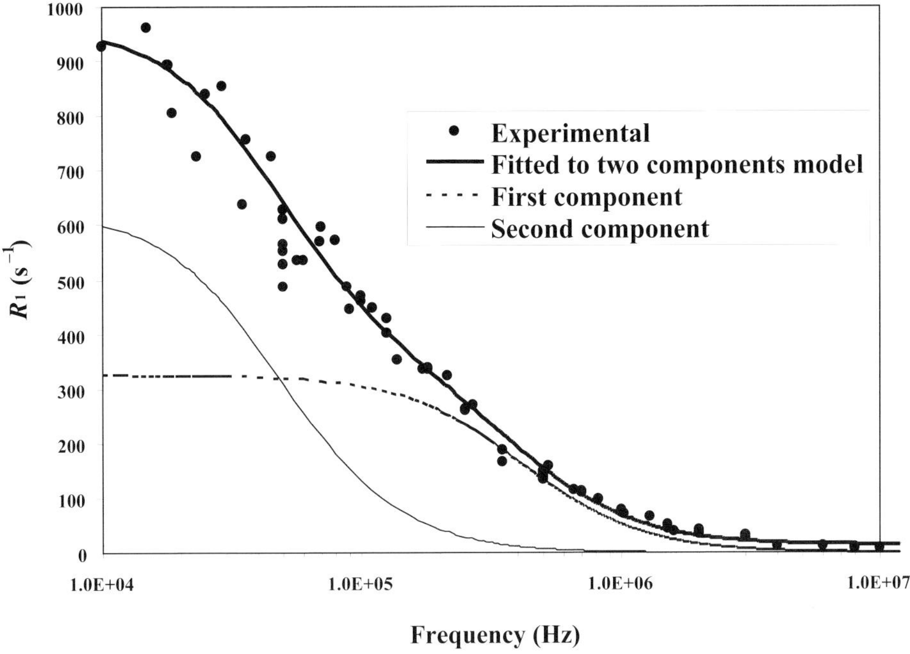

**Fig. 3.9**  The double dispersion for a 90.1% w/w sucrose glass in $H_2O$ at $-42.1°C$. The first component dispersion arises from water, the second from the sucrose lattice. From Hills *et al.*,[58] with permission.

dynamics is perturbed by interaction with a biopolymer surface, and 'structural' or 'interior' water inside the biopolymer molecule. Multistate theory regards each of these states of water as different chemical species and assigns each one a characteristic value of the quantity of interest, such as a water activity, $a_i$, diffusion coefficient, $D_i$ or an NMR relaxation rate, $R_i$ (i = bulk, surface, structural).[1] The Ergodic theorem of statistical mechanics is then invoked to write the time-average property of a particular water molecule exchanging between the various states, i, as the ensemble average over all the water states. Taking the example of the NMR relaxation rate, we can therefore write,

$$R_{av} = \Sigma_{i=1,n} \, x_i R_i \tag{3.44}$$

Here $x_i$ is the mole fraction of water in state i, and $R_i$ is the value of the relaxation rate in state i. Equation 3.44 could, for example, describe the dependence of the NMR water proton relaxation rate, $R_{av}$ on water content in the egg albumin or starch granule microphase. Analogously, we can specify the concentration dependence of the water translational self-diffusion coefficient in each microphase as,

$$D_{av} = \Sigma_{i=1,n} \, x_i D_i \tag{3.45}$$

It could even be used for the calculation of sorption isotherms of each microphase by assigning water activity coefficients to each state of water, in which case,

$$a_{av} = \Sigma_{i=1,n} \, x_i a_i \tag{3.46}$$

Clearly, the major assumption of this approach is that these properties are simply additive over the various states of water in a biopolymer system. However, in most cases this is an untested assumption and great care is needed when implementing this approach. For example, diffusion coefficients may not be additive if the diffusion propagator for water in a particular state is not Gaussian. Likewise, Equation 3.46 assumes that the intrinsic water activities, $a_i$, are constants and independent of the amount of water in each state. Because of changing entropy factors this may not be true. Equation 3.44 also ignores the complications of lifetime broadening and frequency offsets. All these equations are therefore best regarded as semiempirical and approximate relationships whose main value lies in taking explicit account of the changing fractions of multistate water. Despite its weak theoretical basis, multistate theory goes a long way in explaining why plots of water relaxation rates on water content, such as that in Fig. 3.8, resemble sorption isotherms. Multistate theory can also be generalized to multicomponent polymer systems by incorporating empirical 'preference coefficients' $\xi(j)$ for each biopolymer, j, which specify which polymer–water interaction is preferred. This parameter is obviously related in some way to the water–polymer interaction energy, $\chi$, introduced in Section 3.2. The generalized relationships have the form:

$$R_{av} = \Sigma_{ij} \, \xi(j) x_i(j) R_i(j) \tag{3.47}$$

where

$$\Sigma_{ij} \, \xi(j) x_i(j) = 1 \tag{3.48}$$

It is interesting to note that multistate theory predicts hysteresis whenever the desorption or adsorption of water alters matrix structure on the molecular or microscopic distance scales so that the $x_i$ and/or $\xi(j)$ coefficients are different in the two matrices although the total water content is the same. Figure 3.10 shows an example of relaxation and water activity hysteresis for pregelatinized potato starch together with a fit of Equation 3.46 for bulk and surface water. Clearly multistate theory is only a first step towards developing a more rigorous theory of water relations in low-water-content, multicomponent systems and much more theoretical work is required.

The discussion of low-water-content systems has so far focused exclusively on water-soluble biopolymers such as gelatine and BSA. However, many water-insoluble food biopolymers also have properties that are critically affected by small amounts of hydration water. The gluten component in wheat flour, whose viscoelastic properties are so crucial to dough quality, is an interesting example of this class of biopolymer. Gluten is a water-insoluble protein that is nevertheless capable of absorbing large amounts (up to c. 65% w/w) of hydration water. The main effect of this water appears to be in mobilizing (or 'plasticizing') the protein chains, and this enhanced chain mobility permits chain rearrangement that alters the viscoelastic properties and protein functionality. A systematic NMR and FTIR study of a high-molecular-weight subunit of glutenin (a key component of gluten) is instructive.[59] The dependence of the protein chain dynamics on water content was measured indirectly from the changes in the transverse and longitudinal relaxation rates of the nonexchanging

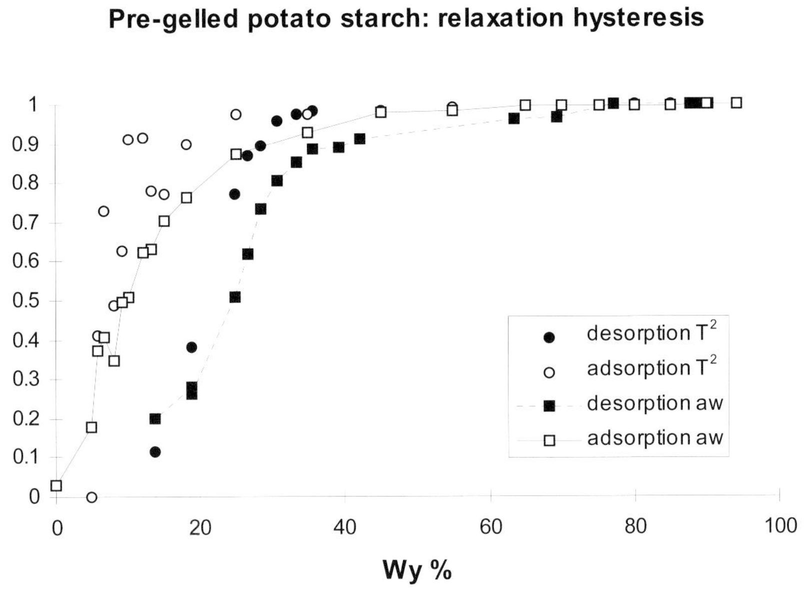

**Pre-gelled potato starch: relaxation hysteresis**



**Fig. 3.10**   The fit of multistate theory to the sorption isotherms of pregelled potato starch during adsorption and desorption hysteresis. Water activity (aw) versus water content by weight (Wy).

protein protons after replacing all exchangeable protons with deuterons and by hydrating with $D_2O$ instead of $H_2O$. These relaxation data were complemented with spectral studies using FTIR and high-resolution measurements of the $^{13}C$ spectral linewidths, which decrease with increasing chain mobility. It was concluded that small amounts of water permit replacement of protein–protein hydrogen bonds with protein–water interactions resulting in chain plasticization and the formation of new intermolecular β-sheet structures in dynamic equilibrium with disordered regions of hydrated chains. It is likely that such hydration-induced conformational changes at the molecular level play an important role in determining dough, and therefore bread-making, quality but they greatly complicate the analysis with multistate theory by changing the number and type of water hydration sites.

### 3.10.2 Nonfreezing water

Despite its shortcomings, multistate theory provides a natural theoretical basis for the phenomenon of 'nonfreezing water'.[60] When a globular protein solution is cooled, ice crystals first form and result in an increase in concentration of the unfrozen protein solution. This continues as the temperature is lowered but there always remains a small fraction of mobile water that is readily detected as a slowly relaxing component in the NMR water proton free induction decay. The reason for this residual 'nonfreezing water' has been a source of controversy over the years. Some have speculated that the nonfreezing water is unable to freeze because it is so viscous and its diffusivity so low that it is in a state of kinetic arrest. However, the molecular dynamics simulations on nanopore water, even at very low water contents, show that the surface water is still highly mobile so this is unlikely. However, kinetic arrest might be the case if ice crystal formation is controlled not by the water dynamics but by the need to rearrange the biopolymers, which could be prohibitively slow. However, it is far more likely that the system is under thermodynamic control and multistate theory provides a ready explanation of nonfreezing water by identifying it with the exchanging populations of surface and internal water. Equation 3.46 shows that the average water activity of surface and internal water is much less than unity, so the phenomenon of nonfreezing water arises whenever this average water activity equals that of ice at a particular subzero temperature.

If this explanation is correct it provides a ready means of measuring and predicting the amount of surface hydration water. Many years ago Kuntz and Kauzmann[61] showed that the amount of nonfreezing water for many water-soluble globular proteins could be estimated by assigning a hydration number to each amino acid in a protein. This ranged from 1 for nonpolar acids to 4 to 7 for charged amino acids. The agreement was even better when account was taken of residues buried inside the globular protein. It would be interesting to try to estimate the mole fraction of surface water in multistate theory using these hydration numbers and thereby predict the dependence of sorption isotherms on biopolymer conformation and chemical modification.

### *3.10.3 Diffusion studies of surface water*

It is difficult to find suitable experimental probes of the dynamics of nonfreezing water (at subzero temperatures) or of surface water (in concentrated, unfrozen, low-water-content systems). In principle it might be possible to use femtosecond spectroscopy; or possibly water oxygen-17 measurements, either in the MRD or triple-quantum filtered relaxation protocols. However, as already mentioned, the short relaxation times of the water oxygen-17 nucleus in frozen and/or dried systems might be prohibitive. As we have seen, proton relaxometry in the transverse mode is useful, but dynamic information can only be extracted by assuming some simple model for the water spectral density. The most direct handle on the water dynamics in concentrated or frozen systems is undoubtedly provided by NMR diffusometry. However, the small values of the water self-diffusion coefficients in these systems means that strong field gradients are needed to observe significant signal attenuation through water diffusion. Care must also be taken when using the stimulated echo pulse sequence to take account of longitudinal cross-relaxation between water and biopolymer (or water and ice) during the diffusion period, and theoretical models have been developed of this effect, at least for the pulsed gradient stimulated echo experiment.[62,63] At the lowest diffusivities it may be necessary to use the very strong, fixed stray (or fringe) field gradient of superconducting magnets, though, of course, they cannot be 'pulsed'. It is then necessary to take account, not only of relaxation and cross-relaxation, but also of slice-selection.[64]

The results of these direct measurements on the diffusivity of nonfreezing water are quite surprising. Using the fringe-field method it was found that the nonfreezing water in frozen solutions of a globular protein (BSA) undergoes unrestricted diffusion in the sense that the root mean square displacement increases linearly with time, such that $<r^2> = 6Dt$ even when diffusion times are long enough for $<r^2>^{1/2}$ to exceed the protein diameter by an order of magnitude. Moreover, the nonfreezing water is still highly mobile, having a self-diffusion coefficient, $D$, that is only about a factor of 2 or 3 less than that of bulk water. It is even more surprising that these results are independent of the initial protein concentration or rate of freezing.[64] This suggests that there is a phase separation during the freezing processes such that a concentrated protein microphase is formed between the ice crystals. Within this unfrozen protein microphase the globular protein molecules must be sufficiently close to each other to allow the surface water to form a connected percolation cluster extending over at least 10 protein diameters.

More recent work has used the pulsed gradient stimulated-echo diffusion protocol to study the mobility of nonfreezing water in packed beds of water-saturated potato starch granules.[65] In contrast to the globular protein solution, the starch is unable to translate and form a separate microphase so we expect, and see, a much smaller diffusion coefficient and stronger temperature dependence. The nonfreezing water is found to undergo unrestricted diffusion with a long-time diffusion coefficient, $D_\infty$, that is only between 1 and 5% of the diffusion coefficient of supercooled bulk water at the same temperature, $D_0$. The nonfreezing water is believed to comprise the unfrozen surface water layer around the amylose and amylopectin polymer chains and is therefore only about a nanometre thick. The unrestricted nature of the diffusion suggests that these channels form a connected network over microscopic distances at least equal to the experimental diffusion distance. The ratio $D_\infty/D_0$ is therefore a measure of the tortuosity of this connected network, and this decreases linearly from about 0.05 to 0.01 with decreasing temperature from –5 to –25°C as the amount of nonfreezing water decreases.[65] If this interpretation is correct, then such measurements might provide a useful probe of the connectivity and tortuosity of the biopolymer matrix in food systems.

Fringe-field diffusion measurements have also been reported[64] on concentrated BSA solutions and gelatine gels as the water content is progressively reduced to 5% w/w at room temperature. The bulk water phase dominates the diffusivity at high water contents above about 50% w/w and the biopolymer matrix serves merely as a geometric barrier to bulk water diffusion. As we saw in Section 3.5, the water self-diffusion propagator is a useful probe of the matrix microstructure in this high-water-content regime. This situation holds until drying has removed most of the bulk water. The diffusion is then mainly determined by fast exchange between residual bulk water and surface (and, to a lesser extent, interior) hydration water. In this regime, multistate theory predicts that, regardless of the effect of geometric restrictions on the diffusion propagator, the water self-diffusion coefficient itself decreases according to Equation 3.45 as the fraction of bulk water decreases. This regime is expected to hold until the connectivity of the surface hydration layer over the experimental diffusion distance is broken. Percolation theory then predicts a dramatic reduction in water diffusivity as the number of connected surface water clusters decreases. In the case of BSA and gelatine, this regime is entered at water contents below about 15% and the diffusion coefficients become too low to be measured even with the fringe field method.[64]

### 3.10.4 Water dynamics under high pressure

It is well known that high pressures can induce conformational changes in biopolymers leading to their aggregation and gelation and that the functional properties of the high-pressure created gel networks differ from those created by heat denaturation.[66] This raises the interesting possibility of creating novel food functionality by appropriate combinations of temperature and high-pressure processing. For this reason the vast majority of research publications on biopolymer systems have focused on the effect of pressure on the structure of the biopolymer component. Comparatively little effort has been spent on the effects of high pressure on water transport and dynamics even in simple protein solutions, let alone in heterogeneous multiphase, multicomponent foods.

The effect of high pressure can be summarized by the thermodynamic relation

$$(\delta\Delta G/\delta P)_{\mathrm{T}} = \Delta V \tag{3.49}$$

which shows that increasing pressure favours a shift in the equilibrium to smaller volume. Because ice has a lower density than water, increased pressure lowers the melting point of ice so it should be possible to study the dynamics of water in complex foods at subzero temperatures under high pressure without ice formation. When microphases exist, as in the starch-egg albumin system, Equation 3.49 predicts that pressure will induce transport of water between microphases whenever the compressibility of one microphase differs from another. For example, pressures of 400 MPa cause damaged wheat starch granules to swell by about 40% by volume, absorbing water from the surrounding microphase.[67] It would therefore be of interest to use techniques such as NMR relaxometry and diffusometry to study pressure-induced water migration between microphases. Work along these lines has just begun with a recent comparison of water distributions in pressure- and heat-generated gels of potato and maize starches using $T_1$–$T_2$ NMR cross-correlation relaxometry.[13]

High pressure can also change the water dynamics within each microphase. The hydration of ions such as $H_3O^+$ and $OH^-$ results in a volume decrease, so high pressure increases the dissociation constant of pure water and results in a pH reduction of –0.3 per 100 MPa pressure. For the same reason ionic groups on biopolymers undergo greater dissociation under high pressure, which suggests that high pressure should increase the proportion of surface hydration water relative to bulk water. However, the situation may not be so straightforward because there are also indications that surface hydration water has a lower compressibility than bulk water, which favours release of surface water as bulk water. This could be accomplished by reducing the surface area through biopolymer aggregation. There is also some evidence that hydrophobic water–polymer interactions lead to increased volume and are therefore destabilized by high pressure. For all these reasons the fraction, $x_i$, of water in each hydration state (see Section 3.10.1) as well as its dynamic state are expected to be pressure-dependent, yet there appear to be no experimental studies of this important phenomenon. One possible reason for this paucity of data is the practical difficulty in undertaking spectroscopic studies such as NMR under high pressure. Jonas[68] has undertaken a systematic study of protein conformational changes under high pressures up to 500 MPa with a purpose-built high-resolution probe operating at 500 MHz, but the biopolymer itself was the focus of this effort rather than the hydration dynamics.

Fortunately, experimental limitations do not apply to molecular dynamics simulations and several have been reported. An 800-ps molecular dynamics simulation investigated the conformational freedom of the protein BPTI in solution under relatively low pressures up to 20 MPa.[69] This demonstrated that the normal structure of water changed from an ice Ih-like to an ice VI-like structure under high pressure, while remaining liquid. This resulted in a higher compressibility of water compared with the BPTI proteins, which led to penetration of water into the hydrophobic core of the protein.

Clearly the pressure dimension has only just begun to be explored and almost all the experimental and theoretical work discussed in previous sections needs to be considered as a function of pressure, as well as temperature.

## 3.11 Conclusion

The flickering three-dimensional network of hydrogen bonds in pure liquid water is an amazingly subtle and unique dynamical structure that has been the subject of experimental and theoretical research for many years and has given rise to numerous models.[70] In this chapter we have seen how this same subtlety and complexity extends into the arena of food science where water plays such an essential role in controlling processing response, storage stability and organoleptic properties.

Given the subtlety of the hydrogen bond network and the complexity of multiphase multicomponent foods it is not surprising that many aspects of the dynamic behaviour of water in food remain a mystery. On the molecular scale, more research is needed to clarify the way in which biopolymer–water interactions alter the dynamic state of surface hydration water. The functional role of surface water also needs to be clarified. On the nanopore scale controversy exists over the existence of stretched nanopore water and the phenomenon of 'microosmosis' as well as its possible role in biological processes. Even on the macroscopic level our knowledge of the factors controlling moisture migration is still largely empirical, which partly explains why the food industry is still searching for effective edible thin-film barriers to prevent moisture transport in multilayer foods. Such practical problems are an added incentive for gaining a deeper understanding of water dynamics and transport on the molecular and microscopic levels.

Although this chapter has focused on water in food, many of the same issues arise in non-food materials. A detailed knowledge of the factors controlling water transport in synthetic hydrogels is essential for the rational development of new contact lens materials and medical implants, where desired mechanical properties have to be combined with specified moisture, gas and salt permeability. At present the research is an empirical trial-and-error exercise involving the labour-intensive synthesis of numerous hydrogels of differing molecular composition and nanopore structure, many of which fail the performance criteria. In like manner, water transport through porous rock plays a vital role in the oil industry where brine is used to extract oil from underground reservoirs. The relationship between water transport and the pore structure and connectivity in rock has therefore been an active research topic for many years. It is a daunting thought that water transport in food is even more complex than in porous rock because, unlike rock, the food biopolymer matrix is not rigid but has a structure and dynamic state that depends on temperature, composition, water content and processing history.

It is to be hoped that, in the not too distant future, the empirical approach to water relations in food will be replaced by rational prediction as computing power increases and our experimental probes become more sophisticated. As we have seen, the problem of compositional and microstructural complexity can, in part, be overcome experimentally by working with higher dimensional spectroscopic techniques; we have seen examples of this in multidimensional relaxation-diffusion NMR correlation spectroscopy, multidimensional high-resolution NMR and in multidimensional femtosecond vibrational spectroscopy, and much remains to be done in refining and exploiting the higher dimensional aspects of these experimental tools. Nor should it be forgotten that techniques such as NMR and vibrational spectroscopy can resolve in spatial dimensions as well as in the time (or frequency) dimensions. NMR k-space microimaging, NMR q-space microscopy as well as infrared microscopy can all give spatial resolutions down to the microscopic distance scale though, to date, the research with these techniques has focused more on determining the matrix structure and composition than on water transport *per se*.

The ever-increasing power of computer processors provides further opportunity for more sophisticated simulations of water dynamics in complex matrices over a range of distance scales. Numerical methods based on finite elements (FEMLAB), random walks (Monte Carlo), cellular automata and neural networks are providing increasingly detailed insights into water–matrix interactions and we have seen several examples in previous sections. It is to be anticipated that many more simulations using reconstructed media and digitized three-dimensional images of real food biopolymer matrices will appear in the near future.

Modelling water dynamics in low-water-content systems and under high pressure remains an outstanding future challenge so the conclusion must be that, despite the deceptive simplicity of the water molecule, its transport and dynamic state in heterogeneous foods will remain a challenging research topic for many years to come.

# 3.12 References


1 Hills, B.P. (1998) *Magnetic Resonance Imaging in Food Science.* John Wiley and Sons, New York.

2 For example, Coulson, J.M. & Richardson, J.F. (1993) *Chemical Engineering, Vol. 1, Fluid Flow, Heat Transfer and Mass Transfer.* Pergamon Press, Oxford.

3 Crank, J. (1993) *The Mathematics of Diffusion.* Oxford Science Publications, Clarendon Press, Oxford.

4 Chatakanonda, P., Chinachoti, P., Sriroth, K. *et al.* (2003) The influence of time and conditions of harvest on the functional behaviour of cassava starch – a proton NMR relaxation study. *Carbohydr. Polymer.* **53**, 233–240.

5 Wright, K.M., Hills, B.P., Hollowood, T.A. *et al.* (2003) Persistence effects in flavour release from liquids in the mouth. *Int. J. Food Sci. Tech.* **38**, 343–350.

6 Wright, K.M., Sprunt, J., Smith, A.C. & Hills, B.P. (2003) Modelling flavour release from a chewed bolus in the mouth. Part I. Mastication. *Int. J. Food Sci. Tech.* **38**, 351–360.

7 Wright, K.M. & Hills, B.P. (2003) Modelling flavour release from a chewed bolus in the mouth. Part II. The release kinetics. *Int. J. Food Sci. Tech.* **38**, 361–368.

8 Flory, P.J. (1953) *Principles of Polymer Chemistry.* Cornell University Press, Ithaca, NY.

9 Tang, H. & Hills, B.P. (2000) The distribution of water in native starch granules – A multinuclear approach. *Carbohydr. Polymer.* **43**, 375–387.

10 Tang, H. & Hills, B.P. (2001) A multinuclear NMR study of the gelatinization and acid hydrolysis of native potato starch. *Carbohydr. Polymer.* **46**, 7–18.

11 Song, Y-Q., Venkataramanan, L., Hurlimann, M.D. *et al.* (2002) $T_1$-$T_2$ correlation spectra obtained using a fast two-dimensional Laplace inversion. *J. Magn. Reson.* **154**, 261–268.

12 Callaghan, P.T., Godefroy, S. & Ryland, B.N. (2003) Use of the second dimension in PGSE NMR studies of porous media. *Magn. Reson. Imaging* **21**, 243–248.

13 Hills, B.P., Costa, A., Marigheto, N. & Wright, K.M. (2005) $T_1$-$T_2$ NMR correlation studies of high pressure processed starch and potato tissue. *Appl. Magn. Reson.* **28**, 13–27.

14 Benamira, S., Wright, K.M., Marigheto, N. & Hills, B.P. (2004) $T_1$-$T_2$ correlation analysis of complex foods. *Appl. Magn. Reson.* **26**, 543–560.

15 Hills, B.P., Wright, K.M. & Belton, P.S. (1989) Proton NMR studies of chemical and diffusive exchange in carbohydrate systems. *Mol. Phys.* **67**, 1309–1326.

16 Callaghan, P.T. (1991) *Principles of NMR Microscopy*. Clarendon Press, Oxford.

17 Karger, J., Pfeifer, H. & Heink, W. (1988) Principles and application of self-diffusion measurements by NMR. *Adv. Magn. Reson.* **12**, 1–30.

18 Adler, P.M. (1992) *Porous Media*. Butterworth-Heinemann, Boston, MA.

19 FEMLAB software for finite element modelling can be found at Comsol. URL www.comsol.com

20 Bear, J. (1972) *Dynamics of Fluids in Porous Media*. Dover Publications, New York.

21 Thomas, N.L. & Windle, A.H. (1982) A theory of case II diffusion. *Polymer* **23,** 529–542.

22 De Groot, S.R. & Mazur, P. (1969) *Non-equilibrium Thermodynamics*. North-Holland Publishing Company, Amsterdam.

23 Bizzarri, A.R. & Cannistraro, S. (2002) Molecular dynamics of water at the protein-solvent interface. *J. Phys. Chem. B*. **106**, 6617–6633.

24 Spohr, E., Hartnig, C., Gallo, P. & Rovere, M. (1999) Water in porous glasses. A computer simulation study. *J. Mol. Liq.* **80**, 165–178.

25 Stapf, S., Kimmich, R. & Seitter, R-O. (1995) Proton and deuteron field-cycling NMR relaxometry of liquids in porous glasses: Evidence for Levy-walk statistics. *Phys. Rev. Letts.* **75**, 2855–2858.

26 Levitz, P.E. (2003) Slow dynamics in colloidal glasses and porous media as probed by NMR relaxometry: assessment of solvent Levy statistics in the strong adsorption regime. *Magn. Reson. Imaging* **21**, 177–184.

27 Murphy, D. & de Pinho, M.N. (1995) An ATR-FTIR study of water in cellulose acetate membranes prepared by phase inversion. *J. Membr. Sci.* **106**, 245–257.

28 Dias, C.R. & de Pinho, M.N. (1999) Water structure and selective permeation of cellulose-based membranes. *J. Mol. Liq.* **80**, 117–132.

29 Wiggins, P.M. (1995) Microosmosis, a chaotic phenomenon of water and solutes in gels. *Langmuir* **11**, 1984–1986.

30 Wiggins, P.M. (1995) Micro-osmosis in gels, cells and enzymes. *Cell Biochem. Funct.* **13**, 165–172.

31 Wiggins, P.M. (1999) Role of water in some biological processes. *Microbiol. Rev.* **54**, 432–449.

32 Grdadolnik, J. & Marechal, Y. (2001) Bovine serum albumin observed by infrared spectroscopy. I. Methodology, structural investigation and water uptake. *Biopolymers* **62**, 40–67.

33 Marechal, Y. (2003) Observing the water molecule in macromolecules and aqueous media using infrared spectroscopy. *J. Mol. Struct.* **648**, 27–47.

34 Ide, M., Mori, T., Ichikawa, K. *et al.* (2003) Structure of water sorbed in Poly(MEA-co-HEMA) films as examined by ATR-IR spectroscopy. *Langmuir* **19**, 429–435.

35 Asplund, M.C., Zanni, M.T. & Hochstrasser, R.M. (2000) Two-dimensional infrared spectroscopy of peptides by phase controlled femtosecond vibrational photon echoes. *Proc. Natl. Acad. Sci. U. S. A.* **97**, 8219–8224.

36 Mattos, C. (2002) Protein-water interactions in a dynamic world. *Trends Biochem. Sci.* **27**, 203–208.

37 Modig, K., Liepinsh, E., Otting, G. & Halle, B. (2004) Dynamics of protein and peptide hydration. *J. Am. Chem. Soc.* **126**, 102–114.

38 Otting, G., Liepinsh, E. & Wuthrich, K. (1991) Protein hydration in aqueous solution. *Science* **254**, 974–980.

39 Halle, B. (2003) Cross relaxation between macromolecular and solvent spins: The role of long range dipole couplings. *J. Chem. Phys.* **119**, 12372–12385.

40 Denisov, V.P. & Halle, B. (1998) Thermal denaturation of ribonuclease A characterised by water $^{17}$O and $^2$H magnetic relaxation dispersion. *Biochemistry* **37**, 9595–9604.

41 Hills, B.P. (1992) The proton exchange-cross relaxation model of water relaxation in biopolymer systems. *Mol. Phys.* **76**, 489–508.

42 Hills, B.P. (1992) The proton exchange-cross relaxation model of water relaxation in biopolymer systems – The sol and gel states of gelatine. *Mol. Phys.* **76**, 509–523.

43 Koenig, S.H., Brown, R.D. & Ugolini, R. (1993) A unified view of relaxation in protein solutions and tissue, including hydration and magnetisation transfer. *Magnet. Reson. Med.* **29**, 77–83.

44 Koenig, S.H. & Brown, R.D. (1991) Field cycling relaxometry of protein solutions and tissue: Implications for MRI. *Progr. NMR Spectroscopy* **22**, 487–567.

45 Halle, B. & Piculell, L. (1986) Water spin relaxation in colloidal systems, Part 3. Interpretation of the low-frequency dispersion. *J. Chem. Soc. Faraday Trans. I* **82**, 415–429.

46 Modig, K., Rademacher, M., Lucke, C. & Halle, B. (2003) Water dynamics in the large cavity of three lipid-binding proteins monitored by $^{17}$O MRD. *J. Mol. Biol.* **332**, 965–977.

47 Baguet, E. & Hennebert, N. (1999) Characterisation by triple-quantum $^{17}$O-NMR of water molecules buried in lysozyme and trapped in a lysozyme-inhibitor complex. *Biophys. Chem.* **77**, 111–121.

48 Baguet, E., Chapman, B.E., Torres, A.M. & Kuchel, P.W. (1996) Determination of the bound water fraction in cells and protein solutions using $^{17}$O-water multiple quantum filtered relaxation analysis. *J. Magn. Reson. Ser. B*, **111**, 1–8.

49 Torres, A., Grieve, S.M., Chapman, B.E. & Kuchel, P.W. (1997) Strong and weak binding of water to proteins studied by NMR triple-quantum filtered relaxation spectroscopy of $^{17}$O water. *Biophys. Chem.* **67**, 187–198.

50 Pal, S.K., Peon, J. & Zewail, A.H. (2002) Biological water at the protein surface: dynamical solvation probed directly with femtosecond resolution. *Proc. Natl. Acad. Sci. U. S. A.* **99**, 1763–1768.

51 Brunne, R.M., Liepinsh, E., Otting, G. *et al.* (1993) A comparison of experimental residence times of water molecules solvating the Bovine Pancreatic Trypsin Inhibitor with theoretical model calculations. *J. Mol. Biol.* **231**, 1040–1048.

52 Sterpone, F., Ceccarelli, M. & Marchi, M. (2001) Dynamics of hydration in Hen Egg White Lysozyme. *J. Mol. Biol.* **311**, 409–419.

53 Russo, D., Hura, G. & Head-Gordan, T. (2004) Hydration dynamics near a model protein surface. *Biophys. J.* **86**, 1852–1862.

54 Wade, R.C. & Goodford, P.J. (1993) Further development of hydrogen-bond functions for use in determining energetically favourable binding sites on molecules of known structure. 2. Ligand probe groups with the ability to form more than two hydrogen bonds. *J. Med Chem.* **36**, 148–156.

55 Ehrlich, L., Reczko, M., Bohr, H. & Wade, R.C. (1998) Prediction of protein hydration sites from sequence by modular neural networks. *Protein Eng.* **11**, 11–19.

56 The RCSB Protein Data Bank. URL www.rcsb.org/pdb

57 Vackier, M-C., Hills, B.P. & Rutledge, D.N. (1999) An NMR relaxation study of the state of water in gelatine gels. *J. Magn Reson.* **138**, 36–42.

58 Hills, B.P., Wang, Y.L. & Tang, H.R. (2001) Molecular dynamics in concentrated sucrose solutions and glasses – an NMR field cycling study. *Mol. Phys.* **19**, 1679–1687.

59 Belton, P.S., Colquhoun, I.J., Grant, A. *et al.* (1995) FTIR and NMR studies on the hydration of a high-$M_r$ subunit of glutenin. *Int. J. Biol. Macromol.* **17**, 74–80.

60 Belton, P.S. (1994) NMR studies of protein hydration. *Prog. Biophys. Molec. Biol.* **61**, 61–79.

61 Kuntz, I.D. & Kauzmann, W. (1974) Hydration of proteins and polypeptides *Adv. Protein Chem.* **28**, 239–345.

62 Peschier, L.J.C., Bouwstra, J.A., De Bleyser, J. *et al.* (1996) Cross-relaxation effects in pulsed-field gradient stimulated echo measurements on water in a macromolecular matrix. *J. Magnet. Reson. Ser. B* **110**, 150–157.

63 Topgaard, D. & Soderman, O. (2001) Diffusion of water adsorbed in cellulose fibers studied with $^1$H-NMR. *Langmuir* **17**, 2694–2702.

64 Klammler, F. & Kimmich, R. (1992) Geometrical restrictions of incoherent transport of water by diffusion in protein and silica fine particle systems and by flow in a sponge. A study of anomalous properties using an NMR field-gradient technique. *Croat. Chem. Acta* **65**, 455–470.

65 Topgaard, D. & Soderman, O. (2002) Self-diffusion of non-freezing water in porous carbohydrate polymer systems studied by NMR. *Biophys. J.* **83**, 3596–3606.

66 Galazka, V.B. & Ledward, D.A. (1998) High pressure effects on biopolymers. In: Hill, S.E., Ledward, D.A. & Mitchell, J.R. (eds) *Functional Properties of Food Macromolecules*, Chapter 7. Aspen Publishers, Gaithersberg, MD.

67 Douzals, J.P., Marechal, P.A., Coquille, J.C. & Gervais, P. (1996) Microscopic study of starch gelatinization under high hydrostatic pressure. *J. Agric. Food Chem.* **44,** 1403–1408.

68 Jonas, J., Ballard, L. & Nash, D. (1998) High-resolution, high-pressure NMR studies of proteins. *Biophys. J.* **75**, 445–452.

69 Marchi, M. & Akasaka, K. (2001) Simulation of hydrated BPTI at high pressure: Changes in hydrogen bonding and its relation with NMR experiments. *J. Phys. Chem.* **105**, 711–714.

70 Ben-Naim, A. (1973) Molecular theories and models of water and of dilute aqueous solutions. In: Franks, F. (ed.) *Water, a Comprehensive Treatise*, Vol. 2, Chapter 11. Plenum Press, New York (see also Frank, H.S., Chapter 14, in Vol. 1 of same series).

# Chapter 4
# Glasses

*Roger Parker and Stephen G. Ring*

## 4.1 Introduction

The glassy state is common in foods. Many dried foods are glassy, and the glassy state is related to their preservation, the foods being relatively stable with respect to physical, chemical and microbiological change. Drying necessarily means that these foods have low water contents and some of the characteristics of glasses originate from molecules being highly concentrated, interacting strongly with one another, the materials taking on the character of the pure unhydrated components. Cooking operations such as frying, baking and toasting can also dry foods into low-water-content glassy materials, typically with crisp textures originating from their brittle fracture behaviour. In addition to preservation, glassy matrices are used for encapsulation. We distinguish two classes: the first class is homogeneous, the encapsulated species being uniformly dispersed throughout the glassy matrix; and the second is inhomogeneous, that is, a two-phase system in which the material encapsulated is dispersed as particles or droplets throughout the glassy matrix. Explanations of the properties of glassy matrices often focus on their slow dynamics; in the context of chemical stability, the reaction rate limit is expressed in terms of diffusion control. While not exclusive to glassy matrices the occurrence and implications of diffusion control in both homogeneous and heterogeneous reaction systems is an important aspect of the chemical physics of glassy food systems.

In this chapter we initially consider the glass transition behaviour of low molecular weight organic liquids and flexible biopolymers. In a food context the most extensively researched organic liquids are the low molecular weight carbohydrates, which are extensively used as food ingredients. The flexible biopolymers include polysaccharides such as starch, and proteins such as gluten and gelatin. Consideration of globular proteins leads to the glass transition behaviour of particulate systems and research on the physics of soft solids, more particularly colloidal glasses and gels. Other food examples could include concentrated emulsions and concentrated starch pastes. Glasses are metastable systems and are subject to time-dependent change. The aspects of stability we consider are mechanical stability and the stability of reactive chemical species encapsulated in glassy matrices.

# 4.2 Glass transitions

## *4.2.1 Low molecular weight liquids and glasses*

As a liquid is cooled, then providing crystallization can be avoided, it will eventually form a glass, which has a liquid-like structure with the mechanical properties of a solid. Of the organic liquids, carbohydrates were amongst the first to be investigated.[1,2] They readily form glasses at modest cooling rates. Crystalline anhydrous $\beta$-D-glucose melts at 150°C. If the melt is cooled at a rate that is rapid compared with the rate of crystallization, then the liquid shows a continuous progressive increase in viscosity, $\eta$, with decreasing temperature (Fig. 4.1). At 26°C the viscosity reaches $\sim 10^{12}$ Pa s. At these viscosities the material behaves as a brittle solid. Mechanical properties of the solid may be characterized through a relaxation time, which describes the response of the system as a function of time following a mechanical perturbation. The shear relaxation time, $\tau$, is given by:

$$\tau = \eta/G_\infty \tag{4.1}$$

where $G_\infty$ is the high frequency limit of the shear modulus, and typically has a value of $\sim 10^{10}$ N m$^{-2}$. For a liquid with a viscosity of $10^{12}$ Pa s, the shear relaxation time, $\tau$, is $\sim 100$ s, which corresponds to solid-like behaviour in everyday experience. The dependence of $\tau$ of the viscous liquid on temperature, $T$, shows a relationship known as the Vogel–Tammann–Fulcher law:

$$\tau \propto \exp(A/T - T_0) \tag{4.2}$$

where $A$ and $T_0$ are constants, with the value of $T_0$ being $\sim 20$–$30°$ below $T_g$. The above relationship describes the relaxation behaviour of low molecular weight organic glass formers. For these materials there is a progressive increase in viscosity on cooling associ-

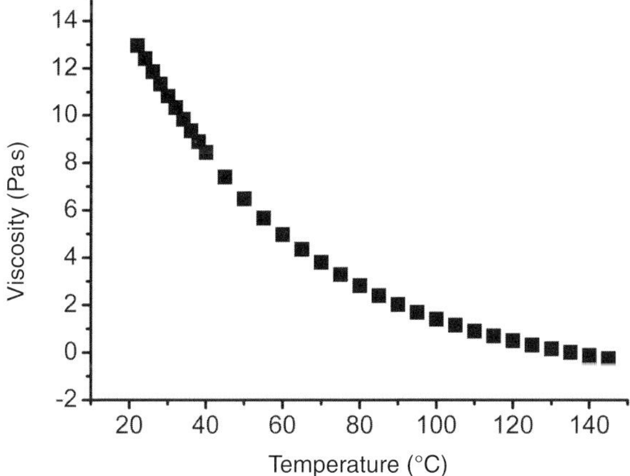

**Fig. 4.1**    Temperature dependence of the viscosity of amorphous glucose, from the melt temperature down to the glass state. After Parks *et al.*[2]

ated with a progressive increase in average size of molecular clusters, which rearrange during viscous flow.[3] For polymeric materials the equivalent relationship is known as the Williams–Landel–Ferry law[4,5] (see later).

At around 26°C the viscosity of glucose shows a marked temperature dependence (Fig. 4.1). If the heat capacity of the amorphous glucose is determined as a function of temperature, then a sharp change in heat capacity is observed at the glass transition temperature, $T_g$,[1] indicating a change from a solid-like heat capacity to a liquid-like heat capacity (Fig. 4.2) over the timescale imposed by the experiment. Calorimetry provides an experimentally convenient method for the determination of $T_g$, as the midpoint of the step change. The precise value of $T_g$ obtained for a particular material depends on the timescale of the calorimetric experiment. Differential scanning calorimetry is routinely employed to determine $T_g$, and in this case $T_g$ will show a dependence on scanning rate. The dependence of $T_g$ on experimental timescale must always be considered in comparing the glass transition behaviour of materials using different techniques. Calorimetry has proved useful for the determination of the glass transition behaviour of low molecular weight organic species such as carbohydrates. Compared with other organic species of similar molecular weight, the carbohydrates have a relatively high $T_g$, which can be attributed to the strength and number of hydrogen-bonded interactions in these liquids.[6] The $T_g$ of monosaccharides shows a small dependence on carbohydrate structure. For a series of oligosaccharides there is a much stronger dependence on degree of polymerization. In the malto-oligomer series, the $T_g$ increases from 26 to 173°C on going from monomer to hexamer.[7] For the higher oligomers it is necessary to examine behaviour in the presence of a diluent, typically water, which depresses $T_g$ and avoids thermal degradation of the carbohydrate.

The preceding discussion was concerned with dry or low-moisture materials. The glass transition may also be relevant to the behaviour of high-moisture materials that are frozen or freeze-dried.[8] Consider an aqueous sucrose solution. If it is cooled below the freezing

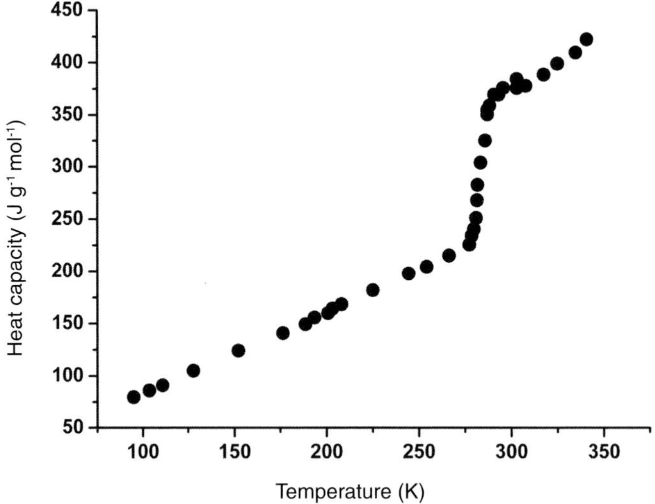

**Fig. 4.2**    Heat capacity of amorphous glucose in the glass transition region.[1]

point of water in the solution, some of the water will crystallize and freeze-concentrate the remaining sucrose solution. The freeze-concentration will further depress the freezing point of water and increase the viscosity of the solution. For a sufficiently deep undercooling, it is found that freeze-concentration is eventually arrested by the vitrification of the concentrated sucrose solution with a concentration, $c_g'$, and a glass transition temperature, $T_g'$. The arrest in crystallization of water may have different physical origins. The enormous viscosity of the sucrose/water glass could slow the transport of water to the growing crystal surface and as a consequence arrest crystallization. Alternatively, the slow structural relaxation of the glassy matrix could physically inhibit the growth of the crystal. The observation[9] that the mobility of water, in carbohydrate/water mixtures, is uncoupled from the bulk shear viscosity and is much more mobile than might be expected, favours the latter explanation. The value of $c_g'$ obtained depends on thermal history and the timescale of the freezing process. In practice, it is found that after the initial freezing and freeze-concentration, there can be a slow crystallization of water during prolonged storage.[10]

### *4.2.2 Biopolymer glasses and plasticization*

For glassy, low molecular weight organic liquids there is a marked change in material properties from a brittle solid to a highly viscous mobile liquid in the vicinity of $T_g$. For polymeric materials, the corresponding change in mechanical properties is generally described as a brittle to rubbery transition.[5] For low molecular weight organic liquids such as carbohydrates, the glass transition is relatively sharp. Even for completely amorphous homobiopolymeric materials such as dextran, the transition may become somewhat more diffuse, reflecting the more complex dynamics of polymeric materials compared with low molecular weight organic compounds.[11] For heterobiopolymers, such as flexible proteins, the transition may become more diffuse as a result of the heterogeneity of chemical structure and the complexity of the molecular dynamics that this introduces.[12] Another effect that can influence behaviour is the tendency of many polymeric materials to associate or crystallize. Even a very limited crystallization, of an otherwise amorphous polymeric material, may effectively cross-link the polymer and reduce the mobility of the polymer chain near the cross-link, with a consequent impact on the characteristics of the glass transition as observed by calorimetry. The fraction of amorphous polymer whose dynamics is slowed as a consequence of the proximity of crystalline regions is called the rigid amorphous fraction.[13,14] This fraction may be considered to have a separate glass transition and may be quantified by temperature-modulated differential scanning calorimetry.[13,14]

The glass transition behaviour of biopolymers is generally examined in the presence of a diluent, typically water, which depresses $T_g$ below the temperature region where thermal degradation would occur.[11,15,16] Figure 4.3 shows the composition dependence of the $T_g$ of a starch-water mixture;[15] addition of water to amorphous starch causes a marked depression in $T_g$, until at 20% w/w water, $T_g$ reaches room temperature. Neutral polysaccharides behave in the same general way, with water being a very effective plasticizing agent, and relatively small effects being observed for different chemical structures. For the glucan polymers – amylose, amylopectin, pullulan and dextran[11] – the range in $T_g$ approached 30°C at a water content of 10% w/w. The branching of some of the glucans is thought to depress $T_g$ as a result of an internal plasticization from the short-chain branches.

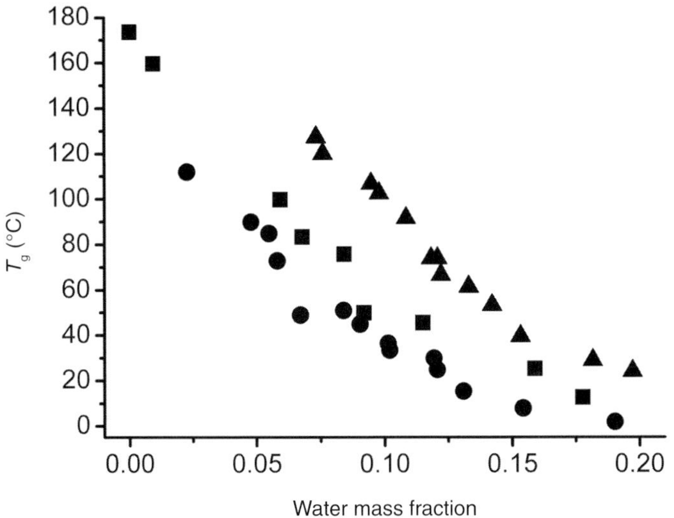

**Fig. 4.3**   Dependence of $T_g$ on water content for starch-water[15] ( ▲ ), and glutenin-water mixtures (Noel *et al.*[16] (■); Micard and Guilbert[18] (●)).

Flexible proteins such as the prolamins (including high molecular weight glutenin and monomeric gliadin),[16–18] elastin[12] and gelatin[19] all show calorimetric glass transitions. The measured $T_g$ of the dry protein is typically in the range 140–195°C. The $T_g$ of a glutenin-water mixture as a function of water content is also shown in Fig. 4.3. As for the polysaccharides, water is an effective diluent and depresses $T_g$. Although all the prolamins show comparable behaviour, small differences are observed depending on the protein examined, its purity and its source.[16] Although the flexible proteins give an obvious calorimetric glass transition, for the globular proteins the observed calorimetric transitions are weak and very diffuse.[20] This is consistent with the globular proteins having a preferred tertiary structure and a somewhat more limited mobility. The glass transition behaviour of an individual globule may be a useful concept in the description of the complex molecular dynamics of a globular protein, and may be used to help interpret aspects of functionality, such as the dependence of enzyme activity on water content in low-moisture systems.[21] For a more complete description, it is also necessary to consider the glassy behaviour of a collection of globules. This latter behaviour should be compared to that of colloidal glasses (see later).

Water is a ubiquitous plasticizer of biopolymeric materials. Its volatility and presence in the atmosphere means that biopolymer-water mixtures are very susceptible to changes in composition and the associated large changes in material properties. The use of other, nonvolatile, plasticizers has also been examined. Most studies have been carried out with low molecular weight hydroxy compounds such as glycerol[22–26] and sorbitol,[27] often in combination with water. There are a number of relationships describing the way that the $T_g$ of mixed binary systems should vary as a function of composition, and these have found use in describing the behaviour of such mixtures. One such is due to Couchman[28,29] and is of the form:

$$T_{gm} = (w_1 \Delta C_{p1} T_{g1} + w_2 \Delta C_{p2} T_{g2})/(w_1 \Delta C_{p1} + w_2 \Delta C_{p2}) \qquad (4.3)$$

The glass transition temperature of the mixture, $T_{gm}$, is related to the glass transition temperatures of the individual components and the heat capacity increment at the glass transition, $\Delta C_{pi}$; $w_i$ is the mass fraction of the $i$th component. The parameters $\Delta C_{pi}$ and $T_{gi}$ are readily determined in a calorimetric experiment, and the above relationship has a practical utility. For example, replacing water, which has a relatively low $T_g$ ($-139°C$),[30,31] with glycerol ($T_g$ $-80°C$)[32] would be expected to lead to an elevation in the $T_g$ of the binary mixture. In more complex mixtures, such as the 'ternary' starch-glycerol-water mixture, other effects need to be considered. Initial additions of glycerol to starch resulted in a depression of a single glass transition. As water was added,[25] $T_g$ was further depressed, and an additional lower glass transition appeared. The lower $T_g$ was close to the $T_g$ of pure glycerol, suggesting that the 'ternary' starch-glycerol-water mixture had phase separated, with the appearance of a glycerol-rich phase.

### 4.2.3 Colloidal glasses

On cooling organic molecular liquids the barriers to thermally activated motions increase, as do the strength of intermolecular interactions and density, and this leads to the progressive increase in viscosity. If a suspension of colloidal particles, with a net repulsive interaction, is concentrated by the application of an osmotic stress, then the viscosity of the suspension will progressively increase with the increasing volume fraction, $\phi$, of particles.[33,34] For suspensions of spherical particles the dependence of relative viscosity, $\eta_r$, on volume fraction is given to a first approximation by:

$$\eta_r = (1 - \phi/\phi_{max})^{-2} \qquad (4.4)$$

where $\eta_r$ is the viscosity of the suspension relative to that of the solvent, and $\phi_{max}$ is the maximum packing fraction, which varies with shear rate. Relationships of the form of Equation 4.4 predict a marked increase in viscosity over a very small range of volume fraction. At high volume fractions there is a sufficient slowing of particle dynamics that liquid-like configurations cannot be explored over practical timescales.[35,36] For small applied stresses the material has the solid-like characteristics of a glass with the particles forming jammed structures that are stress bearing. For a random packing of noninteracting monodisperse hard spheres, these structures may form at volume fractions in the vicinity of 0.6 with a random close packing limit, $\phi_c$, of $\approx 0.644$.[37,38] Ellipsoids can randomly pack more densely up to a $\phi_c$ of 0.68–0.74 depending on aspect ratio.[39,40]

A glass may also be formed if a concentrated colloidal suspension is quenched to a temperature at which the interparticle interaction is sufficiently attractive. With increasing attractive interaction,[41] three dimensional particle networks, with solid-like characteristics (colloidal gels), will form at lower particle volume fractions.[42] Recent research has emphasized the similarities between jammed structures that can form with increasing volume fraction of particles, and those more open structures that form as a result of an increasing attractive interaction between particles. For colloidal glasses, a range of techniques may be used to probe the glass transition region. Mechanical measurements can be used to determine the onset of solid-like behaviour over a defined experimental timescale. Light scattering techniques can give information on particle dynamics,[43] which may also be determined directly using techniques such as confocal microscopy.[44,45] These experimental techniques are often combined with simulation studies on the dynamics of these materials.[43]

The research on colloidal glasses is potentially relevant to many particulate food systems, including the behaviour of concentrated emulsions, suspensions of starch granules and, perhaps more surprisingly, solutions of globular proteins. Globular proteins are particles that are typically a few nanometres in size. While one could consider the polymeric glass transition of the protein chain in the globule, at high volume fractions the packing of protein globules, with a net repulsive interaction, produces disordered solid-like materials with glassy characteristics.[46] The observed onset of glassy behaviour for bovine serum albumin in dilute salt solution occurs at volume fractions in the region of 0.55 (mass fraction ~0.6) and are very different to the compositions at which a flexible protein-water mixture vitrifies (protein mass fraction ~0.8) (cf. Fig. 4.3). This difference has important consequences for the dynamics in these systems and the effects of dynamics on stability. Although at high volume fractions the structural rearrangement of globules may be arrested, the interstitial water and the solutes it contains will be much more mobile. Under some conditions the interaction between protein globules may become attractive. Although this can result in protein crystallization, it can also result in the formation of particle networks or gels, which can be more open structures with solid-like characteristics.

## 4.3 Glassy state dynamics

The relaxation processes and time-dependent material properties of amorphous solids reflect underlying collective molecular motions. The rates of these collective motions are characterized by (mean) relaxation times. For example, a shear-stress relaxation time might be measured in a step-response experiment in which a small instantaneous shear strain is applied to a material, resulting in a jump in shear stress, which subsequently decays (relaxes) with time.[5] For the primary relaxation of amorphous solids, the shear-stress relaxation curve is commonly nonexponential and can be fitted with a stretched exponential function:

$$\phi(t) = \exp[-(t/\tau)^{\beta}] \tag{4.5}$$

to determine the relaxation time, $\tau$. Viscosity measurements and mechanical and dielectric spectroscopy can all be used to characterize relaxation times.[47] The relaxation times are temperature-dependent and conventionally plotted in an Arrhenius plot known as a 'relaxation map'. A particularly comprehensive study of glucitol[47,48] is shown in Fig. 4.4. As liquid glucitol is cooled, the mean relaxation time for the primary relaxation, or $\alpha$-relaxation, process initially increases in a non-Arrhenian manner, with an increasingly large activation energy, before reaching a discontinuity at about $10^4$ s. This discontinuity indicates the glass transition, and in this example the glass transition temperature ($T_g$) was defined as the temperature at which the mean relaxation time is $10^4$ s. At temperatures above $T_g$, the temperature dependence of the various relaxation times can be described by the Vogel–Tammann–Fulcher (VTF) equation. A second, faster relaxation process, termed

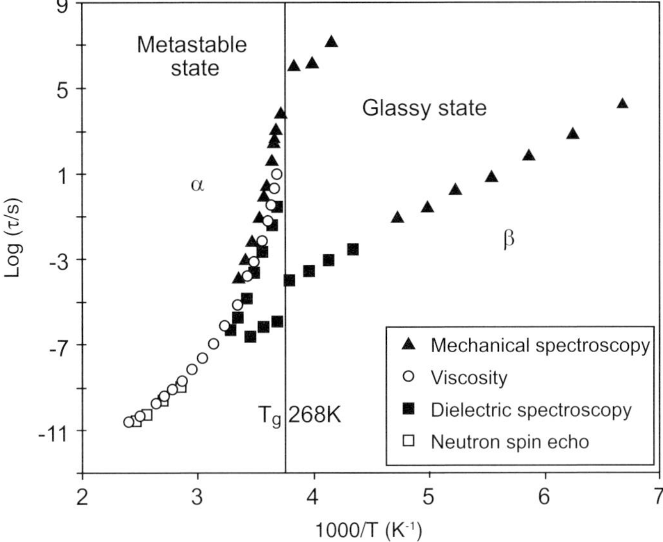


**Fig. 4.4**  Relaxation time plotted against inverse absolute temperature (relaxation map) for glucitol (sorbitol) above and below its glass transition. Based upon Faivre,[48] with permission.

the secondary relaxation, or β-relaxation, is observed at lower temperatures; it shows an Arrhenius temperature dependence. Although this relaxation is observed predominantly in the glassy state, it can be observed at temperatures above $T_g$, where it ultimately merges with the α-relaxation process. A detailed assignment of the molecular motions giving rise to these relaxations has not yet been achieved. Roughly speaking, the β-relaxation has some of the characteristics of a localized, short-time, cage-rattling motion,[49] whereas the α-relaxation, the main structural relaxation, occurs over longer timescales and corresponds to the cages formed by nearest neighbours relaxing and allowing diffusive motions. During structural relaxation, the system is exploring its energy landscape, and for this reason when the structural (primary) relaxation time exceeds experimental timescales, the system becomes nonequilibrium, that is, in the glassy state the system no longer fully explores its energy landscape during an experiment. It is this effect that is responsible for the observed discontinuity in Fig. 4.4 (and Fig. 4.2).

In complementary studies on amorphous glucose[50] and glucose-water mixtures,[51] specialized NMR techniques have probed the slow rotational motions associated with the primary relaxation. Moran and Jeffrey[51] found that at temperatures close to $T_g$ conventional NMR relaxation time measurements probed processes that were faster than the primary relaxation, and at temperatures below $T_g$ these faster relaxations had lower activation energies than those of secondary relaxations observed by dielectric techniques. Even single-component glass-formers have complicated dynamics that require specialized techniques to probe them.[50] Dielectric spectroscopic studies of secondary relaxations in carbohydrates and their water mixtures show variation with molecular structure.[52] Figure 4.5 shows temperature scans of the dielectric tan δ (= dielectric loss, ε″/dielectric constant, ε′, the ratio of the imaginary and real parts of the complex permittivity) at 1 kHz, for a range of dry amorphous carbohydrates. The main peak in tan δ is due to the primary relaxation, and as these measurements are at

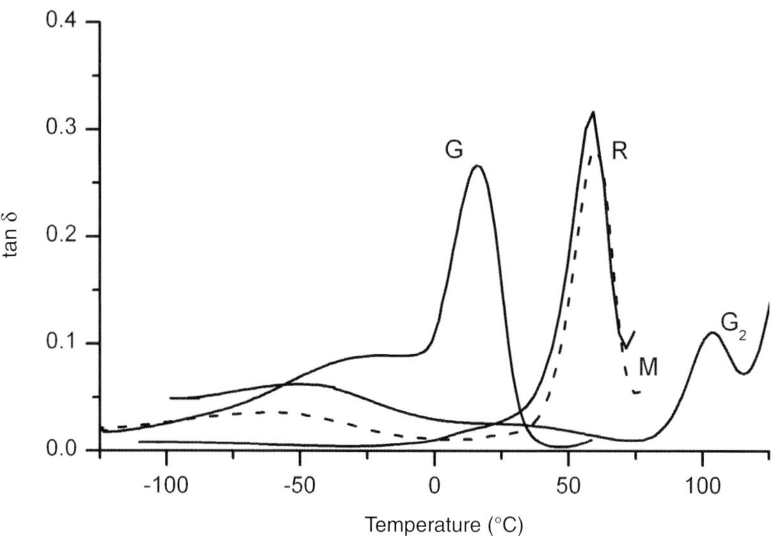



**Fig. 4.5**   Primary and secondary dielectric relaxations for a range of dry amorphous carbohydrates at 1kHz. G, glucitol; R, L-rhamnose; M, D-mannose; $G_2$, maltose.

1 kHz, it peaks at a temperature above the calorimetrically determined $T_g$. In these scans, $T_g$ corresponds to the position of the low-temperature shoulder of the primary relaxation peak. Peaks in tan δ at temperatures below the primary relaxation peak are due to secondary relaxation processes. In glucitol, an acyclic carbohydrate, the secondary relaxation is strong and occurs at temperatures not far below the primary relaxation, so that the two relaxations overlap, even at 1 kHz. A comparison of the secondary relaxations in the monosaccharide D-mannose, which has a hydroxymethyl group at C-6, and its deoxy-sugar L-rhamnose, which has a methyl group at C-6, indicates that the presence of a polar exocyclic group has a strong influence on the secondary relaxation. In the disaccharide maltose, the primary relaxation is relatively weak, and the secondary relaxation peak is deep in the glassy state (−50°C). The full significance of secondary relaxation processes to functionality is not yet understood, although polymer scientists associate strong secondary relaxations with enhanced transport properties, such as high gas permeability.[53]

Glassy state dynamics has an extensive and fast-growing literature. For access to this, the reader is directed to other recent reviews.[54–58] One topical issue is the nonexponential nature of the primary relaxation. The origins of this behaviour could arise from the intrinsically nonexponential nature of the process, or from the dynamic heterogeneity of the liquid[57,58] with the observed nonexponentiality being a consequence of the different dynamic environments that are present. NMR studies[59] and single-molecule spectroscopy confirm the latter view.[60,61] The glassy liquid may be characterized by domains that each have a particular relaxation time. These domains can rearrange through large collective motions to form new domains with different relaxation times. The observed relaxation in the bulk is therefore an average over all these relaxations. Recent DSC-based and NMR techniques give a length scale of 3–5 nm for the size of these regions in glucitol at $T_g$, which corresponds to about 100 molecules rearranging cooperatively.[62] As temperature is increased the length scale decreases.

## 4.4 Structural relaxation in low molecular weight organic liquids and biopolymers

Amorphous materials that are stored at a temperature below the melting temperature of the crystalline solid can be subject to time-dependent change in mechanical properties as a result of a crystallization process. With the enormous viscosity of glassy materials, it is to be expected that crystallization within the glass is relatively slow. There is another process that can lead to time-dependent change in material properties of glassy materials that is associated with structural relaxation within the glass.[63,64]

For most amorphous organic materials – low molecular weight organic liquids, synthetic polymers, and biopolymers (including flexible proteins such as gluten and polysaccharides) – the expected dependence of structural relaxation time on temperature is broadly similar. An expression that has been shown to be widely applicable in describing this temperature dependence for fully relaxed, synthetic polymers is the Williams–Landel–Ferry (WLF) relationship:[4,5]

$$\log a_{\mathrm{T}} = -c_{1,0}(T - T_0)/(c_{2,0} + T - T_0) \tag{4.6}$$

where $a_{\mathrm{T}}$ is the ratio of relaxation times, $\tau_{\mathrm{T}}/\tau_0$, at a temperature $T$, and a reference temperature, $T_0$. If $T_g$ becomes the reference temperature then:

$$\log a_{\mathrm{T}} = -c_{1g}(T - T_g)/(c_{2g} + T - T_g) \tag{4.7}$$

with values of the coefficients $c_{1g}$ and $c_{2g}$ obtained from fitting data on a range of synthetic polymers, being 17.44 $K^{-1}$ and 51.6 K, respectively. The temperature and composition dependence of the relaxation time, $\tau$, is shown schematically in Fig. 4.6 for an amorphous starch-water mixture. As the glass transition is approached, either through reducing temperature or water content, there is predicted to be a very marked change in relaxation behaviour. For a more detailed discussion on the use of Equation 4.6 to describe the behaviour of synthetic polymers the reader is referred to the work of Ferry.[5]

If the structural relaxation is sufficiently slow, it can lead to time-dependent changes in the properties of glassy materials over practical timescales of hours to weeks,[63–71] and is potentially relevant to the observed ageing of low-water-content products.[72–76] The origin of this physical ageing is the dependence of the 'equilibrium' structure of an undercooled liquid on temperature, and the timescale required to achieve it. For example, it is generally expected that reducing temperature would increase the density of a material. As a liquid is undercooled toward the glass transition its viscosity increases, as does the associated structural relaxation time. If an amorphous material is rapidly quenched into the glass state, the structural relaxation time may be so high that the amorphous structure, and resulting density, is effectively 'frozen'. The structure will gradually evolve and at very long times it will have a fully relaxed, 'equilibrium' structure. If it is cooled again, further structural relaxations and rearrangements within the undercooled liquid will occur until, given sufficient time, a new 'equilibrium' structure is obtained. The structure of the undercooled amorphous liquid can therefore show a dependence on time and thermal history. This relaxation behaviour has been extensively studied, particularly for polymers and inorganic glasses. The temperature-dependent

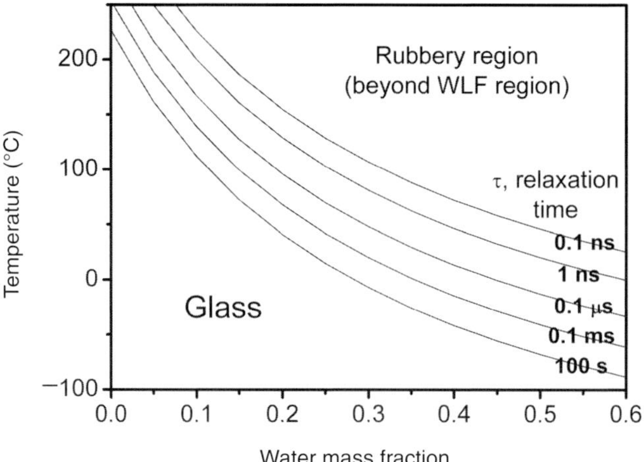


**Fig. 4.6**  Williams–Landel–Ferry (WLF) predictions of the dependence of structural relaxation time on water content for a starch-water mixture. Constructed using data from Fig. 4.3 and Equations 4.3 and 4.7.

changes in bonding between molecules and their configuration are associated with changes in volume, enthalpy, heat capacity and material properties, including mechanical behaviour and diffusivity. During ageing the material becomes stiffer and less compliant, with increasing tendency to fracture.

Although the increase in density of a liquid with time is one way of probing this structural relaxation, its use is rather restricted. As densification also affects the energetics of interaction between molecules, and the accessibility of liquid configurations, it can be probed in a calorimetric experiment where structural relaxation is observed as a peak in heat capacity preceding $T_g$ or an overshoot at $T_g$, rather than the simple step change shown in Fig. 4.2. There is often a requirement to be able to predict the change in the material properties of a glassy product with time. Common questions might be 'at what temperature do I need to store the product to minimize this time-dependent change?', and 'how might fluctuations in water content affect the observed behaviour?' Fortunately there are various phenomenological approaches for describing the observed time-dependent behaviour; a widely applied one, which has a useful predictive capability, is the Tool–Narayanaswamy (TN) method,[70] which has been applied to polymeric systems. The dependence of structural relaxation on time, $t$, can be described by an empirical relaxation function, $\phi$, of the form of Equation 4.5, where $\beta$ $(0 < \beta \leq 1)$ is a measure of the nonexponentiality of the relaxation, which is dependent on both temperature, $T$, and, to an extent, liquid structure (characterized on a temperature scale through the notion of a fictive temperature, $T_f$ – the temperature at which a particular structure would be fully relaxed). The relaxation time, $\tau$, may be obtained using the expression:

$$\tau = A\,\exp[x\Delta h^*/RT + (1-x)Dh^*/RT_f] \qquad (4.8)$$

where $A$, $x$ $(0 < x \leq 1)$ and $\Delta h^*$ are constants. The term $\Delta h^*$ can be determined from the dependence of the calorimetric $T_g$ on scanning rate. These relationships can be used to calculate the time dependence of $T_f$ following a temperature step, and from this the heat capacity change with temperature can be predicted. By appropriate selection of the constants $A$, $x$ and $\beta$ the experimentally observed behaviour may be modelled. For a simple carbohydrate, such as maltose,[76] it was found that a single set of constants described the dependence of ageing on time and temperature and had a useful predictive utility. More recently it was found that the above approach was useful in describing the ageing of a plasticized starchy material.[75]

## 4.5 Mechanical stability – colloidal systems

A jamming phase diagram has been suggested,[77] and subsequently confirmed experimentally,[78] which describes the effect of temperature, particle volume fraction, and stress on the stability of jammed structures. Jamming can only occur at a sufficiently high volume fraction of particles, and jammed structures may be disrupted by raising temperature or applying a stress. Colloidal systems can therefore display a range of ageing phenomena. One of the simpler systems studied is neutrally buoyant concentrated suspension with a net repulsive interaction. In this case the system was initially disrupted by stirring and its relaxation back to its 'equilibrium' state observed by confocal microscopy.[44] The motion of several thousand particles was followed as a function of time. It was found that particle motion significantly slowed and the characteristic relaxation time increased with increasing age of the sample. The ageing process was spatially and temporally heterogeneous.

Systems with a net attraction also show time-dependent behaviour.[79–81] In many practical situations the density difference between particle and suspending medium and the effects of gravity lead to additional effects. In a food context this might involve the creaming of emulsion droplets or the sedimentation of starch granules. There have been a number of studies where an attractive interaction between particles has been produced by depletion flocculation through the addition of a soluble polymer.[82] The combination of sedimentation and an attractive interaction produces jammed sediments with an open gel structure. On ageing, the structure of the gel coarsens and eventually collapses under the influence of gravity. Increasing the strength of the interparticle interaction slows the evolution of structural change.

## 4.6 Chemical stability

In this section we describe some theoretical approaches connecting chemical reaction kinetics and glassy state dynamics[83] together with relevant experimental studies probing translational mobility in glassy and near glassy systems. Reactions in homogeneous and heterogeneous multiphase systems[84] are considered.

Seeking a link between biostabilization and glassy state dynamics is one way in which the 'vitrification hypothesis'[85] can be elaborated. Coupling to the glassy state dynamics is not the only potential mechanism by which glasses may achieve their preservative properties. Another complementary hypothesis for the preservative action of amorphous carbohydrate matrices, first applied to anhydrobiotic organisms, is the 'water replacement hypothesis'.[86] A further potential role of carbohydrates in biostabilization is as a nonvolatile solvent and diluent, a role that is clearly related to the 'water replacement hypothesis'. This aspect of the behaviour of amorphous carbohydrates is described in sections on encapsulation, and the solvent properties of carbohydrates.

### 4.6.1 Chemical kinetics and the glassy state in single-phase systems

Preservation can be characterized in terms of the timescale of deteriorative reactions. Typical timescales for preservation are in the range $10^6$ s (11.6 days) to $10^8$ s (3.2 years). In a glass, the timescale of the main structural relaxation (see Fig. 4.4) can be comparable to that of preservation, but would the chemical reactions of species dispersed in a glass be expected to occur over a similar timescale? Distinct approaches to unimolecular and bimolecular reactions must be taken.[83] Whereas a unimolecular reaction can be modelled as a process of passing over an energy barrier, a bimolecular reaction also involves a diffusive step, in which the reactants diffuse together prior to reaction. Thus, the very high viscosity of a glass may affect a reaction by decreasing the frequency of diffusive encounters of molecules and by slowing the rate at which reactive molecules pass over energy barriers.

For diffusion-controlled bimolecular reactions, Smoluchowski's theory[87,88] can be used to estimate reaction rate. This predicts that the second-order rate constant $k_2 = 8RT/3\eta$, where $R$ is the gas constant, $T$ the absolute temperature, and $\eta$ the viscosity. At $T_g$, $\eta$ is about $10^{12}$ Pa s. The half-life for the reaction is $t_{1/2} = 1/k_2[A]_0$,[89] where $[A]_0$ is the initial reactant concentration. The half-lives for some typical reactant concentrations are shown in Table 4.1. The half-life varies between systems, simply because $[A]_0$ is varying. This has a dramatic effect, although the emulsion and protein are predicted to be stable (with respect to aggregation) for timescales greater than years; the molecular reactant is predicted to react over a timescale of 17.5 days, an unacceptably short half-life for many applications. There are, however, a number of assumptions implicit in these predictions, which require further examination.

One assumption in Smoluchowski theory is that of angular-independent reactivity. While this is acceptable for applications to emulsions and, possibly, to proteins, it is inappropriate for most chemical reactions that are subject to steric constraints. Solc and Stockmayer[90,91] developed a theory in which the reactivity was angular-dependent; it depended upon the size of circular, reactive, patches. Reactive patch sizes can plausibly be estimated using simple geometric arguments. This effect is estimated to reduce the diffusion-controlled rate by a factor of $10^2$ to $10^3$ for small molecular reactants.[92]

Another potential shortcoming of Smoluchowski theory is the assumption of diffusion control. In aqueous solution, diffusion-controlled reactions appear to be unusual, the few

**Table 4.1**    Half-lives for diffusion-controlled reaction at $T_g$, predicted using Smoluchowski theory.

| Material | Concentration | $t_{1/2}$ |
|---|---|---|
| Emulsion | 20% v/v | $7.5 \times 10^6$ years |
| | 1 μm diameter | |
| Protein | 10% w/w | 2.3 years |
| | 0.75 cm$^3$ g$^{-1}$ | |
| | 5 nm diameter | |
| Molecule | 100 mM | 17.5 days |

examples being well-documented,[87] and most reactions are reaction- or activation-controlled. Collins and Kimball[93] modified the Smoluchowski theory to include the effects of finite chemical reactivity, that is, in contrast to the Smoluchowski approach, on molecular encounter there is a possibility that molecules may simply diffuse apart again without reaction. At steady state, the overall bimolecular rate constant, $k$, for the reaction is predicted to be:

$$k = \frac{4\pi\, r_c D k_{act}}{4\pi\, r_c D + k_{act}} \tag{4.9}$$

where $4\pi r_c D$ is the diffusion-controlled rate, $r_c$ is the collision diameter, $D$ is the relative diffusivity, and $k_{act}$ is the reaction-controlled rate constant. This theory predicts that, as the relative diffusive mobility of reacting species is reduced, there is a crossover from reaction- to diffusion-control, as shown in Fig. 4.7. The crossover occurs when $k_{act} = 4\pi r_c D$. An outcome of this theory is that slow reactions will remain reaction-controlled, until the diffusive mobility is sufficiently small for the crossover to occur. It may be that, even at the glass transition, there is sufficient mobility for a reaction to remain reaction-controlled, and under these conditions, it would be expected to exhibit classical Arrhenius temperature-dependence.[89,92]

A comparison of the initial rate of the Maillard reaction between glucose and lysine in an amorphous sucrose-trehalose-water matrix with the Smoluchowski theory prediction is shown in Fig. 4.8. Above $T_g$ the reaction rate has an Arrhenian temperature dependence with an activation energy of 140 kJ mol$^{-1}$, lower than that predicted by Smoluchowski theory. This suggests reaction-control at temperatures above $T_g$. Finite reaction rates are measured in the glassy state; however, the Smoluchowski theory predictions show that these are not unexpected. In order to measure reaction over a practical timescale the reactants are highly concentrated, with glucose and lysine concentrations of 5 and 10% w/w, respectively (i.e. about 0.4 and 0.8 M). Assuming this reaction is bimolecular and second order, the inverse

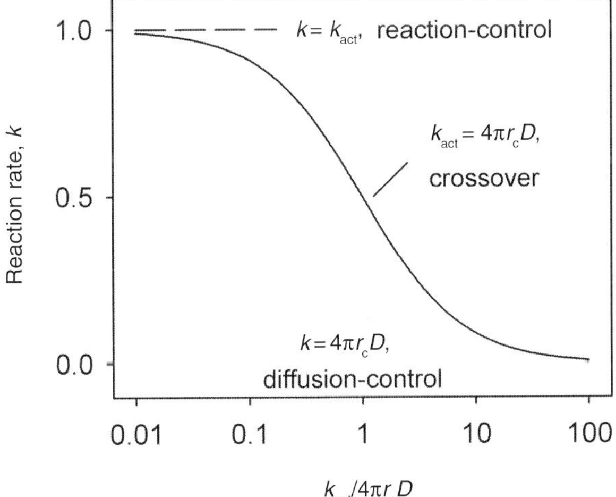


**Fig. 4.7**   Crossover between reaction-control and diffusion-control for a bimolecular reaction with a finite rate.[93]

dependence on reactant concentration means that reducing these concentrations 100-fold would give a corresponding increase in reaction half-life, that is, a practical preservation timescale of 4–8 years.[92]

For reactions of small molecules, the most serious shortcoming of the Smoluchowski prediction of reaction rate is the use of the Stokes–Einstein (SE) relationship,[94] $D = kT/6\pi\eta r_h$, where $k$ is Boltzmann's constant and $r_h$ a hydrodynamic radius (modelling the molecule as a spherical particle). The SE relationship underestimates the diffusivity, particularly in the case of a small molecule population in a mixture or when the viscosity is high. A test of the SE relationship was carried out for fluorescein diffusing in sucrose-water mixtures, measured using fluorescence recovery after photobleaching (FRAP).[95] The ratio $T/D\eta$ ($\propto$ hydrodynamic radius) was constant at temperatures well above the glass transition ($T \geq T_g/0.86$), as predicted by the SE relationship; however, at temperatures close to the glass transition ($T < T_g/0.86$), the relationship broke down. At $T_g$, the diffusivity was about $10^7$ faster than that predicted by the SE relationship. For small molecules, the factor by which the SE relationship underestimates diffusion can be larger,[96] for example, for ethanol $\sim 10^7$, and water $\sim 10^9$. Conductivity measurements can also be applied to measure translational mobility in near-glassy amorphous carbohydrates.[97] The molar conductivity, $\Lambda_m$, of a symmetrical electrolyte (ion charge $ze$, where $e$ is the magnitude of the charge on the electron) is related to the self-diffusion coefficients of the ions, $D_+$ and $D_-$, through the Nernst–Einstein relationship, $\Lambda_m = z^2F^2(D_+ + D_-)/RT$, where $F$ is the Faraday constant. Figure 4.9a shows a $T_g$-scaled Arrhenius plot of the molar conductivity of KCl in a series of amorphous carbohydrate-10% w/w water mixtures. At $T_g$, the molar conductivity varies in the order monosaccharides < disaccharide <

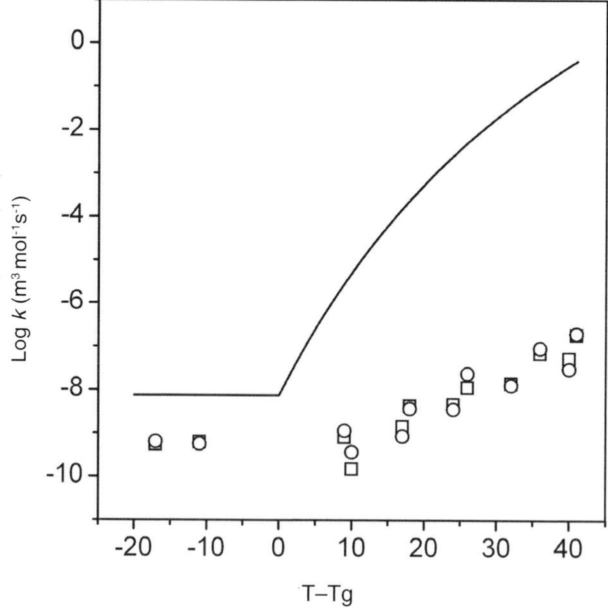



**Fig. 4.8**  Comparison of second-order rate constants for the consumption of lysine ($\circ$) and glucose ($\square$) with predicted diffusion-controlled rates for reaction in the glass transition region.[92]

trisaccharides. Some care needs to be taken in interpreting the change in molar conductivity, because, in addition to the changing ion mobility, there may potentially be a change in ion pairing. Calculations based upon Bjerrum theory[98] suggest that the ion-pairing effect is small, but this is worthy of further investigation. Figure 4.9b shows the same data as in Fig. 4.9a but plotted as a conventional Arrhenius plot. In Fig. 4.9b, the order is reversed; at constant temperature, molar conductivity varies in the order trisaccharides and disaccharide < monosaccharides. Thus, using conductivity as an index of mobility, it would be predicted that, at $T_g$, preservation in a glassy monosaccharide would minimize mobility, whereas at constant temperature, preservation in a di- or trisaccharide would minimize mobility. However, the magnitude of the conductivities indicates considerable uncoupling of the ionic motions from the main structural relaxation and diffusion coefficients ($D_+$ and $D_-$), which would indicate that there are many ion–ion encounters over preservation timescales.

Hagen *et al.*[83] argued that the effects of glassy state dynamics on the rates of barrier crossing can be predicted using Kramers' theory:

$$k = \frac{A}{\eta}\exp(-E_0/RT) \tag{4.10}$$

where $A$ is a constant and $E_0$ an energy barrier. This equation (slightly modified) was successfully applied to the rearrangements of electronically excited myoglobin in a trehalose glass, a unimolecular reaction.[99] It could also be applied to the reaction step of bimolecular reactions ($k_{act}$ in Equation 4.9). We are not aware whether the theory has been tested for reactions of small molecules. Kramers' theory is a model for diffusion over an energy barrier, and the inverse viscosity dependence originates from the SE relationship, and so is likely

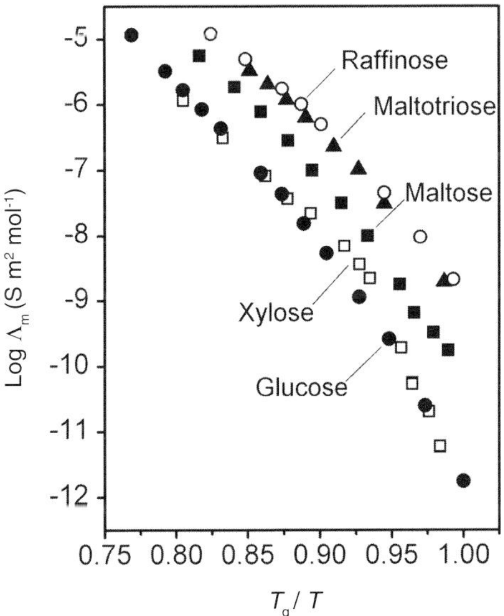



**Fig. 4.9**  Molar conductivity of KCl in amorphous carbohydrate-10% w/w water mixtures at temperatures above $T_g$. (a) $T_g$-scaled Arrhenius plot. (*Continued.*)

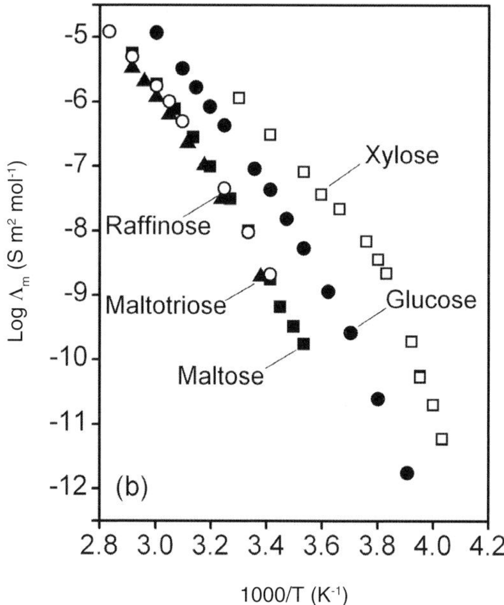



**Fig. 4.9**    (*Continued.*) (b) Arrhenius plot.

to suffer from similar problems of SE breakdown as the above Smoluchowski predictions. Furthermore, the breakdown is likely to be more severe, because the SE relationship is being applied to more localized subdiffusive motions, as the reacting system passes through the transition state, motions that occur over a shorter length scale than that characterizing the primary relaxations.

In conclusion, the SE breakdown and limitations in our theoretical understanding of the connection between glassy state dynamics and chemical kinetics mean that we are some way from a comprehensive physicochemical basis for the effect of vitrification on deteriorative reactions of interest to food scientists. It should be stressed that diffusion-controlled bimolecular reactions and 'WLF kinetics' are well-founded in the physical chemistry literature. For example, in the 1970s, Dainton *et al.*[100] performed photochemical studies of diffusion-controlled energy transfer kinetics in glass-forming solvents at near-glassy temperatures. The diffusion-controlled kinetics showed 'WLF' dependence, and further analysis revealed the SE breakdown described above. The experimentally measured second-order rate constants indicated that, depending upon reactant concentration, diffusion-controlled reactions can have half-lives of the order of hours in near-glassy solvents.

On the basis of this discussion, what advice can be given for controlling reaction rate through vitrification? The answer to this question can be no more than our best guess. Firstly, minimize mobility by putting the system into a glass.[85] This can be achieved by mixing the material with a carbohydrate in aqueous solution and then drying to a low-water-content amorphous glass. Because ageing reduces the diffusivity of gases through glassy polymers, annealing the glass to achieve maximal ageing/density should be performed. This can be optimized using TN models.[76] Minimize mobility by optimizing storage temperature and the molecular structure of the carbohydrate. If storage at ambient temperatures is a constraint, then use higher molecular weight carbohydrates, for example, at least di- or trisaccharides.[101] In order to minimize mobility, it is simpler if only a single phase is present, that is, partially miscible mixtures lead to the complication of minimizing mobility in multiple phases. There is now a body of literature showing that chemical reactions can occur in the glassy state,[92,102,103] from which we conclude that vitrification technology does not replace the need to understand the chemistry of the deteriorative reactions in a system. Acid–base catalysis is common, and so pH should be optimized to minimize reaction rate. If the reaction is bimolecular, dilution of the reactants with amorphous carbohydrate increases reaction half-life. If chemical deterioration persists, then analyse the chemistry and look for chemical preservation strategies (there are recent examples of faster deterioration upon encapsulation[104]).

### 4.6.2 Chemical kinetics and the glassy state in multiphase systems

Some essential oils and lipids, which are susceptible to oxidation, are encapsulated in glassy matrices as vitrified emulsions by processes such as spray drying,[84] freeze drying[105] and extrusion.[106] Oxidation of oils and lipids can result in the consumption of oxygen and so these systems represent heterogeneous reaction systems.[84] While the initial conditions in these diffusion-reaction systems depend on details of processing (e.g. the initial oxygen levels in the oil and matrix[107]) and chemistry (e.g. identity and concentrations of free radical initiators[105]), subsequent reaction ultimately depends on the passage of oxygen through the matrix to the reacting species.[84,108] In these multiphase systems the distance characterizing the diffusion path length is determined by the size of the particles. In order fully to understand these systems both the reaction chemistry and the oxygen permeation need to be understood. Recent progress has been made[105,107] using oxygen-sensitive nitroxyl spin probes, and electron spin resonance (ESR) spectroscopy, to measure the oxygen permeation through glassy encapsulation matrices. In this study the authors found that oxygen, initially in the oil droplets, was not removed during freeze drying, a further indication of the impermeability of the matrix to oxygen and emphasizing the importance of initial oxygen levels.

## 4.7 Glassy carbohydrates as encapsulation matrices and solvents

### 4.7.1 Flavour encapsulation in glassy carbohydrates

Liquid flavours are encapsulated in glassy carbohydrates to prevent their evaporation, to protect them from adverse chemical reactions, and to convert them into easily handled free-flowing powders. Carbohydrate-encapsulated flavours are prepared by processes such as spray-drying[84] and extrusion.[106] To act as encapsulation matrices, glassy carbohydrates must be, to a large extent, impermeable to the components of a flavour. Impermeability could result from flavour components having a low solubility in the matrix, a limited diffusive mobility in the matrix, or a combination of both. Furthermore, the flavour must not plasticize the matrix, making it sticky and liable to caking.[101] Thus, miscibility and component partitioning in flavour-carbohydrate-water mixtures are important to understanding the function of encapsulation systems.

Flavours contain components of varying hydrophobicity, a property that can be characterized using octanol-water partition coefficients, $P$. Table 4.2 shows values of log $P$ for components present in a cherry flavour; they vary from hydrophobic components that are poorly miscible with water, such as benzaldehyde and benzyl alcohol, to relatively hydrophilic components, such as diacetyl, which is fully miscible with water.[106,109–110] Comparison of partition coefficients in benzyl alcohol-glucitol (a model of the flavour-carbohydrate matrix system) with those in octanol-water (Fig. 4.10) shows that the effect of substituting octanol for benzyl alcohol, and water for glucitol, is to increase the partitioning of species into the hydrophobic phase by at least an order of magnitude. Overall, only the most hydrophilic components show significant partitioning into the amorphous carbohydrate matrix, and, from the solubility viewpoint, it is only these components that will permeate the matrix. As these components are at low concentrations, and there are small amounts of unencapsulated flavour oil, leakage rates through matrices are difficult to detect. However, it is straightforward to measure release of encapsulated flavours, in response to changes in water content (and temperature).[106] This can be achieved by conditioning samples in different humid atmospheres, followed by sealing samples in headspace vials and analysing the headspace concentration after 24 hours. Figure 4.11 shows that release increases in response to both increases and decreases in the 'as prepared' water content (3.5% w/w). After an increase in water content, microscopy showed that the sucrose in the matrix had crystallized, causing droplets of flavour oil to break the surface of the matrix, thus allowing release into the headspace. Release upon reduction of matrix water content was thought to be due to the glassy matrix cracking under drying stresses. Even the 'as prepared' material showed small levels of release that were attributed to the evaporation of small amounts of surface oil and subsurface oil in cracks. Overall, there was no unambiguous evidence for the matrix being permeable to any of the flavour components. The release of all components followed essentially the same pattern, and could be explained in terms of processes by which the flavour phase gained direct access to the headspace. Experiments using single-phase mixtures would aid unambiguous identification of matrix permeability.

### *4.7.2 Solvent properties of amorphous carbohydrates*

The above partitioning studies show that, although amorphous carbohydrates, like water, are extensively hydrogen-bonded, their solvent properties are not identical. Maltotriose is a convenient material to use to study the general solvent behaviour of a pure amorphous carbohydrate, because, to our knowledge, it has never been observed to crystallize (unless

**Table 4.2**   Octanol–water partition coefficients, $P$, for components of a cherry flavour.

| Component | Log $P$ |
| --- | --- |
| Diacetyl | −1.3 |
| Acetaldehyde | −0.22 |
| Ethyl acetate | 1.04 |
| Benzyl alcohol | 1.10 |
| Benzaldehyde | 1.48 |

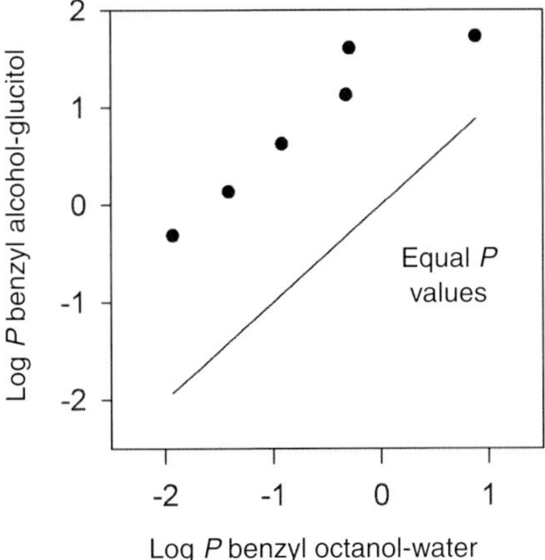



**Fig. 4.10**  Correlation between benzyl alcohol-glucitol partition coefficients (70°C) and octanol-water partition coefficients (20°C) for a series of alcohols.

derivatized[111]), and so only amorphous states appear in its state diagram. An interesting range of phenomena is shown by 50:50 w/w mixtures of dry glassy maltotriose ($T_g = 134°C$) with water, dry methanol, ethanol, *n*-propanol and *n*-butanol, at ambient temperature. Whereas maltotriose and water are completely miscible and form a single solution phase, the alcohol mixtures are only partially miscible and form two separate phases. Methanol is sufficiently miscible to swell the maltotriose particles, plasticizing the mixture through its glass transition and causing the viscous, maltotriose-rich, liquid phase to collapse. In contrast, the ethanol, *n*-propanol and *n*-butanol mixtures are insufficiently miscible to plasticize the maltotriose-rich phase through the glass transition. On stirring these mixtures, the maltotriose-rich particles remain discrete, with no apparent stickiness. Analysis of the alcohol-rich phase does reveal some solubility of the glassy maltotriose-rich particles in the alcohols; it is not

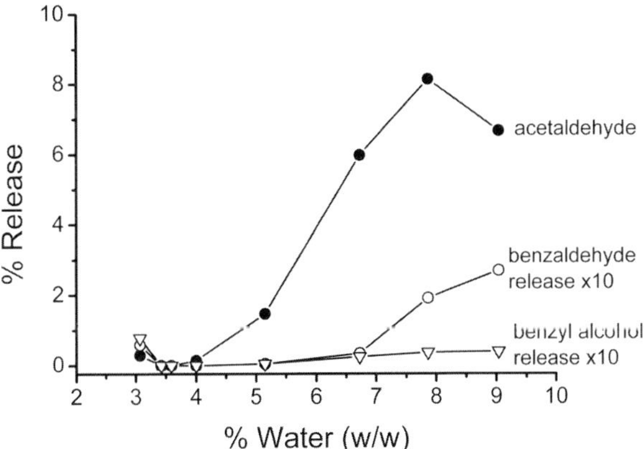



**Fig. 4.11**  Effect of water content on headspace release of a model cherry flavour encapsulated in an initially amorphous sucrose-maltodextrin matrix.

yet known whether this nonequilibrium solubility is sensitive to ageing. The physics of the situation is similar to water sorption isotherms of carbohydrate glasses, wherein one phase is in equilibrium, while the other is not. By variations in composition in ternary maltotriose-alcohol-water mixtures, behaviour can be varied continuously. Figure 4.12 shows a ternary state diagram for maltotriose-ethanol-water mixtures. Phase separation only occurs below about 25% w/w water, and below this limit the water content has a strong influence on the miscibility of the two phases and on the viscosity of the maltotriose-rich phase. Water and ethanol plasticize maltotriose similarly, so that when the combined water and ethanol content is less than 4.8% w/w, the mixture is glassy.

At higher temperatures ethanol and maltotriose become more miscible and, by appropriate choice of composition, it is possible to quench single-phase solutions into the two-phase region of the state diagram.[112] Figure 4.13a and b show the structures that result when the composition crosses from the liquid–liquid region (composition A in Fig. 4.12) to the liquid–glass region (composition B in Fig. 4.12). In Fig. 4.13a the precipitating maltotriose-rich particles are above their glass transition temperature, have stuck together and are undergoing a sintering-like coalescence process. Figure 4.13b shows an aggregate in which the constituent particles are glassy and show limited coalescence. The particles are about 2 μm in diameter and it is clear that the phase separation has proceeded to a large extent before being arrested by the glass transition. Although these systems are models, they are not far removed from some practical systems, and they show a rich variety of phenomena involving phase separation and glass transitions (e.g. critical points, spinodals, sintering, 'glassy precipitation').

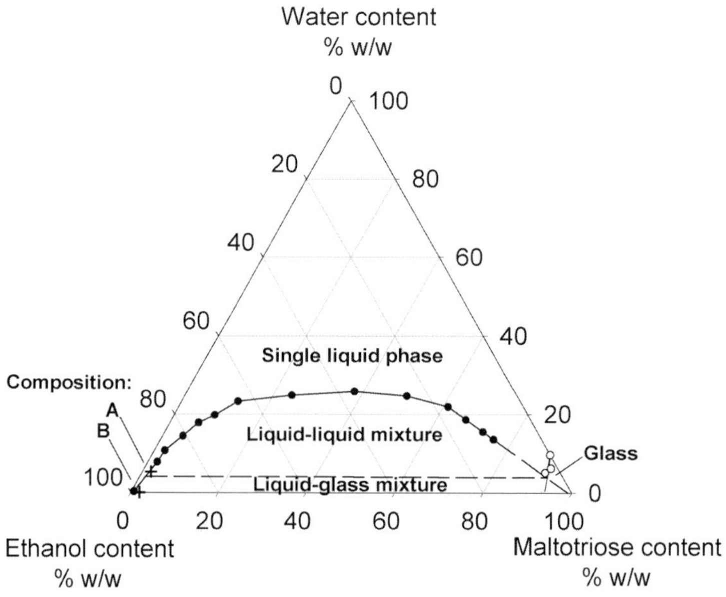


**Fig. 4.12**  Maltotriose-ethanol-water state diagram at 20°C. Liquid–liquid boundary (●); glass transition curve (○); composition before quenching to study the microstructure (see Fig. 4.13) (+).

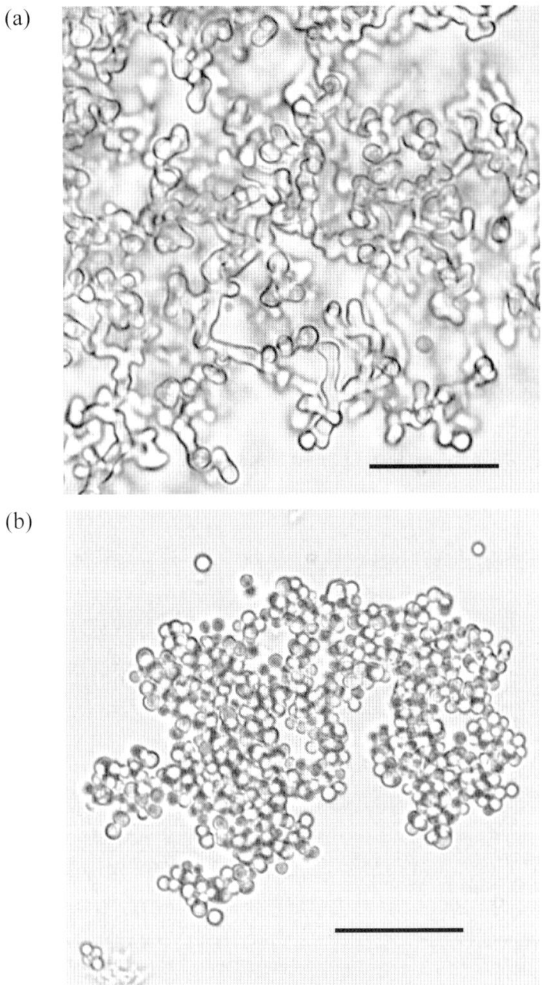

**Fig. 4.13**   Structure of maltotriose-rich precipitate in a quenched ethanol-water-maltotriose mixture. (a) Composition A highly viscous, (b) composition B glass, as shown in Fig. 4.12. Scale bars represent 20 μm.

## 4.8 Concluding remarks

As a result of recent research, there is a much wider recognition of the importance of the glass transition in influencing the properties of low-moisture food materials, and the importance of moisture management in the control of product quality. Developing areas of research include the role of glasses in preservation and their effect on chemical stability. The notion that all chemical reaction was arrested in the glassy state has proved too simplistic a view and is in need of revision. While the glass transition was thought to be relevant mainly to low-moisture materials (or frozen systems), research on the physics of colloidal glasses is showing that useful insight may be gained into the behaviour of other types of food material including concentrated emulsions, and particulate systems such as starch pastes and globular protein gels.

## Acknowledgements

This research was supported by the core strategic grant of the Biotechnology and Biological Sciences Research Council.

## 4.9 References

1 Parks, G.S., Huffman, H.M. & Cattoir, F.R. (1928) Studies on glass – the transition between the liquid and glassy states of glucose. *J. Phys. Chem.* **32**, 1366–1379.

2 Parks, G.S., Barton, L.E., Spaght, M.E. & Richardson, J.W. (1934) The viscosity of undercooled liquid glucose. *Physics* **8**, 193–199.

3 Green, J.L., Ito, K., Xu, K. & Angell, C.A. (1999) Fragility in liquids and polymers: New, simple quantifications and interpretations. *J. Phys. Chem. B* **103**, 3991–3996.

4 Williams, M.L., Landel, R.F. & Ferry, J.D. (1955) The temperature dependence of relaxation mechanisms in amorphous polymers and other glass-forming liquids. *J. Am. Chem. Soc.* **77**, 3701–3707.

5 Ferry, J.D. (1980) *Viscoelastic Properties of Polymers.* Wiley, New York.

6 Tromp, R.H., Parker, R. & Ring, S.G. (1997) A neutron scattering study of the structure of amorphous glucose. *J. Chem. Phys.* **107**, 6038–6049.

7 Orford, P.D., Parker, R., Ring, S.G. & Smith, A.C. (1989) Effect of water as a diluent on the glass-transition behavior of malto-oligosaccharides, amylose and amylopectin. *Int. J. Biol. Macromol.* **11**, 91–96.

8 Franks, F. (1985) *Biophysics and Biochemistry at Low Temperatures.* Cambridge University Press, Cambridge.

9 Parker, R. & Ring, S.G. (1995) Diffusion in maltose-water mixtures at temperatures close to the glass transition. *Carbohydr. Res.* **273**, 147–155.

10 Ablett, S, Izzard, M.J. & Lillford, P.J. (1992) Differential scanning calorimetric study of frozen sucrose and glycerol solutions. *J. Chem. Soc. Faraday Trans.* **88**, 789–794.

11 Bizot, H., Lebail, P., Leroux, B. *et al.* (1997) Calorimetric evaluation of the glass transition in hydrated, linear and branched polyanhydroglucose compounds. *Carbohydr. Polym.* **32**, 33–50.

12 Lillie, M.A. & Gosline, J.M. (1990) The effects of hydration on the dynamical mechanical properties of elastin. *Biopolymers* **29**, 1147–1160.

13 Pak, J., Pyda, M. & Wunderlich, B. (2003) Rigid amorphous fractions and glass transitions in poly(oxy-2,6-dimethyl-1,4-phenylene). *Macromolecules* **36**, 495–499.

14 Wunderlich, B. (2003) Reversible crystallization and the rigid-amorphous phase in semicrystalline macromolecules. *Prog. Polym. Sci.* **28**, 383–450.

15 Zeleznak, K.J. & Hoseney, R.C. (1987) The glass transition in starch. *Cereal Chem.* **64**, 121–124.

16 Noel, T.R., Parker, R., Ring, S.G. & Tatham, A.S. (1995) The glass-transition behavior of wheat gluten proteins. *Int. J. Biol. Macromol.* **17**, 81–85.

17 Ferrari, C. & Johari, G.P. (1997) Thermodynamic behaviour of gliadins mixture and the glass-softening transition of its dried state. *Int. J. Biol. Macromol.* **21**, 231–241.

18 Micard, V. & Guilbert, S. (2000) Thermal behavior of native and hydrophobized wheat gluten, gliadin and glutenin-rich fractions by modulated DSC. *Int. J. Biol. Macromol.* **27**, 229–236.

19 Kasapis, S., Al Marhoobi, I.M. & Mitchell, J.R. (2003) Molecular weight effects on the glass transition of gelatin/cosolute mixtures. *Biopolymers* **70**, 169–185.

20 Sartor, G., Mayer, E. & Johari, G.P. (1994) Calorimetric studies of the kinetic unfreezing of molecular motions in hydrated lysozyme, hemoglobin, and myoglobin. *Biophys. J.* **66**, 249–258.

21 Rupley, J.A. & Careri, G. (1997) Protein hydration and function. *Adv. Protein Chem.* **41**, 37–172.

22 van Soest, J.J.G. & Knooren, N. (1997) Influence of glycerol and water content on the structure and properties of extruded starch plastic sheets during aging. *J. Appl. Polym. Sci.* **64**, 1411–1422.

23 Lourdin, D., Bizot, H. & Colonna, P. (1997) "Antiplasticization" in starch-glycerol films? *J. Appl. Polym. Sci.* **63**, 1047–1053.

24 Lourdin, D., Ring, S.G. & Colonna, P. (1998) Study of plasticizer-oligomer and plasticizer-polymer interactions by dielectric analysis: maltose-glycerol and amylose-glycerol-water systems. *Carbohydr. Res.* **306**, 551–558.

25 Forssell, P., Mikkila, J., Suortti, T. *et al.* (1996) Plasticization of barley starch with glycerol and water. *J. Macromol. Sci. Pure A* **33**, 703–715.

26 Forssell, P.M., Mikkila, J.M., Moates, G.K. & Parker, R. (1997) Phase and glass transition behaviour of concentrated barley starch-glycerol-water mixtures, a model for thermoplastic starch. *Carbohydr. Polym.* **34**, 275–282.

27 Gaudin, S., Lourdin, D., Forssell, P.M. & Colonna, P. (2000) Antiplasticisation and oxygen permeability of starch-sorbitol films. *Carbohydr. Polym.* **43**, 33–37.

28 Couchman, P.R. & Karasz, F.E. (1978) A classical thermodynamic discussion of the effect of composition on glass-transition temperatures. *Macromolecules* **11**, 117–119.

29 Couchman, P.R. (1978) Compositional variation of glass-transition temperatures. 2. Application of the thermodynamic theory to compatible polymer blends. *Macromolecules* **11**, 1156–1161.

30 Sugisaki, M., Suga, H. & Seki, S. (1968) Calorimetric study of the glassy state. IV Heat capacity of glassy water and cubic ice. *Bull. Chem. Soc. Jpn.* **41**, 2591–2599.

31 Johari, G.P., Hallbrucker, A. & Mayer, E. (1987) The glass liquid transition of hyperquenched water. *Nature* **330**, 552–553.

32 Champeney, D.C. & Ould-Kaddour, F. (1987) Conductance of KCl-glycerol solutions – test of the Fuoss paired ion model. *Phys. Chem. Liquids* **16**, 239–247.

33 Phan, S.E., Russel, W.B., Cheng, Z.D. *et al.* (1996) Phase transition, equation of state, and limiting shear viscosities of hard sphere dispersions. *Phys. Rev. E* **54**, 6633–6645.

34 Krieger, I.M. (1972) Rheology of monodisperse lattices. *Adv. Colloid Interface Sci.* **3**, 111–136.

35 Mason, T.G., Lacasse, M.D., Grest, G.S. *et al.* Osmotic pressure and viscoelastic shear moduli of concentrated emulsions. *Phys. Rev. E* **56**, 3150–3166.

36 Mason, T.G. & Weitz, D.A. (1995) Linear viscoelasticity of colloidal hard-sphere suspensions near the glass transition. *Phys. Rev. Lett.* **75**, 2770–2773.

37 Torquato, S., Truskett, T.M. & Debenedetti, P.G. (2000) Is random close packing of spheres well defined? *Phys. Rev. Lett.* **84**, 2064–2067.

38 Rintoul, M.D. & Torquato, S. (1998) Hard-sphere statistics along the metastable amorphous branch. *Phys. Rev. E* **58**, 532–537.

39 Donev, A., Cisse, I., Sachs, D. *et al.* (2004) Improving the density of jammed disordered packings using ellipsoids. *Science* **303**, 990–993.

40 Donev, A., Torquato, S., Stillinger, F.H. & Connelly, R. (2004) Jamming in hard sphere and disk packings. *J. Appl. Physics* **95**, 989–999.

41 Trappe, V., Prasad, V., Cipelletti, L. *et al.* (2001) Jamming phase diagram for attractive particles. *Nature* **411**, 772–775.

42 Segre, P.N., Prasad, V., Schofield, A.B. & Weitz, D.A. (2001) Glasslike kinetic arrest at the colloidal-gelation transition. *Phys. Rev. Lett.* **86**, 6042–6045.

43 Pham, K.N., Puertas, A.M., Bergenholtz, J. *et al.* (2002) Multiple glassy states in a simple model system. *Science* **296**, 104–106.

44 Courtland, R.E. & Weeks, E.R. (2003) Direct visualization of ageing in colloidal glasses. *J. Phys. Condensed Matter* **15**, S359–S365.

45 Weeks, E.R., Crocker, J.C., Levitt, A.C. *et al.* (2000) Three-dimensional direct imaging of structural relaxation near the colloidal glass transition. *Science* **287**, 627–631.

46 Brownsey, G.J., Noel, T.R., Parker, R. & Ring, S.G. (2003) The glass transition behavior of the globular protein bovine serum albumin. *Biophys. J.* **85**, 3943–3950.

47  Faivre, A., Niquet, G., Maglione, M. *et al.* (1999) Dynamics of sorbitol and maltitol over a wide time-temperature range. *Eur. Polym. J. B* **10**, 277–286.

48  Faivre, A. (1997) *Étude des Phenomes de Relaxation Associés à la Transition Vitreuse.* Ph.D thesis, INSA de Lyon, France.

49  Angell, C.A. (2000) Ten questions on glassformers, and a real space "excitations" model with some answers on fragility and phase transitions. *J. Phys. Condens. Mat.* **12**, 6463–6475.

50  van Dusschoten, D., Tracht, U., Heuer, A. & Speiss, H.W. (1999) Site specific rotational mobility of anhydrous glucose near the glass transition as studied by 2D echo decay 13C NMR. *J. Phys. Chem. A* **103**, 8359–8364.

51  Moran, G.R. & Jeffrey, K.R. (1999) A study of the molecular motion in glucose/water mixtures using deuterium nuclear magnetic resonance. *J. Chem. Phys.* **110**, 3472–3483.

52  Noel, T.R., Parker, R. & Ring, S.G. (2000) Effect of molecular structure and water content on the dielectric relaxation behaviour of amorphous low molecular weight carbohydrates above and below their glass transition. *Carbohydr. Res.* **329**, 839–845.

53  Freeman, B.D. & Hill, A.J. (1998) Free volume and transport properties of barrier and membrane polymers. *ACS Symp. Ser.* **710**, 306.

54  Debenedetti, P.G. (1996) *Metastable Liquids – Concepts and Principles.* Princeton University Press, Princeton, NJ.

55  Debenedetti, P.G. & Stillinger, F.H (2001) Supercooled liquids and the glass transition. *Nature* **410**, 259–267.

56  Sillescu, H. (1999) Heterogeneity at the glass transition: a review. *J. Non-Cryst. Solids* **243**, 81–108.

57  Ediger, M.D. (2000) Spatially heterogeneous dynamics in supercooled liquids. *Ann. Rev. Phys. Chem.* **51**, 99–128.

58  Richert, R. (2002) Heterogeneous dynamics in liquids: fluctuations in space and time. *J. Phys.-Cond. Mat.* **14**, R703-R738.

59  Bohmer, R., Hinze, G., Diezemann, G. *et al.* (1996) Dynamic heterogeneity in supercooled ortho-terphenyl studied by multidimensional deuteron NMR. *Europhys. Lett.* **36**, 55–60.

60  Deschenes, L.A. & Bout, D.A.V. (2002) Heterogeneous dynamics and domains in supercooled o-terphenyl: A single molecule study. *J. Phys. Chem. B* **106**, 11438–11445.

61  Deschenes, L.A. & Vanden Bout, D.A. (2001) Single-molecule studies of heterogeneous dynamics in polymer melts near the glass transition. *Science* **292**, 255–258.

62  Hempel, E., Hempel, G., Hensel, A. *et al.* (2000) Characteristic length of dynamic glass transition near $T_g$ for a wide assortment of glass-forming substances. *J. Phys. Chem. B* **104**, 2460–2466.

63  Hodge, I.M. (1995) Physical ageing in polymer glasses. *Science* **267**, 1945–1947.

64  Moynihan, C.T., Crichton, S.N. and Opalka, S.M. (1991) Linear and non-linear structural relaxation. *J. Non-Cryst. Solids* **131**, 420–434.

65  Hodge, I.M. & Berens, A.R. (1985) Effects of annealing and prior history on the enthalpy relaxation in glassy polymers. 5. Mathematical modeling of nonthermal preageing perturbations. *Macromolecules* **18**, 1980–1984.

66  Hodge, I.M. (1994) Enthalpy relaxation and recovery in amorphous materials. *J. Non-Cryst. Solids* **169**, 211–266.

67  Berens, A.R. & Hodge, I.M. (1982) Effects of annealing and prior history on enthalpy relaxation in glassy polymers. 1. Experimental study on polyvinyl-chloride. *Macromolecules* **15**, 756–761.

68  Hodge, I.M. & Berens, A.R. (1981) Calculation of the effects of annealing on sub-Tg endotherms. *Macromolecules* **14**, 1598–1599.

69  Hodge, I.M. (1983) Effects of annealing and prior history on enthalpy relaxation in glassy polymers. 4. Comparison of 5 polymers. *Macromolecules* **16**, 898–902.

70  Hodge, I.M. & Berens, A.R. (1982) Effects of annealing and prior history on enthalpy relaxation in glassy polymers. 2. Mathematical modeling. *Macromolecules* **15**, 762–770.

71  Lee, M., Saha, S.K., Moynihan, C.T. & Schroeder, J. (1997) Non-exponential structural relaxation, anomalous light scattering and nanoscale inhomogeneities in glasses. *J. Non-Cryst. Solids* **222**, 369–375.

72  Lammert, A.M., Lammert, R.M. & Schmidt, S.J. (1999) Physical aging of maltose glasses as measured by standard and modulated differential scanning calorimetry. *J. Thermal Anal. Calorimetry* **55**, 949–975.

73  Schmidt, S.J. & Lammert, A.M. (1996) Physical aging of maltose glasses. *J. Food Sci.* **61**, 870–875.

74  Shogren, R.L. (1992) Effect of moisture content on the melting and subsequent physical aging of cornstarch. *Carbohydr. Polym.* **19**, 83–90.

75  Lourdin, D., Colonna, P., Brownsey, G.J. *et al.* (2002) Structural relaxation and physical ageing of starchy materials. *Carbohydr. Res.* **337**, 827–833.

76  Noel, T.R., Parker, R., Ring, S.M. & Ring, S.G. (1999) A calorimetric study of structural relaxation in a maltose glass. *Carbohydr. Res.* **319**, 166–171.

77  Liu, A.J. & Nagel, S.R. (1998) Nonlinear dynamics – Jamming is not just cool any more. *Nature* **396**, 21–22.

78  Trappe, V., Prasad, V., Cipelletti, L. *et al.* (2001) Jamming phase diagram for attractive particles. *Nature* **411**, 772–775.

79  Dawson, K.A. (2002) The glass paradigm for colloidal glasses, gels, and other arrested states driven by attractive interactions. *Curr. Opin. Colloid Interface Sci.* **7**, 218–227.

80  Cipelletti, L., Ramos, L., Manley, S. *et al.* (2003) Universal non-diffusive slow dynamics in aging soft matter. *Faraday Discussions* **123**, 237–251.

81  Prasad, V., Trappe, V., Dinsmore, A.D. *et al.* (2003) Universal features of the fluid to solid transition for attractive colloidal particles. *Faraday Discussions* **123**, 1–12.

82  Kilfoil, M.L., Pashkovski, E.E., Masters, J.A. & Weitz, D.A. (2003) Dynamics of weakly aggregated colloidal particles. *Philos. T. Roy. Soc. A* **361**, 753–764.

83  Hagen, S.J., Hofrichter, H.J., Bunn, H.F. & Eaton, W.A. (1995) Comments on the physics and chemistry of trehalose as a storage medium for hemoglobin-based blood substitutes: "From Kramers theory to the battlefield". *Transfus. Clin. Biol.* **2**, 423–426.

84  Anandaraman, S. & Reineccius, G.A. (1986) Stability of encapsulated orange peel oil. *Food Technol.* **40**, 88–93.

85  Franks, F. (1994) Long-term stabilization of biologicals. *Biotechnology* **12**, 253–256.

86  Crowe, J.H., Carpenter, J.F. & Crowe, L.M. (1998) The role of vitrification in anhydrobiosis. *Ann. Rev. Physiol.* **60**, 73–103.

87  Rice, S.A. (1985) *Diffusion Limited Reactions.* Elsevier, Amsterdam.

88  Russel, W.B., Saville, D.A. & Schowalter, W.R. (1989) *Colloidal Dispersions.* Cambridge University Press, Cambridge.

89  Atkins, P.W. (1978) *Physical Chemistry.* Oxford University Press, Oxford.

90  Solc, K. & Stockmayer, W.H. (1971) Kinetics of diffusion-controlled reaction between chemically asymmetric molecules. I. General theory. *J. Chem. Phys.* **54**, 2981–2988.

91  Zhou, H.X. (1993) Brownian dynamics study of the influences of electrostatic interaction and diffusion on protein-protein association kinetics. *Biophys. J.* **64**, 1711–1726.

92  Craig, I.D., Parker, R., Rigby, N.M. *et al.* (2001) Maillard reaction kinetics in model preservation systems in the vicinity of the glass transition: Experiment and theory. *J. Agric. Food Chem.* **49**, 4706–4712.

93  Collins, F.C. & Kimball, G.E. (1949) Diffusion-controlled reaction rates. *J. Colloid Sci.* **4**, 425–437.

94  Tyrrell, H.J.V. & Harris, K.R. (1984) *Diffusion in Liquids.* Butterworths, London.

95  Champion, D., Hervet, H., Blond, G. *et al.* (1997) Translational diffusion in sucrose solutions in the vicinity of their glass transition temperature. *J. Phys. Chem. B* **101**, 10674–10679.

96  Gunning, Y.M., Parker, R. & Ring, S.G. (2000) Diffusion of short chain alcohols from amorphous maltose-water mixtures above and below their glass transition temperature. *Carbohydr. Res.* **329**, 377–385.

97  Noel, T.R., Parker, R. & Ring, S.G. (1996) Conductivity of maltose-water-KCl mixtures in the supercooled liquid and glassy states. *J. Chem. Soc. Faraday Trans.* **92**, 1921–1926.

98  Robinson, R.A. & Stokes, R.H. (1959) *Electrolyte Solutions.* Butterworths, London.

99 Hagen, S.J., Hofrichter, J. & Eaton, W.A. (1995) Protein reaction kinetics in a room-temperature glass. *Science* **269**, 959–962.

100 Dainton, F.S., Henry, M.S., Pilling, M.J. & Spencer, P.C. (1977) Viscosity dependence of diffusion-controlled triplet energy transfer in 2-methylpentan 2-,4-diol. *J. Chem. Soc. Faraday Trans.* **73**, 243–249.

101 Levine, H. & Slade, L. (1986) A polymer physicochemical approach to the study of commercial starch hydrolysis products (SHPS). *Carbohydr. Polym.* **6**, 213–244.

102 Streefland, L., Auffret, A.D. & Franks, F. (1998) Bond cleavage reactions in solid aqueous carbohydrate solutions. *Pharm. Res.* **15**, 843–849.

103 Schebor, C., Buera, M.D., Karel, M. & Chirife, J. (1999) Color formation due to non-enzymatic browning in amorphous, glassy, anhydrous, model systems. *Food Chem.* **65**, 427–432.

104 Lai, M.C., Schowen, R.L., Borchardt, R.T. & Topp, E.M. (2000) Deamidation of a model hexapeptide in poly(vinyl alcohol) hydrogels and xerogels. *J. Peptide Res.* **55**, 93–101.

105 Orlien, V., Andersen, A.B., Sinkko, T. & Skibsted, L.H. (2000) Hydroperoxide formation in rapeseed oil encapsulated in a glassy food model as influenced by hydrophilic and lipophilic radicals. *Food Chem.* **68**, 191–199.

106 Gunning, Y.M., Gunning, P.A., Kemsley, E. *et al.* (1999) Factors affecting the release of flavor encapsulated in carbohydrate matrixes. *J. Agric. Food Chem.* **47**, 5198–5205.

107 Andersen, A.B., Risbo, J., Andersen, M.L. & Skibsted, L.H. (2000) Oxygen permeation through an oil-encapsulating glassy food matrix studied by ESR line broadening using a nitroxyl spin probe. *Food Chem.* **70**, 499–508.

108 Imagi, J., Muraya, K., Yamashita, D. *et al.* (1992) Retarded oxidation of liquid lipids entrapped in matrices of saccharides or proteins. *Biosci. Biotech. Biochem.* **56**, 1236–1240.

109 Leo, A.J., Hansch, C. & Elkins, D. (1971) Partition coefficients and their uses. *Chem. Rev.* **71**(6), 525–616.

110 Gunning, Y.M., Parker, R., Ring, S.G. *et al.* (2000) Phase behavior and component partitioning in low water content amorphous carbohydrates and their potential impact on encapsulation of flavors. *J. Agric. Food Chem.* **48**, 395–399.

111 Pangborn, W., Langs, D. & Perez, S. (1985) Regular left-handed fragment of amylose: crystal and molecular structure of methyl-α-maltotrioside, $4H_2O$. *Int. J. Biol. Macromol.* **7**, 363–369.

112 Gunning, Y.M., Lalloue, B., Noel, T.R. *et al.* (2004) Phase behaviour and microstructure of partially-miscible, glass-forming alcohol-water-maltotriose mixtures. *J. Materials Sci.* **39**, 6945–6950.

# Chapter 5
# Powders and Granular Materials

*Gary C. Barker*

## 5.1 Introduction

Powders and granular materials can be found in all phases of food manufacture. Raw materials, processed materials, ingredients and many finished food products are particulates. These range from fundamental materials like flour and salt to complex, engineered substances like powdered flavours, granulated beverages and branded breakfast cereals. Jenike and Johanson (http://www.jenike.com/pages/experience/food_ind/food.html) list nearly 200 common food materials that have particulate properties and are used in typical solids handling operations. The UK flour milling industry alone has a turnover of £1 billion (Flour Advisory Bureau; URL http://www.fabflour.co.uk), and coffee, in all its many particulate forms, is the second most traded commodity on world markets (after petroleum). Even common liquid food materials, like milk or wine, are often turned into particulate solids for storage or transportation. Statistics from the UK show that approximately 15% of all milk utilized by dairies in 2002–03 (~2 × 10$^9$ litres) was turned into either full cream or skim milk powders[1] (fluid cow's milk is about 87% water). These powders are shelf stable, convenient and easily transported worldwide.

A precise definition of granular material does not exist. The term 'granular material' describes any large, dense collection of discrete, macroscopic solid particles. This description applies equally to mustard seeds and cornflakes (typically mustard seeds used in commercial preparation of spices are spherical and 95% are smaller than 400 μm in diameter). Sometimes granular materials that have a typical particle size that is smaller than a threshold size, ~100 μm, are called powders, as opposed to granular solids, but this distinction is somewhat arbitrary.[2] Another classification scheme defines particulates as either free-flowing or cohesive. Usually free-flowing means that there are no attractive forces between individual particles whereas a powder is cohesive if particles stick together. For particularly small solid particles van der Waals forces are significant compared with gravity and cause cohesion. Alternatively moisture within the surrounding air (or other fluid) can exist as small, liquid bridges between particles. In this configuration the surface energy of the fluid acts to attract particles to their neighbours. In some food materials particle sizes are such that surface forces are not important. Flour and fine sugar particles are significant exceptions. Fine icing sugars may include particles in the 3–100 μm range (British Sugar, URL http://www.britishsugar.co.uk) where van der Waals forces are significant. Commercial icing sugar products therefore often include additional, inert species of particles, called free-flow agents, to reduce cohesion and assist with handling or processing operations.

The mechanical and flow properties of granular materials are very complex and in many current food manufacturing operations all but the most basic understanding is based on empiricism. It is well known that granular material may appear to flow like a liquid in some situations and still behave like a traditional solid, supporting external stress, in others. However, it is difficult to predict when the solid-like to liquid-like change will occur for a particular material, for example, as the result of an increasing shear stress. Similarly it is difficult to predict when an initially loose packing of grains in a container will become solid and impenetrable from extended storage and ageing. A few fundamental principles of powder mechanics, including static-friction and dilatancy, were first identified, more than a century ago, by classical physicists like Coulomb, Faraday and Reynolds. However, these principles are difficult to apply in real situations and are easily obscured by the wide variety of materials and processes that are commonly encountered in solids handling. For many years soil science was the only area in which principles of powder mechanics were used regularly but, more recently, there has been new interest from physicists.[3–5] This interest stems only partially from a drive for better understanding and improved control in areas of significant commercial interest like food manufacturing. Additionally, academic research into powder mechanics has mushroomed because newly identified analogies, between granular dynamics and motions in other slow-moving systems like glasses, indicate that granular materials might provide ideal opportunities for understanding nonequilibrium systems at a fundamental level.[6]

Dissipation is one fundamental aspect of granular dynamics. Every time that two grains of powder collide, or move while they are in contact, some of the kinetic energy is lost to internal degrees of freedom, heat, sound, etc. This means that, as a collection, grains slow down rapidly unless there is a constant source of kinetic energy (driving). When powder is poured from one container into another the grains quickly come to rest and form a static bed or a pile. The individual powder grains are trapped in a fixed configuration or packing. Dissipative dynamics is the origin of the frequent statement that a powder has temperature $T = 0$. The molecules that form the solid matter of each particle remain fully mobilized but this molecular motion has no effect on the configuration of the powder as a whole, at the grain level, which is trapped when all the kinetic energy disappears. The thermodynamic temperature may as well be zero with regard to its effect on grain packings. This behaviour is in complete contrast to that for a poured liquid, which, when the pouring motion has died away, has a state that is determined by the motion of the constituent molecules, that is, by the absolute temperature. Dissipative dynamics has profound consequences. Raising the thermodynamic temperature does not cause a powder to explore different configurations within its configuration space; only the addition of kinetic energy can cause the grains to find new packings. In turn, classical thermodynamics cannot be applied, simply, to grain configurations so that traditional concepts like entropy do not have a clear meaning. In this case grain configurations that include large amounts of order can be favoured over disordered configurations if they can be achieved more easily while the grains are in motion, that is, if they are favoured by kinetics. Many effects called powder demixing or segregation illustrate this phenomenon.

Rapid dissipation ensures that static, close-packed, granular configurations are metastable. This means that packings with lower (potential) energy could be found but, because $T = 0$, there are no random fluctuations to enable granular materials to swap rapidly between different (closely related) states. As a consequence the densities, connectivities and stresses that exist in one metastable state, after a period of external driving, are often strongly related to those that existed previously. This means that the properties and behaviour of granular materials are fundamentally process history dependent. Knowledge about the current state of a powdered material is often insufficient to predict how it will respond to further driving. This aspect has a crucial impact on industrial solids handling where sources of materials, factory-scale processes and operational conditions are not always assured or repeatable.

The decoupling of grain configurations from the thermodynamic environment initiates a new scale of energy that dominates granular behaviour. The role of kinetic energy in exploring different grain configurations ensures that the relevant energy scale is $\sim mgd$, the potential energy required to raise a single grain through its own diameter, $d$, where $m$ is a typical grain mass and $g$ is gravitational acceleration. For typical granular materials this energy is many orders of magnitude larger than the thermodynamic scale and it determines the practical scale for input energy, in stirrers, agitators, etc., for use in powder handling.

Static friction is another fundamental element of powder mechanics. Static friction, $\mathbf{F}$, is the force that resists the relative sliding motion for two solid surfaces and is often expressed as $\mathbf{F} \leq \mu \mathbf{N}$ where $\mathbf{N}$ is the normal contact force and $\mu$ the limiting value of the friction coefficient. For solid objects the most familiar experimental technique for measuring the limiting friction coefficient involves placing a block of material on a horizontal surface and then tilting until the object begins to slide. The limiting coefficient can be expressed as the tangent of the limiting angle (the friction angle). For granular materials, where there are an enormous number of solid–solid contacts, the role of static friction is much more complex but when grains are poured onto a flat surface they form a pile that has a fixed maximum slope. Stability of the pile at the maximum slope, resisting the gravitational stress, originates from solid–solid friction at grain–grain and grain–floor contacts. The angle made by the free surface to the horizontal is called an angle of repose. From Coulomb onwards scientists have failed to find a simple relationship that links the friction angle with the angle of repose, and this remains an area of active research. In addition to static friction the maximum stable angle of the free surface of a pile depends on details of the particle shapes, because interlocking asperities can act like friction, and on the way that the pile is made. The 'poured' angle of repose is slightly different from the 'drained' angle of repose.

The relationship between friction and repose has far more than an academic interest. Wall friction, the friction between the particles and the material used for constructing a container, determines the optimum angle required for a conical-shaped hopper to deliver desirable mass flows. If the hopper angle is too shallow the discharge may be intermittent and if the angle is too steep an internal 'funnel' may form leaving static material in a zone at the periphery of the discharge cone. Funnel flows, and associated long residence times, are particularly undesirable for food powders. Currently, appropriate hopper angles for each material are found by incorporating the results from (expensive) laboratory tests, for each material, into continuum mechanics models that describe stresses in hopper geometries (Powder and Bulk, URL http://www.powderandbulk.com). Although this approach has strong empirical validity the cost of shear tests is often limiting and additional testing is required every time that powders are mixed, get damp or just get older. An understanding of static friction at the particulate level would have a huge impact on solids handling and would bring significant commercial advantages.

Wall friction is also responsible for the nonlinear variation of pressure in a vertical cylinder containing granular material – in solids handling these vessels are usually called bins. In contrast to the hydrostatic pressure developed by liquids the vertical stress in packed granular material saturates over distances $\sim D/\mu$ where $D$ is the diameter of the cylinder and $\mu$ is the friction coefficient for the granular material and the walls. For tall, narrow bins this effect ensures a constant flow rate for granular discharge (independent of the head height as in an egg timer) but it also means that the walls of the silo or bin must support part of the load. There are numerous engineering standards to ensure that silo walls have sufficient strength to cope with a broad range of conditions but occasionally unpredicted, eccentric or dynamic, loads arise and cause catastrophic failure (Jenike & Johanson, URL http://www.jenike.com/pages/education/cases/pdf_docs/02jjcase.pdf). Many current estimates of the wall loads used for silo designs are based on assumptions, originally proposed by Janssen more than a century ago, concerning the stress in ideal (continuum) granular material.[2] Although modern viewpoints provide a critique of this model,[7] there is still insufficient understanding of the friction forces in particulate materials to provide improved, alternative, inputs into most silo designs.

A further fundamental property of granular materials is called dilatancy. This term was first used by Reynolds to indicate that static granular material must expand, at least momentarily, before it can flow.[2] The phenomenon has simple geometrical origins and can be observed for a wide variety of disordered materials that are composed from close packed, nondeformable, particles. Sometimes the effects of dilatancy are quite dramatic. When ground coffee is vacuum packed in foil, so that air pressure forces the packaging material tightly against the coffee grains, the package becomes completely rigid – called a 'brick pack'. In these conditions there is no free volume and therefore no possibility for the granular material to expand. Even though the material in the packet is made from loose individual grains it cannot flow or deform. Once the vacuum is released the flexible pack provides room for expansion and the material becomes soft and free flowing. Removing oxygen from the packet in this way, and preventing oxidation of the aromatic compounds, can extend the shelf life of freshly roasted coffee by approximately three times. However, there is some market resistance to the brick packs, which look and feel unnatural, and in some cases flushing the packet with nitrogen gas is a preferable means for reducing the oxidation and staling of ground coffee.

Dilatancy is particularly relevant in relation to sheared granular materials. The structural deformations that precede flow for sheared particulate media are inhomogeneous and are often localized in the form of shear bands.[8] In these zones, which are typically a few particles wide lying parallel to the shear plane, the volume of the void space increases in response to shear stress. In turn this expansion releases some packing constraints and initiates granular flow. Relatively few structural changes occur in the surrounding material and the shear bands persist in the velocity profile of the flowing material. In this regime the dilatancy acts as a relationship between the density and the shear stress for dense granular material. The numbers, widths and locations of shear bands depend on the particulate structure, in addition to the operating configuration, and, therefore, they depend on the process history of the material. Powders that flow along chutes, through valves and out of bins are subject to shear forces so that an improved appreciation of dilatancy has a significant impact on many areas of powder handling.

In many commercial situations, which include food manufacture, dissipation, friction and dilatancy dominate the behaviour of powders and granular solids. However, it is often apparent that inability to apply a fundamental understanding of these complex phenomena means that the development of new materials and novel processes is impeded. Within the food industry developments involving granular materials often require time-consuming and expensive testing and pilot operations. A recent European Union report,[9] concerning powder research within the food industry, stressed that knowledge of powder processes is far behind that of liquid processes and, therefore, there remain a great many practical problems that current methods do not address effectively. In its conclusion the EU report points to issues, like stabilizing functionality and control of contamination, that are particularly important for food powders, but also stresses that particulate science and technology, developed in other fields, may prove beneficial to improved food manufacturing. Opportunities for innovations in food materials science are increasing in line with the consumer-driven desire for higher quality, higher value and convenient products. In the remainder of this chapter I will highlight some of the fundamental science of granular materials that has emerged in the last decade and indicate the relevance to modern commercial food powder handling operations.

## 5.2 Packing

Powders and granular materials do not have fixed or unique densities. Firstly the dilatancy of granular material causes a local reduction of the density whenever flow is initiated, but additionally the metastability of close-packed grain configurations ensures that even static granular materials have variable density. A sample of granular material can easily be prepared with a range of densities; for example, gently tapping the side of a coffee jar causes the upper level of the grains to sink and the coffee density increases. This variability alone has a huge impact on material processing operations and, for example, to ensure the delivery of uniform quantities, with low tolerance, it may be necessary to replace inexpensive volumetric feeding by an expensive gravimetric option. Additionally, many secondary operations in food manufacturing, such as redispersion or coating, rely on predictable and reproducible powder densities and are compromised by spontaneous variability. For example, the powder density is a significant factor contributing to the nonelectrostatic transfer efficiency of powdered flavours onto savoury snacks using a tumble drum operation.[10]

In quantitative analyses of granular materials the fractional volume, $\phi$, occupied by the grains is more significant than the absolute (bulk) density (which involves properties of both the solid material and the surrounding fluid). The volume fraction facilitates comparison between different granular materials and emphasizes the important role played by particle packing in the determination of material properties. For materials made from porous particles even the volume fraction is open to interpretation. The volume fraction for static granular material depends, in a complex way, on the particle geometries, on the interparticle interactions and, as illustrated above, on the process history. However, although the values for the volume fraction vary considerably from material to material and from one condition to another, in general $\phi$ has two easily expressed limits. For each material the loosest granular structures, with volume fraction $\phi_{min}$, occur following high-energy processes such as pouring material from one container into another or spraying material from a nozzle into a bin. The densest packings, with volume fraction $\phi_{max}$, occur following a sustained period of low-energy excitation such as gentle tapping. These volume fractions are sometimes called the random loose packing fraction and the random close packing fraction, and for noncohesive, monodisperse, hard sphere materials they have values ~0.55 and ~0.64. The Hausner ratio,[11] $H = \phi_{max}/\phi_{min}$, is used to quantify the range of densities for a particular material and large values are considered indicative of cohesivity and poor flowability for powdered materials. For many food powders[12,13] this ratio falls in the range $1.2 < H < 1.6$.

In industrial situations it is almost impossible to shield granular materials from background vibrations so that slow densification, or consolidation, is a fundamental aspect of material handling. The rate and extent of the densification is usually unpredictable and this equates to a significant loss of control. Ultimately strongly consolidated materials are undesirable because they resist flow and hinder mixing operations. Preventative or remedial actions are invariably nonincremental and involve large energy inputs in the form of agitation or fluidization. It is almost impossible to return material directly to a uniform loose packing, with density $\phi_{min}$, following consolidation so that as part of natural ageing processes granular material quality is invariably reduced.

Consolidation of granular material is made up of many small changes in the particle positions, biased by downward movements, which are driven by a slow input of kinetic energy. Reorganization of the packing is very complex and includes both the independent motion of individual particles as well as collective motion of groups or clusters. Recently the nonequilibrium statistical mechanics of this relaxation has been the subject of intense investigations. Nowak *et al.*[14] studied the density of a packing of monodisperse spherical particles in a tall cylindrical tube that was subjected to a series of discrete, vertical shakes or 'taps'. In response to tapping there are density fluctuations throughout the tube but the overall trend is a slow densification of the packing with a logarithmic dependence of density on time described by:

$$\phi(t) = \phi(\infty) - \frac{\phi(\infty) - \phi(0)}{1 + B\ln(1 + t/\tau)} \tag{5.1}$$

where $t$ is the time and $B$, $\tau$ are parameters. The slow increase of the volume fraction does not depend fundamentally on the initial value, $\phi(0)$. However the final, steady-state, fraction, $\phi(\infty)$, decreases monotonically as the tapping intensity is increased. The model parameters also depend on the tapping intensity and, in particular, the rate of approach to the steady-state volume fraction, $\tau^{-1}$, slows as the tapping gets weaker. For especially weak tapping, below a threshold strength, the packing becomes trapped in metastable configurations and progress towards a steady state is interrupted.

The threshold behaviour, which appears during the controlled shaking of granular material, leads to a novel view of the compaction process. In the presence of the threshold the only way to reach a steady state for low-intensity excitation involves a careful pretreatment of the sample with stronger vibrations – an annealing process. Nowak *et al.*[14] used a ramped variation of the tapping intensity with time to examine the low-intensity-shaking steady state.

Starting with low-density material tapping is applied with slowly increasing intensity. The material initially consolidates but, when the shaking strength exceeds the threshold strength, the steady packing fraction decreases monotonically with increasing intensity. When the ramp on the shaking intensity is reversed the density of the material rises as the intensity decreases but, now, it continues to rise when the shaking strength dips below the threshold strength. In this regime consolidation is reversible along a continuous, monotonic curve relating the packing density and the excitation strength. The initial evolution, including an apparent compaction with increasing intensity, exists as a separate, irreversible, branch of the density-shaking relationship.

In addition to the slow evolution, described above, the volume fraction of granular material fluctuates in response to tapping. The spectra of density fluctuations in the steady state regime reveal details of the micromechanics of structural relaxations and provide a direct way to compare the quasistatic dynamics of granular materials with other, usually thermal, systems that relax very slowly. Time series analysis by Nowak *et al.*[14] reveals stationary Gaussian fluctuations of the density, for a range of tapping intensities, although some deviations from this pattern are still the subject of debate. Power spectra for density fluctuations show at least two characteristic timescales for structural relaxations. These times depend on the excitation intensity and can be aligned with single particle and collective reorganization processes. The density fluctuations provide an appealing picture of granular dynamics involving a crossover between individual and collective 'equilibrium' responses to tapping as the packing density increases. As the system moves to higher density a growing number of particles have to be rearranged simultaneously to reduce the volume (locally) so that the timescale for density relaxation grows to make the approach to a steady state logarithmically slow.

Several other approaches support this complex picture of granular dynamics. Three-dimensional computer simulations of shaken hard spheres[15] have identified detailed patterns of individual particle and collective relaxations and also describe a monotonic decay of the steady-state volume fraction with shaking intensity. The crucial role of local free volume has been included into several theoretical models that examine the transition to reversible behaviour[16] and that pursue the similarity with glass-like systems.[17] In particular Edwards and his colleagues[18] have developed a well-posed analogy between the athermal statistical mechanics of powders and classical equilibrium thermodynamics. In this analogy, which was one of the drivers for the controlled experimental studies in Chicago,[14] all granular configurations with the same volume have identical probability. In this way the volume takes the place of thermal energy but otherwise the counting and weighting of grain configurations, subject to constraints imposed by particle overlaps and stability, etc., follows the rules of molecular thermodynamics. In this scheme the partial derivative of volume with respect to the entropy is a new parameter, called the 'compactivity', which corresponds, in the analogy, to the thermodynamic temperature. Compactivity quantifies the state of a static powder in the same way that temperature quantifies the state of a fluid. The equilibrium density fluctuations observed by Nowak *et al.* provide a measure of the compactivity. Together these results provide a novel picture of granular systems that are quantified not just by their density but also by the statistics of the configurations. In consequence macroscopic behaviour, such as consolidation, can be understood in terms of statistical processes at the grain level and emergent phenomena, such as the reversibility of the densification process, can be developed into industrially significant strategies for improved materials handling.

Interparticle cohesive forces influence the values of the bulk packing fraction for granular materials, particularly $\phi_{min}$, but do not prevent slow densification.[19] Cohesive materials develop open structures when they are initially poured into a bin so that their loose packing fractions, $\phi_{min}$, are relatively small. This behaviour underlies much of the practical significance that is attributed to the Hausner ratio. Experiments[20] using magnetic particles, with variable interaction strength, support a very simple relationship between the loose packing fraction and the strength of cohesive forces:

$$\phi_{min}^{0} - \phi_{min} \sim \frac{F_c}{mg} \tag{5.2}$$

where $F_c$ is the interparticle force, $mg$ is the particle weight and $\phi_{min}^{0}$ is the limiting value of $\phi_{min}$ for noncohesive particles ($\phi_{min}^{0} \sim 0.56$ for spheres). The loose packing fraction does not depend explicitly on the particle size or on the nature of the cohesive interactions.

This view of granular materials, and of the compaction process, is not restricted to idealized materials composed from hard spheres. Experiments[21] indicate that materials made from hard rod-shaped particles behave like those made from spheres when they are excited by tapping. For rod-shaped particulates the limiting values of volume fraction, $0.49 < \phi < 0.72$, are different from those for spheres but, nonetheless, packings of cylinders have a logarithmically slow densification and a reversible branch of the excitation curve. This indicates that the macroscopic responses of particulate materials to shaking are qualitatively independent of the shapes of the constituent particles. However, rod-shaped particles have additional, orientational, degrees of freedom and this ensures that there is also a relatively fast, shaking-driven, transition to a dense system with orientational order (nematic). For packings of cylinders the transition to an ordered system occurs in addition to the slow consolidation response. Additional, shape, degrees of freedom appear to have considerable impact on the limiting packing fraction even in the absence of an ordering process. Donev *et al.*[22] have measured packing densities for consolidated, ellipsoidal, candy-coated chocolates and found, somewhat counterintuitively, that the maximum packing fraction is higher than the random close packing fraction for spheres. The experiments show that materials made from ellipsoid-shaped particles may achieve random packings with volume fractions $\sim 0.72$. Donev *et al.* suggest that the additional degrees of freedom for nonspherical particles require additional particle contacts in order to maintain mechanical stability and, in turn, this requires higher densities. This integration of particle shape degrees of freedom into models of the density response of granular materials is not complete but it opens additional options for the development and control of industrial powder processing; particularly in situations where optimal performance occurs at high density. Discussions of particle packing and of the relationship between density and coordination have gone on for centuries[23] – and will certainly continue.

## 5.3 Segregation

Free-flowing granular materials segregate. Operations like shaking, stirring or pouring cause initially homogeneous particulate materials to develop regions that are rich in particles with one size, or one shape, or one density, etc. Materials that contain particles with a large range of sizes, such as breakfast cereals, provide particularly visible examples of segregation[24] but the phenomenon is widespread and mixtures that have only a small range in particle properties still separate. Segregation amounts to an ordering process so that the appearance of spontaneous demixing illustrates, starkly, the nonequilibrium nature of granular dynamics. Additionally the appearance of large-scale order contrasts with the random nature of common driving forces, most of which would be thought to encourage mixing, so that granular segregation is often counterintuitive and clearly indicates complex behaviour. Currently this complexity is such that detailed analyses of particular, large-scale, segregation behaviour are not well established.

The industrial impact of particulate segregation is enormous. Many dry food materials are particulate mixtures or blends and in-line segregation leads to a direct loss of product quality or impaired process performance. The individual components of a food powder may be flavours, colours, thickeners or flow agents so that segregation can lead to inconsistent taste, poor appearance, imperfect rehydration, etc. Jenike & Johanson (http://www.jenike. com/pages/experience/food_ind/food.html) list several case studies of food particulate segregation phenomena, which occur in real operational conditions, and the actions that have been used to alleviate the problems. These examples include the demixing of citric acid powder in a drink mix that caused fluctuations in tartness of the final product, the sifting of particulates in a stuffing mix that led to unacceptable variation of the particle size distribution of the bread crumbs, and the separation of the components of a cake mix during pneumatic conveying. Solutions to practical demixing problems usually involve specific, ingenious engineering as well as specialist designs for agitators, chamber geometry, baffles and fill levels in mixers, etc. There are some general rules, built from a strong history of operational experience, but few fundamental principles that promote generic, transferable developments to control segregation.

Particle size differences are a dominant driver for demixing. It is very difficult to maintain a proper mix of unequal particles if the material is subject to fluctuating mechanical forces such as those that arise from discharging hoppers, conveying operations and unshielded vibrations. In 1987 Rosato *et al.*[25] used a very simple Monte Carlo computer simulation model to illustrate the micromechanics of particle size segregation. In the simulations isolated particles, which were larger than their neighbours, rose to the top of a particle bed in response to simple vertical shaking. Similarly shaken mixtures, with two distinct particle sizes, developed a clear separation with a layer of larger particles resting on a bed of smaller ones. The simulations, although abstract, provided a clear picture of the separation phenomenon and gave detailed information about particle positions during the shaking process. Rosato *et al.* identified a statistical mechanism, which favours the downward motion of smaller particles, as the underlying driver of 'The Brazil Nut' effect. This mechanism drives separation for any shaken mixture as long as there is a disparity of particle sizes. The rate of separation increases with the size difference.

Subsequently, further simulations, as well as advanced imaging experiments, identified additional patterns of particulate dynamics that may drive vibration-induced size separation in granular media. Internal avalanches, arching by large particles and mass convection all provide mechanisms that, under appropriate conditions, might cause large particles to rise to the top of shaken granular mixtures. The convective mechanism that leads to size segregation is particularly significant and was described by Knight *et al.*[26] The convective flow of shaken granular material, although well known, was observed,[27] noninvasively and with high resolution, using magnetic resonance imaging to explore the possibilities for segregation. These experiments confirmed that for typical shaken granular materials the downward flow region of a convective cell may be very narrow and so there is a limit on the participation of large particles in the complete flow cycle. This limit amounts to size segregation; in a convective flow pattern large particles remain on the surface because they cannot fit into the region of the downward flow. Friction at the container walls is essential for the initiation of a convective pattern in granular material.[26]

Although the convection mechanism implies mass flow, and therefore no explicit mechanism to distinguish between different particle properties, the separation proceeds with a rate that depends, in a complex fashion, on both particle size and particle density.[28] The large particles disturb the smooth flow pattern during upward convection and move at a different rate to their surroundings. The separation rate does not depend monotonically on size or density and this complex picture culminates in 'reverse Brazil nuts' – that is, large, light particles that actually sink to the bottom of a shaken mixture of grains.[29] This behaviour is still the subject of great debate.[30] The flow of interstitial fluid during the shaking process is also an important variable.[28] Some of the parameter spaces for shaking-induced separation of particular granular materials have been mapped but a complete picture does not exist. However, because it is clear that, at least in the convection regime, changes in density can be used to reverse the direction of separation, there are indications that developments in particle fabrication methods may provide options for improved process control.

Segregation processes in granular materials are not restricted to simple configurations like vertical shaking. Pouring a particulate mixture to make a conical pile can cause segregation of particle sizes; typically large particles are found predominantly at the base of the pile but sometimes particles with distinct sizes form a stratified pattern as the pile grows.[31] The extent of the size separation depends on the deposition rate[32] and on several particulate properties including shape.

Homogeneous mixtures of particulates are particularly difficult to obtain from constituents with very distinct particle sizes or from constituents with very distinct number fractions. Additionally, changing attractive interactions can lead to cohesion and cause changes in the segregation response.[33] However, granular materials with relatively similar, free-flowing particles may also resist mixing. Many commercial mixing operations, for dry granular materials, use a circular bin, or drum, that rotates about a horizontal axis. The granular material in a drum mixer remains static while it follows the rotation of the bin for some of the spin cycle but then becomes unstable and either flows down the free surface, by continuous rolling or as a series of avalanches, or falls freely under gravity and is deposited at the base of the drum. Contrary to most expectations it is impossible to guarantee better blending, and greater homogeneity, simply by increasing the speed, or extending the residence time, for a tumble mixer. A wide range of particulate flow regimes and steady, nonequilibrium, patterns have been observed in rotating drums.[34] The patterns include both radial and axial banding as well as apparently chaotic formations. Noninvasive observations, such as nuclear magnetic resonance (NMR) imaging,[35] show that distinct microscopic particulate transport regimes, like percolation, convection, avalanching or diffusion, are always in competition during driven granular flows. Banding occurs when rapid mixing, for example resulting from a convective mechanism, is restricted by static cores or noninteracting segments of material that are set up by the flow. Because the effective 'granular temperatures' are very low the diffusion mechanism in particulate materials is very inefficient at breaking the barriers to mixing that are imposed by the flow. In industrial configurations, baffles, additional agitators or complex 'double-cone' geometries are usually added to restrict the onset of organized flows but this process is largely empirical and may result in other drawbacks or unforeseen consequences.

The rational design of powder mixing processes, and in particular the ability to scale up observations made in a bench-top blender to industrial sizes, is still a major objective for the food and other process industries.[36] For analogous fluid-phase systems it is clear that organized behaviour at an intermediate scale, between molecular dynamics and continuum flows, is an essential element of scale-up. In this case fluid dynamics provides a catalogue of well-defined instabilities and transitions that dictate the path between the behaviours observed on different scales so that process design is often routine. In contrast, current models of granular dynamics do not provide a simple way to step between different scales of organization in particulate flows and the intermediate scale is largely unexplored. For dense, slow-moving, granular materials the numerous metastable trapped states act as a barrier for the systematic observation of ordered flows at an intermediate scale. However, by removing the jamming from a granular medium, using weak fluidization, Conway *et al.*[37] have identified a series of well-defined driven instabilities in granular flows. The instabilities are observed in a classical Couette geometry and arise from the development of vortices, at intermediate scales, that are similar to the primary Taylor instability for fluids. Transitions are controlled by the shear stress and a characteristic hierarchy of banded patterns is observed. The vortices observed by Conway *et al.* are accompanied by novel segregation transitions and also spawn additional vortices that modify the scale of the kinetic interactions. Although these observations are still incomplete they indicate that control of convective mixing of particles, and prediction based on fundamental understanding, may become possible.

## 5.4 Jamming

Nothing typifies the unpredictability of granular materials behaviour better than the sudden failure of apparently loose granular material to flow out of a hopper or bin. The cessation in flow is usually due to arching, or bridging, of the material across the outlet and, in turn, it is the source of considerable frustration in many granular materials processing situations. Remedial actions, which alleviate bridging in hoppers, are often rather crude and their effects have become the subject of many amusing descriptions such as 'hammer rash'[38] (see Ajax Equipment, the 'Society for the Prevention of Cruelty to Hoppers', URL http://www.ajax. co.uk/spoch.htm). However, bridging, and the associated effects that arise from loss of control and lost production, also have a serious economic impact on many real materials processing operations.[39]

There are at least two types of arching phenomena that occur in granular materials. Firstly, 'cohesive' arches can occur when dense particulates, with attractive interactions, enter a convergent flow pattern such as that at the approach to the outlet of a conical or wedge-shaped hopper. Secondly 'interlocking' arches, or bridges, can be formed when a group of hard particles become trapped, that is, their relative motion is impossible, because of multiple, direct, particle–particle contacts. Cohesive arches have been analysed extensively because they are a major industrial concern. The analysis of cohesive arches combines a continuum model of granular material with extensive flow property measurements and gives an established methodology, developed by Jenike, for the design of a hopper (see D. Schulze, Storage of powders and bulk solids in silos, URL http://www.dietmar-schulze.de/storagepr.html). The angle of the hopper that ensures a mass flow regime is calculated from the friction coefficients of the granular material and of the hopper wall. Then the minimum opening size that prevents arching is established from the yield strength of the material (which varies with density and hence with the distance from the opening) and from an expression for the variation of the vertical stress with height in the particular hopper geometry – cones and wedges can be very different. While the vertical stress exceeds the material strength cohesive arches do not interrupt the discharge. Although the Jenike methodology is an accepted industry standard it inevitably involves time-consuming and expensive shear cell testing. Currently it is impossible to develop theoretical expressions that allow flow properties to be predicted, *a priori,* from simple physical properties of granular materials like particle size, temperature, moisture content and time of storage at rest. For many food powders this relationship, between material properties and flowability, does not even present a clear pattern.[40]

In commercial applications interlocking arches are rarer than cohesive arches but they may prove much more significant as a basis for understanding the properties of dense particulate materials. Interlocking arches are built at the particulate scale and they correspond to configurations that involve several particles that are mutually constrained by their hard contacts. The interlocking configurations are geometrical in nature and do not rely strongly on the details of the particle interactions. The simplest visualization of an interlocking arch is a chain of particles, bowed slightly upwards, that bridge over the outlet of a hopper. In this configuration the particles in the bridge are jammed together tightly and none of them can move downwards; this structure can clearly hinder hopper discharge. In two dimensions jamming bridges have been examined by To *et al.*[41] These experiments showed that the statistics of the bridges, formed by discs at the outlet of a two-dimensional hopper, have a relatively simple form (constrained, self-avoiding, random walks) and can be used to predict the probability of a hopper-jamming event. For example, the random walk model showed that two-dimensional hoppers, with openings that are between three and four particles wide, are most often jammed by five-particle bridges although six-, seven- and eight-particle bridges jam the outlet with smaller frequencies.

Pneumatic devices, just hi-tech hammers, are often used with hoppers to reduce the probability of the jamming caused by interlocking arches. However, particle bridges are far more general and occur throughout static close-packed granular structures. Mutual stabilizations identify the particles that make a bridge and these can always be found even though the network of contacts formed in a close-packed set of particles is not fully constrained. Mehta and Barker[15,42] have identified bridge formation as an essential result of gravity stabilization for particulate materials. Pouring, stirring and shaking operations for particulate materials all end with static granular structures that include bridges. In turn the bridge structures are then influential on further, slow, dynamics of particulates such as structural relaxations and consolidation. In three dimensions the statistical geometry of the bridges is complex but it is still possible to identify an important subset of structures that have string-like shapes. Pugnaloni *et al.*[43] have shown that the three-dimensional bridges, in computer-simulated sphere deposits, are very common and have a well-defined scaling behaviour so that bridges built from $n$ spheres occur with probability $p(n) \sim n^{-\alpha}$ with $\alpha \sim 1$. This scaling does not appear to depend strongly on the way a deposit is formed or on the packing density.

The particle bridges in granular material have irregular shapes and have a wide range of lateral extensions. The distribution of bridge extensions (the cumulative form is related to the jamming probability for a hopper) has a very characteristic shape; the maximum probability occurs for bridges with extensions that are slightly smaller than the mean extension, and bridges with large extensions occur with probabilities that decrease exponentially. This form is provocative because it is similar to that for the distribution of normal contact forces in granular solids. In a close-packed bed of particles the normal forces, at the particle–particle contacts, do not develop uniformly throughout the material but conspire to form force chains.[44] Force chains concentrate the stress along a few well-defined paths and leave other regions of material bearing very little load; the inhomogeneous stress is reflected by the exponential tail of the distribution function and by the high probability for small contact forces. Force chains are responsible for anomalous behaviour in granular materials such as the stress minimum underneath a conical pile[45] or the large rise in the resisting force when granular material is probed near the bottom of a container.[46] The similarity of the two distribution functions, for bridge extensions and normal contact forces, and their independence from other parameters, like the amount of disorder included in a packing, indicate that force chains may be related to bridging – possibly bridges provide zones that accommodate the numerous small contact forces. In turn, in this case, the presence of bridges in granular materials must be reflected in the form of the continuum constitutive relations that relate the stress and fabric properties of granular solids. These relations are needed to develop quantitative descriptions of granular media and, potentially, they lead to a description of arching that is based on first principles.

The microscopic processes that cause bridges to form in granular deposits have generated considerable interest. The grains that form a bridge do so because their motion is stopped by a local crowding effect before they can sink lower in the gravitational field but, as part of this process, the interlocked grains give the granular material a large-scale mechanical integrity or yield stress. This phenomenon has become known as 'jamming' and it is believed to have several universal features that make granular materials research relevant in other fields. Colloidal materials,[47] supercooled liquids,[48] foams and even football crowds[49] experience jamming. A unifying organizational scheme, which highlights the universality of the nonequilibrium transitions in slow dynamical systems, called the jamming phase diagram, has been suggested.[50] The space of configurations is spanned by the density, the temperature and the applied load. Granular systems are athermal and occupy one plane of the diagram, with $T = 0$, whereas molecular systems, with no external load, fit into a perpendicular plane. However, crucially, the transitions between the states in this space are believed to follow a single set of rules. Paramount in this shared understanding is the possibility that the slow heterogeneous dynamics that occurs in and around interlocking bridges in granular deposits may reveal details concerning the molecular motions that occur in liquids as they approach the glass transition – information that is otherwise notoriously difficult to obtain.

## 5.5 Discussion

The number and variety of food powders is set to increase. Powdered food materials and ingredients improve the efficiency, flexibility and hygiene of many food processing operations. New powdered materials, like flavours, active ingredients or stabilizers, promote novel foods and can give added value to established products. In this short review we have pursued a very limited range of science that relates to food powders; we have not addressed powder fabrication methods like spraying, drying and freezing fracture. We have not examined powder functionality like dispersion, rehydration, coating or encapsulation and release. We have not explored aspects of safety like dust formation or allergenic effects, and we have not examined the chemistry of finely divided solid food materials. However, this review does indicate that robust science has a role to play in the developing use of food powders. Increasingly, a detailed understanding of particle technology is leading the drive to replace empiricism by rational design[36] in powder processing.

The profile of granular materials research is changing. Understanding brick packs, Brazil nuts and jamming has projected research on particulate systems to a wider audience and so has attracted greater interest. Current research in granular materials embraces problems concerning complexity, slow dynamics, nonequilibrium phase transitions and self-organization so that it has significant academic appeal; a mechanics of granular materials experiment, exploring the slow dynamics of particulates in low gravity, has been included in several recent NASA space shuttle missions (NASA, Using microgravity to understand soil behavior, URL http://spaceresearch.nasa.gov/sts-107/107_mgm.pdf). It is clear that fundamental research at the particulate level, supported by advanced imaging and visualization techniques and by sophisticated computer simulations, provides a powerful addition to traditional continuum considerations for granular solids.

Research and development requirements, to promote competitive manufacturing of powdered food materials and ingredients, have been identified by several organizations.[9,51] In the future it is likely that these requirements will merge with others from commensurate businesses like pharmaceuticals, agrochemicals and the management and use of biological waste materials. It is clear that several of the challenges identified for improved handling of granular materials, including advanced and hygienic food processing, are within reach of fundamental science. Predictive strategies for mixing, prevention of the ageing caused by consolidation, controlled discharge from bulk storage and many other real problems will soon be addressed by ongoing shared research in the chemical physics of granular materials and powders.

## 5.6 References


1 DEFRA (2006) *Milk Statistics*. URL http://statistics.defra.gov.uk/esg/statnot/milk.pdf
2 Brown, R.L. & Richards, J.C. (1966) *Principles of Powder Mechanics*. Pergamon, Oxford.
3 Jaeger, H.M. & Nagel, S.R. (1992) Physics of the granular state. *Science* **255**, 1523–1531.
4 Mehta, A. (ed.) (1994) *Granular Matter: An Interdisciplinary Approach*. Springer-Verlag, New York.
5 Kadanoff, L.P. (1999) Built upon sand: Theoretical ideas inspired by granular flows. *Rev. Mod. Phys.* **71**, 435–444.

6 Trappe, V., Prasad, V., Cipelletti, L. *et al.* (2001) Jamming phase diagram for attractive particles. *Nature* **411**, 772–775.

7 de Gennes, P.G. (1999) Granular matter: a tentative view. *Rev. Mod. Phys.* **71**, S374–S382.

8 Mueth, D.M., Debregeas, G.F., Karczmar, G.S. *et al.* (2000) Signatures of granular microstructure in dense shear flows. *Nature* **406**, 385–389.

9 European Commission (2003) Powder research to promote competitive manufacture of added-value food ingredients. Accompany Measure No. QLK1-CT-2001–30172. URL www.foodpowders.net

10 Biehl, H.L. & Barringer, S.A. (2003) Physical properties important to electrostatic and nonelectrostatic powder transfer efficiency in a tumble drum. *J. Food Sci.* **68**, 2512–2515.

11 Mohammadi, M.S. & Harnby, N. (1997) Bulk density modelling as a means of typifying the microstructure and flow characteristics of cohesive powders. *Powder Tech.* **92**, 1–8.

12 Ilari, J.L. (2002) Flow properties of industrial dairy powders. *Lait* **82**, 383–399.

13 Malave, J., Barbosa-Canovas, G.V. & Peleg, M. (1985) Comparison of the compaction characteristics of selected food powders by vibration, tapping and mechanical compression. *J. Food Sci.* **50**, 1473–1476.

14 Nowak, E.R., Knight, J.B., Ben-Naim, E. *et al.* (1998) Density fluctuations in vibrated granular materials. *Phys. Rev. E.* **57**, 1971–1982.

15 Barker, G.C. & Mehta, A. (1991) Vibrated powders: A microscopic approach. *Phys. Rev. Lett.* **67**, 394–397.

16 Edwards, S.F. & Grinev, D.V. (1998) Statistical mechanics of vibration-induced compaction of powders. *Phys. Rev. E* **58**, 4758–4762.

17 de Gennes, P.G. (2000) Tapping of granular packs: A model based on local two-level systems. *J. Coll. Int. Sci.* **226**, 1–4.

18 Mehta, A. & Edwards, S.F. (1989) Statistical mechanics of powder mixtures. *Physica A* **157**, 1091–1100.

19 Gioia, G., Cuitino, A.M., Zheng, S. & Uribe, T. (2002) Two-phase densification of cohesive granular aggregates. *Phys. Rev. Lett.* **88**, 204302(4).

20 Forsyth, A.J., Hutton, S.R., Osborne, C.F. & Rhodes, M.J. (2001) Effects of interparticle force on the packing of spherical granular material. *Phys. Rev. Lett.* **87**, 244301(3).

21 Villarruel, F.X., Lauderdale, B.E., Meuth, D.M. & Jaeger, H.M. (2000) Compaction of rods: relaxation and ordering in vibrated, anisotropic granular material. *Phys. Rev. E* **61**, 6914–6921.

22 Donev, A., Cisse, L., Sachs, D. *et al.* (2004) Improving the density of jammed disordered packings using ellipsoids. *Science* **303**, 990–993.

23 Aste, T. & Weaire, D. (2000) *The Pursuit of Perfect Packing*. Institute of Physics, Bristol, UK.

24 Barker, G.C. & Grimson, M.J. (1990) The physics of muesli. *New Scientist* No. 1718, 37–40.

25 Rosato, A., Strandburg, K.J., Prinz, F. & Swendsen, R.H. (1987) Why the Brazil nuts are on top: size segregation of particulate matter by shaking. *Phys. Rev. Lett.* **58**, 1038–1040.

26 Knight, J.B., Jaeger, H.M. & Nagel, S.R. (1993) Vibration-induced size separation in granular media: the convection connection. *Phys. Rev. Lett.* **70**, 3728–3731.

27 Ehrichs, E.E., Jaeger, H.M., Karczmar, G.S. *et al.* (1995) Granular convection observed by magnetic resonance imaging. *Science* **267**, 1632–1634.

28 Mobius, M.E., Lauderdale, B.E., Nagel, S.R. & Jaeger, H.M. (2001) Size separation of granular particles. *Nature* **414**, 270.

29 Hong, D.C., Quinn, P.V. & Luding, S. (2001) Reverse Brazil nut problem: competition between percolation and condensation. *Phys. Rev. Lett.* **86**, 3423–3426.

30 Shinbrot, T. (2004) The Brazil nut effect – in reverse. *Nature* **429**, 352–353.

31 Maske, H A., Havlin, S., King, P.R. & Stanley, H.E. (1997) Spontaneous stratification in granular mixtures. *Nature* **386**, 379–382.

32 Baxter, J., Tuzun, U., Heyes, D. *et al.* (1998) Stratification in poured granular heaps. *Nature* **391**, 136.

33 Li, H. & McCarthy, J.J. (2003) Controlling cohesive particle mixing and segregation. *Phys. Rev. Lett.* **90**, 184301.

34 Shinbrot, T. & Muzzio, F.J. (2000) Nonequilibrium patterns in granular mixing and segregation. *Physics Today* March, 25–30.

35 Seymour, J.D., Caprihan, A., Altobelli, S.A. & Fukushima, E. (2000) Pulsed gradient spin echo nuclear magnetic resonance imaging of diffusion in granular flow. *Phys. Rev. Lett.* **84**, 266–269.

36 Michaels, J.N. (2003) Toward rational design of powder processes. *Powder Tech.* **138**, 1–6.

37 Conway, S.L., Shinbrot, T. & Glasser, B.J. (2004) A Taylor vortex analogy in granular flows. *Nature* **431**, 433–437.

38 Clow, A. (2004) Powder flowability measurement. *FoodLINK News* **49**, 12–13.

39 Cooper, P. (1988) Carnation Foods solve bridging problems. *Bulk Solids Handling* **8**, 162.

40 Fitzpatrick, J.J., Barringer, S.A. & Iqbal, T. (2004) Flow property measurement of food powders and sensitivity of Jenike's hopper design methodology to measured values. *J. Food Eng.* **61**, 399–405.

41 To, K., Lai, P-K. & Pak, H.K. (2001) Jamming of granular flow in a two-dimensional hopper. *Phys. Rev. Lett.* **86**, 71–74.

42 Barker, G.C. & Mehta, A. (1992) Vibrated powders: structure correlations and dynamics. *Phys. Rev. A.* **45**, 3435–3446.

43 Pugnaloni, L.A., Barker, G.C. & Mehta, A. (2001) Multi-particle structures in non-sequentially reorganised hard sphere deposits. *Adv. Complex Systems* **4**, 289–297.

44 Liu, C-h., Nagel, S.R., Schecter, D.A. *et al.* (1995) Force fluctuations in bead packs. *Science* **269**, 513–514.

45 Vanel, L., Howell, D., Clark, D. *et al.* (1999) Memories in sand: experimental tests of construction history on stress distribution under sandpiles. *Phys. Rev. E.* **60**, R5040–R5043.

46 Stone, M.B., Bernstein, D.P., Barry, R. *et al.* (2004) Getting to the bottom of a granular medium. *Nature* **427**, 503–504.

47 Haw, M.D. (2004) Jamming, two-fluid behaviour and self-filtration in concentrated particulate suspensions. *Phys. Rev. Lett.* **92**, 185506.

48 O'Hern, C.S., Langer, S.A., Liu, A.J. & Nagel, S.R. (2001) Force distributions near the jamming and glass transition. *Phys. Rev. Lett.* **86**, 111–114.

49 Helbing, D., Farkas, I. & Vicsek, T. (2000) Simulating dynamical features of escape panic. *Nature* **407**, 487–490.

50 Liu, A.J. & Nagel, S.R. (1998) Jamming is not just cool any more. *Nature* **396**, 21–22.

51 Fitzpatrick, J.J. & Ahrne, L. (2005) Food powder handling and processing: Industry problems, knowledge barriers and research opportunities. *Chem. Eng. Proc.* **44**, 209–214.

Chapter 6
# Gels

*Victor J. Morris*

## 6.1 Introduction

Food gels are more easily recognized than defined. Familiar examples are dessert jellies, aspics, jams, hard-boiled eggs, custard and yoghurts. These are all examples of biopolymer gels, and this chapter will focus on this type of material. Most proteins and some polysaccharides (carbohydrates) will form these types of gels. A common feature is that once formed (set), usually within a container, the gels retain their shape, even when the container is removed. If the gel is compressed or twisted it deforms, storing energy elastically and, when the stress is removed, it recovers its original shape. These materials are called strong gels. At high applied stresses they may fail or break, and the material remains fragmented upon removing the applied stress. At moderate applied stresses, above a threshold value, certain biopolymer gels may flow. In this case, upon removal of the applied stress, the material remains deformed but, when twisted or compressed again, below this threshold value, they respond elastically. These materials are usually termed weak or fluid gels.

Gels are usually prepared from solutions of biopolymers. It seems fairly well established that these sols contain individual biopolymers that, on setting, associate to form the space-filling networks that generate the 'solid-like' consistency of the gel. The types of network formed, and the method of formation, are often peculiar to the particular biopolymer or, at least, to particular classes of biopolymers. For the same biopolymer different methods of gelation can yield different types of networks and different types of gel. The type of network structure formed will determine certain characteristic properties of the formed gels. It will determine their elastic moduli: the extent to which the gels will deform when subjected to small twists or compression. At larger stresses the network structure will determine whether the gels flow or fail. The mode and extent of aggregation will also determine the opacity of the gels. Most sols are transparent. Provided the polymers remain fairly uniformly distributed and the level of aggregation is small then the gels will also be transparent. High degrees of aggregation and a nonuniform distribution of biopolymers will lead to opaque gels.

Although the nature of the biopolymers, and the way they associate, determine the appearance and texture of the gels it is not possible to infer from rheological, mechanical or turbidity measurements alone the nature of the network structures within the gels. Thus, with the aim of eventually being able to upgrade raw materials, or rationally to control processing, it is necessary to understand the structural changes and form of association of gelling biopolymers. This research has chiefly involved the use of spectroscopic or physicochemical methods, often on dilute solutions of the biopolymers, to define the structural changes accompanying gelation and the nature of the junction zones, the regions of associated biopolymers that hold the networks together. Models for junction zones can be tested using small-angle neutron scattering, small-angle X-ray scattering and/or wide-angle X-ray scattering methods. In the case of fibrous proteins and polysaccharides X-ray diffraction of oriented fibres prepared from gels has allowed molecular models for junction zones to be proposed and tested at atomic resolution. Light, X-ray and neutron scattering methods have been used to follow the early stages of association through studies on gel precursors and to probe the kinetics of structure formation in intact gels. Less is known about the long-range structure in biopolymer gels with most information being obtained from electron microscopy and, more recently, atomic force microscopy. This literature on the molecular basis of gelation is extensive and has been reviewed elsewhere.[1–11] It is not planned to duplicate this information in the present short review. Rather, it is intended to emphasize the general physical chemical basis of the gelation of proteins and polysaccharides, and to illustrate certain polymer-specific features of gelation with specific examples.

Historically the approach to studying gelation has been to investigate the properties of individual biopolymers and to infer their behaviour within foods. In this chapter we will be concerned with the use of biopolymers to introduce and control structure and texture in foods. Even in this fairly simple situation the basic raw materials are often not single polymers. The ingredients are often extracts from plant or animal tissue that contain polymers with variations in their structure, or mixtures of different polymers. The extracts are usually used in the presence of other biopolymers or ingredients that can influence their behaviour. Recently there has been a growth of interest in the study of simple biopolymer mixtures. This is partly in order to obtain a better understanding of the actual structures present in complex foods, but more importantly because it is possible to generate a whole new range of structures and textures by controlled blending and gelation of biopolymer mixtures. It might be thought, given the uniqueness of many of the gelling biopolymers, the large number of possible mixtures, and the range of conditions for blending and setting the materials, that it would not be possible to draw generic conclusions about the types of structures that could be formed. In fact it is possible to classify such mixtures and to identify characteristic patterns of behaviour for these classes of material. This area of research is still new and the basic ideas and concepts will be described.

A feature of mixed polymers is the prospect of phase separation, which leads to pockets of one structure present in a matrix with a different composition. This composite, or particulate, structure is also found in starch-based foods and adds an extra layer of complexity to the description of the structure and properties. Other particulate structures can be created by processing, and the main examples in the food area are emulsions and foams. For oil-in-water emulsions the aim is to generate droplets of oil in an aqueous phase and to stabilize these structures to prolong the shelf life of the products. Gelling hydrocolloids play several roles in this scenario. The first role is to modify the properties of the aqueous phase to inhibit association (coalescence) or segregation (creaming) of the oil droplets. This can be done by increasing the viscosity of the aqueous phase. A better solution is to use a weak gel. This would inhibit motion of the oil droplets but, if the product is shaken, stirred or poured the network breaks, and the product flows. The consistency of the product is retained once the stress is removed. Another cause of instability is the merging, or coalescence, of droplets when they meet. Coalescence can be controlled by creating structures at the oil–water interface. The main biopolymers used for this application are proteins, and the assembly and breakdown of what are often termed two-dimensional protein gels are major factors determining emulsion stability. As well as controlling the surface rheology at the oil–water interface the surface-adsorbed polymers can additionally prevent coalescence in the way in which they interact with similar structures on the surfaces of other droplets, or by means of coupling to the polymeric structure of the bulk aqueous phase. In the latter case this can lead to highly stable structures called emulsion gels. Proteins play a similar role in the stabilization of the air–water interfaces generated in the creation of foams. The molecular understanding of the structure and role of these networks is still fairly new and will be discussed in some detail. Perhaps the most complex processed gel structure is that found in whipped foods and ice creams. These are essentially solidified foamed emulsions. The protein networks play roles in controlling the stability of the air–water and oil–water interfaces and control limited association (partial coalescence) of solidifying fat droplets. In partial coalescence the droplets retain their shape on association forming aggregates and/or networks. If the droplets adsorb to the air–water interface they can lock the foam structure into the complex multicomponent gel network. The new molecular understanding of the interfacial structures is suggesting rational approaches to the manipulation of product quality. An emerging concept is the potential for engineering and design of interfacial structures and these concepts will be discussed.

The emphasis in this chapter is on the new and emerging ideas about the nature and role of gelling biopolymers. Such newly emerging areas are always controversial, and the views expressed in this chapter are personal. The reader is advised to consult the cited reviews and literature for alternative views on these areas. It is hoped that focusing on the frontiers of this subject will spur further research and understanding of these areas.

## 6.2 Polysaccharide gels

### 6.2.1 What are polysaccharides?

The basic building blocks of polysaccharides are sugars (Fig. 6.1). Polysaccharides that can function as gelling agents are high molecular weight fibrous materials. In the sol state the polysaccharides can generally be described as semiflexible coils with the stiffness (persistence length) of the polymers being dependent on the structure of the polymer and, for charged polymers, the ionic strength of the aqueous medium.

The structure of the polysaccharides depends on the source and method of extraction. Both factors are crucial in determining the nature of gelation[12] and the quality of the gels produced. Only a few of the gelling polysaccharides are homopolymers (Fig. 6.2a). Examples are the starch polysaccharides or cellulose derivatives.

Two polysaccharides can be extracted from starch.[1,13] Amylose is an essentially linear polymer of $\alpha(1\rightarrow4)$-D-glucose (Fig. 6.1b). The molecular weight will depend on the source and the method of extraction, but typical polymers will contain several thousand glucose units. Solutions of amylose in water are unstable at room temperature but the polymer can

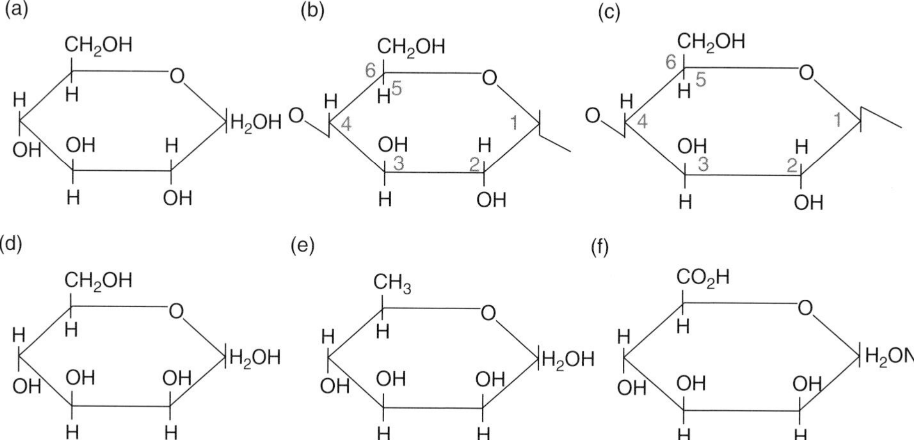

**Fig. 6.1** The figure shows examples of sugar building blocks in gel-forming polysaccharides. (a) Glucose is a hexose containing six carbon atoms. The positions of the carbon in the sugar ring are numbered 1–6 (b,c). The distribution of hydroxyl groups around the ring determines the type of sugar. For example mannose (d) and glucose (a) differ in the positioning of the hydroxyl group at C2. Linkages between sugars are formed by elimination of water and (b) and (c) show (1→4) linked glucose rings. For individual sugars there are two anomeric ($\alpha$ and $\beta$) forms corresponding to the two possible locations of the hydroxyl group at C1. Thus two types of linkages can be formed as illustrated for $\alpha(1\rightarrow4)$ linked (b) and $\beta(1\rightarrow4)$ linked (c) glucose. In addition to the normal sugars there are deoxysugars, such as 6-deoxymannose, or rhamnose (e), and uronic acids such as mannuronic acid (f). The sugars shown are D optical isomers. The L isomers are mirror-image structures. The influence of glycosidic linkages on polysaccharide structure and gelation has been discussed by Rees.[12]

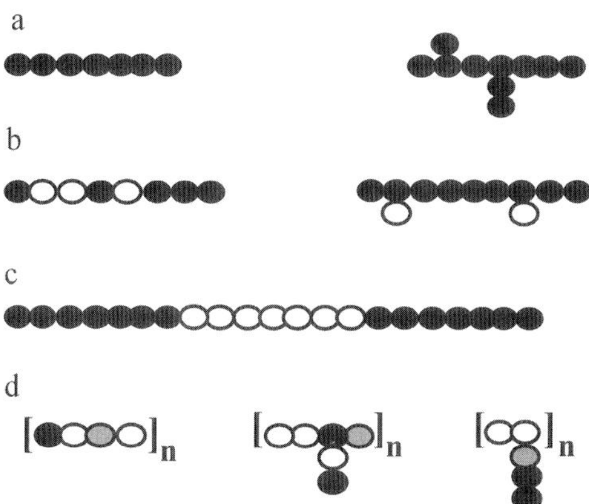

**Fig. 6.2** Different types of polysaccharide structures. (a) Homopolymers are polysaccharides that contain just one type of sugar. They may be linear or branched structures. Heteropolysaccharides contain more than one type of sugar unit. The structures may be (b) irregular linear or branched structures, (c) block copolymers or (d) contain discrete chemical repeat units. The influence of glycosidic linkages on polysaccharide structure and gelation has been discussed by Rees.[12]

be stabilized by the formation of suitable helical inclusion complexes.[14] This allows the polymeric structure to be visualized (Fig. 6.3). The atomic force microscopy (AFM) image shows the flexible fibrous structure of the molecule. The other polysaccharide extractable from starch is amylopectin.[1,13] This is also a homopolymer but it is a highly multiply branched

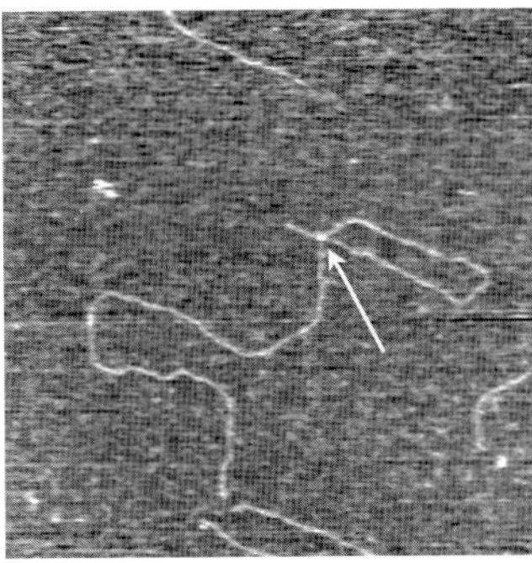

**Fig. 6.3**   An atomic force microscopy image of an amylose molecule. The image size is 638 × 638 nm. Note the bright spot (arrowed) at the point where two chains cross one another. The images are three-dimensional images and the height (brightness) doubles at the crossover point. The amylose chains have been solubilized by forming complexes[14] with iodine and a nonionic surfactant (Tween 20).

structure, with $\alpha(1\rightarrow6)$-linked amylosic chains. Amylopectin molecules can contain several hundreds of thousands of glucose units. Within the native starch granule the branches of the amylopectin molecules are present as crystalline lamellae. Thus the branched structure of amylopectin molecules is not random but reflects the location and type of crystalline structure within the granule.

Cellulose is also a homopolymer of glucose[15] but the sugar monomers are joined by $\beta(1\rightarrow4)$ linkages (Fig. 6.1c). Cellulose is also insoluble and the soluble derivatives are prepared by introducing charge, or substituents that can block hydrogen bonding between molecules, thus inhibiting aggregation and crystallization.

Most gelling polysaccharides are heteropolysaccharides containing more than one type of sugar unit (Fig. 6.2b–d). The types of polymers are varied, comprising irregular unbranched structures, regular or irregular branched structures, and block copolymers.

The bacterial polysaccharides xanthan and gellan are complex structures but they contain regular repeat units (Fig. 6.2d), a pentasaccharide for xanthan[1,16] (Fig. 6.4a) and a tetrasaccharide for gellan[1,16] (Fig. 6.4b). Even for these systems the noncarbohydrate substitution is variable. It is not clear whether the substitution is irregular and incomplete for all the polymers, or whether the extract is a mixture of fully substituted and unsubstituted polymers. The main commercial form of gellan is subjected to an alkaline treatment at high temperature during extraction, which de-esterifies the material, leaving a regular polymer structure.

The algal polysaccharides known as agar and the carrageenans from red seaweeds (*Rhodophyceae*) also show structures[1,17,18] (Fig. 6.5) that approximate to simple repeat units (disaccharides). The carrageenans can be subdivided into three types of gelling polymers (κ-carrageenan, ι-carrageenan and furcellaran) based on the site and level of sulfation (Fig. 6.5a–c). Almost all carrageenans and related polysaccharides are heterogeneous, with structural variation occurring both within and between polysaccharides. Choice of seaweed and extraction procedures can be used to improve the structural regularity and enhance the gelling ability of the polymers. For commercially important carrageenans a particular type of disaccharide repeat unit will represent the predominant chemical structure (Fig. 6.5a–c). An important type of structural defect for both agar and the carrageenans is the replacement of $(1\rightarrow4)$-linked anhydrogalactose by galactose or galactose-6-sulfate (Fig. 6.5d). These

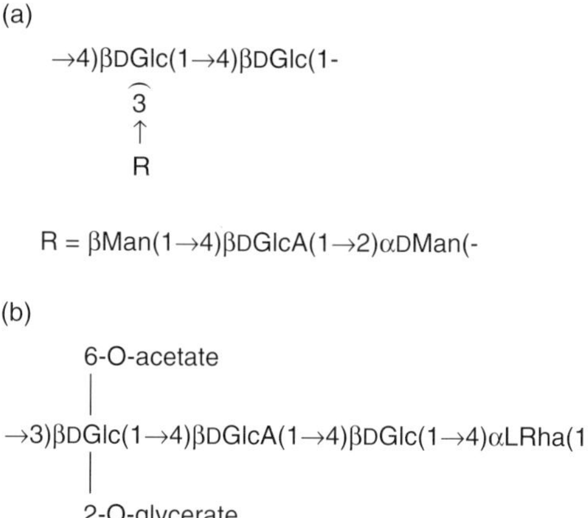



**Fig. 6.4**    The chemical repeat unit structures for (a) xanthan and (b) gellan gum. The noncarbohydrate structure of xanthan is not shown in the figure. The commercially available polymer normally contains an acetate substituent on the first mannose residue on the side chain and the terminal mannose contains a pyruvate substituent. This substitution is normally nonstoichiometric. The noncarbohydrate substituents are shown for gellan gum but in commercially available material this substitution is normally incomplete. Note that the abbreviations stand for the following sugars: Glc, glucose; Man, mannose; Rha, rhamnose; GlcA, glucuronic acid.

residues, believed to be biological precursors to the anhydride residues, interfere with the formation of a regular helical structure and have been termed kinks.[12] Kinking residues are believed to be concentrated in certain regions of the chains. Gelation is dependent on helix formation and as little as 1 kink per 200 normal residues is believed to seriously impair gelation. Alkali treatment during extraction can convert galactose-6-sulfate residues to anhydrogalactose residues and this enhances gelation. Although extracts from certain seaweeds such as *Eucheuma spinosa* and *Eucheuma cottoni* yield nearly pure ι- and κ-carrageenans the extracts are generally considered to be mixtures of the two forms. The use of ι- and κ-carrageenases is providing evidence to suggest these samples may actually be intramolecular hybrid structures.[19,20] This is probably also true of furcellaran, which is a hybrid of κ-residues and unsubstituted disaccharides. Agar also possesses a non-sulfated disaccharide repeat unit,[1,17] but it contains the 3,6-anhydro-α-L-galactose unit rather than the 3,6-anhydro-α-D-galactose found in the carrageenans.

Examples of irregularly substituted linear or branched heteropolysaccharides are glucomannans (konjac mannan)[21] and galactomannans[22,23] (carob, tara and guar). Typical structures for these polymers are shown in Fig. 6.2b. Konjac mannan is a predominantly linear molecule containing β(1→4)-D-mannose and β(1→4)-D-glucose in the ratio ~1:6. Detailed structural analysis suggests that the polysaccharide does not contain a regular repeating sequence of large blocks of glucose (cellulosic blocks) or blocks of mannose (mannan blocks). Approximately 5–10% of the residues are acetylated but the substitution pattern is not known. The only difference between glucose and mannose is the positioning of the hydroxyl group at C2 (Fig. 6.1a,d). Thus konjac mannan can be viewed as similar to a derivatized cellulose. Galactomannans consist of a mannan backbone (β(1→4)-D-mannose) irregularly substituted with α(1→6)-D-galactose residues. The side chains solubilize the otherwise insoluble mannan backbone. Galactomannans extracted from different plant

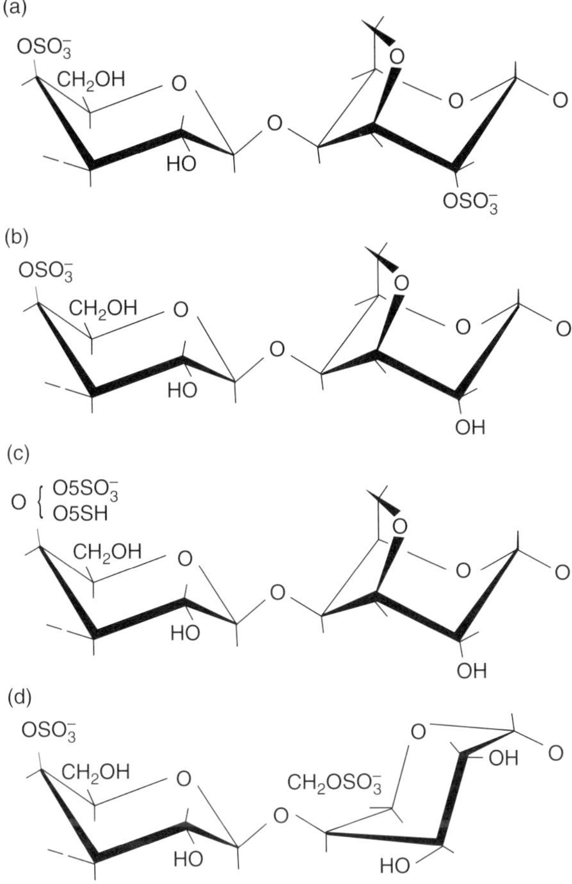

**Fig. 6.5**  Idealized chemical repeat unit structures for the members of the carrageenan family: (a) ι-carrageenan, (b) κ-carrageenan and (c) furcellaran. The basic carbohydrate repeat unit is a disaccharide containing 3-linked β-D-galactose and 4-linked 3,6-anhydro-β-D-galactose. Note galactose differs from glucose in the location of the hydroxyl at C4. The site and level of sulfation dictates the type of carrageenan. A common irregularity in the structure, known as kinking residues, is the presence of 4-linked galactose 6-sulfate instead of the anhydro-sugar.

sources under different experimental conditions contain different levels of galactose side chains. Commercially available samples of carob, tara and guar gum have mannose/galactose (M/G) ratios of 3.55, 3.0 and 1.56 respectively.

There are two examples of gelling polysaccharides that are block copolymers (Fig. 6.2c): alginate and pectin. Alginate from brown seaweed (*Phaeophyceae*) has the simplest chemical structure[1,24] containing two sugars: α(1→4)-linked D-mannuronic acid (M) and α(1→4)-linked L-guluronic acid (G). The polymer essentially consist of blocks of M residues (M blocks) and sequences of G residues (G blocks) interspersed with irregular sequences containing both M and G residues. The M/G ratio and the sequence of residues within the chain depend on the algal source, growth conditions, age and location within the plant. The gelation of alginates requires the presence of G blocks of >20 residues,[24] and hence high M/G alginates are poor gelling agents. Pectins are extracted from the cell walls of land plants and the chief

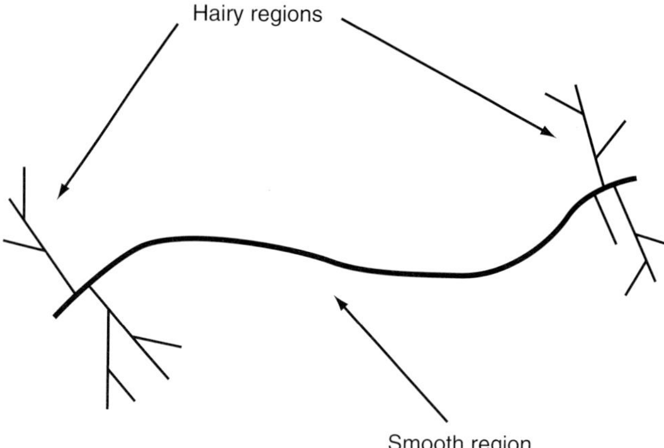



**Fig. 6.6**   A schematic picture of pectin structure. The polymer is considered to consist of 'smooth regions' composed of galacturonic acid or its methyl ester, contained between 'hairy regions' containing the neutral sugars.

commercial sources are citrus peel and apple pumice from the drinks and cider industries. Sugar beet pulp also provides a commercial source for pectins. The structures[25,26] can be complex and will vary with plant source and the method of extraction. Pectins approximate to a structure mainly composed of poly-$\alpha$(1→4)-linked D-galacturonic acid containing small quantities of neutral sugars, namely rhamnose, galactose and arabinose. Detailed structural analysis suggests that the neutral sugars are concentrated in so-called 'hairy regions' of the polymers composed of alternating sequences of $\alpha$(1→2)-linked L-rhamnose and $\alpha$(1→4)-linked D-galacturonic acid decorated with arabinose and galactose as side chains (Fig. 6.6). Pectins contain elongated sequences of $\alpha$(1→4)-linked D-galacturonic acid (smooth regions) interspersed between the hairy regions. The galacturonic acid can be methyl esterified and the degree of esterification (DE) depends on the age and location of the pectin within the plant cell wall and the method of extraction. The modes of gelation are different for high-DE (55–80%) pectins and low-DE pectins. Gelation of low-DE pectins (DE < 40%) requires the presence of unesterified blocks of galacturonic acid of >15–20 residues.[25] Acid and enzyme de-esterification leads to different distributions of methyl substituents and different gelling profiles for the same DE value. Sugar beet pectins may contain acetyl substituents and phenolic esters. The acetate inhibits gelation but can be selectively removed by commercially available acetyl esterases. The phenolic esters provide an additional means of cross-linking and gelling the pectins.[27,28] Amide groups can be introduced by the action of ammonia in alcoholic suspensions of pectins. The amide groups can be used to modify the gelling profiles of low methoxy (LM) pectins.

### 6.2.2 How do polysaccharides form networks?

In the sol state the gelling polysaccharides behave as stiff coils in solution. The stiffness of the polymers and the volume occupied in solution depend on the chemical structure of the polymers. If the concentration of the sol is increased then, at a concentration $c^*$, the volumes occupied by individual polysaccharides will begin to interpenetrate. Above this overlap concentration the polysaccharides form entangled solutions that restrict the flow of the polymers. These structures can be thought of as temporary networks. Gelation involves inducing the polymers to associate and form a more substantial network structure. There is quite strong evidence to suggest that for most polysaccharides the critical concentration for gelation $c_0 < c^*$. This would seem to imply that inducing aggregation leads to some degree of liquid–liquid phase separation as a prelude to formation of the gel network. As the polymers associate this will trap the polymers in a nonequilibrium structure. Further long-term shrinkage (demixing) and further consolidation of the network may account for the release of water (syneresis) from certain gels prepared under different conditions.

The modes of gelation tend to be different for different polysaccharides[1] but certain general principles can be described. The methods of gelation will lead to gels with different properties. The most obvious of these are the production of cold-setting gels, gels that set on heating, and those that set on cooling the sol. The nature of the linkage will determine whether the gel is strong or weak, and whether the sol–gel transition is reversible or irreversible.

### 6.2.2.1 Point cross-links

The simplest type of cross-linkage that can be introduced is a point cross-link (Fig. 6.7a). The best reported example of such a structure is the cross-linking of sugar beet pectin through the generation of diferulic acid linkages.[27,28] This type of structure is likely to be the closest approximation to a classic rubber structure with the shear modulus of the gel determined by the number of cross-links per unit volume, and hence determined by the length of the polysaccharide chain between cross-links. The gels can be set at room temperature and are irreversible. The linkages should lead to strong gels and, if the level of cross-linking is low, the gels should be transparent. These types of structure and the mechanism of association are not common in food systems, although cross-linking via phenolics is considered to be important in the association of water-soluble pentosans[29] (arabinoxylans) and this may occur during baking.

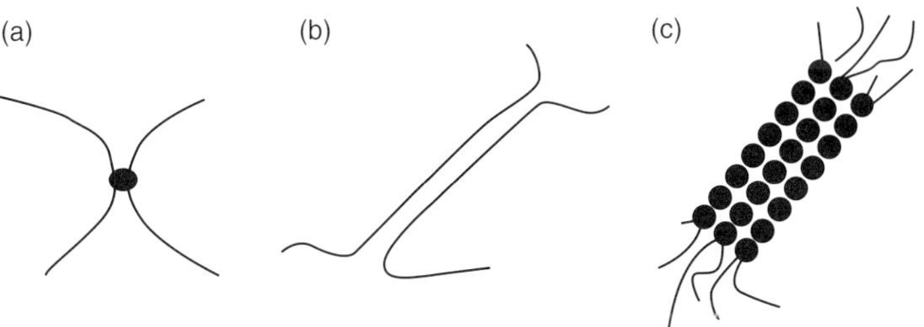


**Fig. 6.7**   Examples of the types of junction zones that can be found in gels. (a) Point cross-links. (b) Extended cross-links. These linkages bind at least two chains together. This could involve intermolecular cooperative binding between segments on adjacent chains, or the formation of a multiple helical structure. (c) In this case the fundamental units that can link at least two chains undergo further association, via a different binding mechanism, to form more complex cross-links.

### *6.2.2.2 Block structures*

Instead of linking points on different chains together it is possible to cause blocks of structure on different polymers to associate leading to network formation (Fig. 6.7b). This type of cross-link is quite common in food gels, although the mechanism for inducing association is different for different polysaccharides.

The best example of this type of gelation is alkali-induced gelation of konjac mannan.[30] Alkali treatment deacetylates the glucomannan. X-ray fibre diffraction studies have shown[31] that konjac mannan can form crystallites with a structure related to that of mannan II, the crystal structure for pure mannan. Molecular modelling studies have shown that isomorphic replacement of mannose by glucose within the mannan II lattice is sterically feasible. This approach has been adapted to examine the effects of deacetylation of konjac mannan.[31] It was found that not all possible sites of acetylation are stereochemically allowed: these substituents would inhibit crystallization. It is proposed that alkali treatment causes gelation by generating blocks of insoluble mannan regions that associate, cross-linking the chains into a gel network.

A related effect can be seen for galactomannans. In solution galactomannans behave as semiflexible coils. However, their rheological properties show notable departures from classical coil-like behaviour. This is normally attributed to intermolecular association[32] at concentrations above $c^*$. This effect depends on the M/G ratio, and lower galactose content favours reduced solubility, aggregation and gelation.[22] Gelation is sensitive not only to the M/G ratio but also to the distribution of galactose side chains along the mannan backbone. Galactomannans with the same M/G ratio may show different solubilities and different tendencies to gel. Gelation of less-soluble galactomannans may occur on standing, can be induced by freeze-thaw cycles, or by the addition of solutes that are claimed to lower water activity.[22] As the galactose content is reduced, the statistical probability of bare mannan blocks appearing increases, and it is generally accepted that association of these blocks is responsible for gelation.

The insolubility of cellulose blocks is probably responsible for the gelation of cellulose derivatives. The most important gelling derivatives are the ether derivatives methylcellulose (MC) or hydroxypropylmethylcellulose (HPMC). Aqueous sols of MC or HPMC (>1.5%) form gels on heating and reform sols on cooling.[15] The nature of the gelation process is still poorly understood. However, it is generally considered that these materials have a lower critical solution temperature and, when heated above this temperature, form gels due to precipitation from solution.

The presence and association of blocks is also considered to be responsible for the gelation of alginates[1,24] and LM pectins.[1,25] In both these cases the blocks are charged and neutralization of the charge plays an important role in promoting association of the blocks. For alginates the important block structures are polyguluronic acid. There is strong physical chemical evidence[33] showing that, provided the blocks are above a particular length (>20 sugars), then cooperative binding of cations occurs, and leads to aggregation of the blocks. The ion binding is sensitive to ion type, and calcium binding is favoured and used to gel the polymers.[1,24] So why does calcium bind to G blocks and not M blocks? The answer to this depends on the detailed structure of the polymer. Polarized infrared and X-ray fibre diffraction studies[34–36] of polyuronic acids suggested that clefts or cavities in the buckled ribbon structure of polyguluronic acid could accommodate water molecules by intermolecular hydrogen bonding, and it was proposed that calcium ions could replace these water molecules in alginate gels. Studies on calcium alginates suggested similar structures. Biologically G units are produced from M units by a C5 epimerase, which inverts the position of the uronic acid within the sugar ring. The result is to change the shape of the sugar, converting M blocks from flat ribbons to more buckled ribbon structures.[12] It is this change in shape that explains the preferential accommodation and binding of calcium ions. On the basis of circular dichroism and conformational studies[37] an 'egg box' model (Fig. 6.8) was proposed in which the calcium ions (eggs) were contained within associated G blocks (the egg box). This general concept of the mode of gelation seems to be fairly well accepted, although there is still some dispute about the number of blocks involved in the junction zones, a factor that may in fact depend on the detailed conditions used to prepare or study the gels. The sol–gel transition is reversible to the introduction or removal of calcium ions and the gels, when prepared at neutral pH, are normally thermoirreversible.

The gelation of LM pectin is considered to be analogous to that of alginate. De-esterification of pectin will lead to the appearance of blocks of polygalacturonic acid. Physical chemical studies[33] suggest that for block lengths above 15–20 residues cooperative binding of calcium occurs, and that this can lead to association of the blocks. On the basis of the near mirror-image relationship between polygalacturonic acid and polyguluronic acid, and the interpretation of changes in circular dichroism during gelation, an egg-box model for the gelation of LM pectin has been proposed[37,38] (Fig. 6.8). The charge on the galacturonic acid can be neutralized by lowering the pH, and LM pectins can be set at low pH. It is assumed this also arises due to the association of blocks with water molecules occupying the cavities. The gel structures are reversible to the introduction and removal of calcium and to lowering and raising the pH. Calcium-set LM pectin gels prepared at neutral pH are thermally irreversible, but calcium-set gels prepared at lower pH are thermally reversible. This has been attributed[39,40] to two modes of association (ionic and nonionic) in low pH calcium-set gels. Calcium binding is sensitive to DE and the detailed block distribution within the pectin. Given samples will preferentially gel at characteristic calcium concentrations and this makes gelation sensitive to the hardness of the water. Broader calcium sensitivity for gelation can be induced by amidation. Quite how this works is not clear. Perhaps amidation randomizes the distribution of charged blocks available for calcium binding. There are suggestions that blocks of amidated regions may be present in the polymers, and that these can lead to the formation

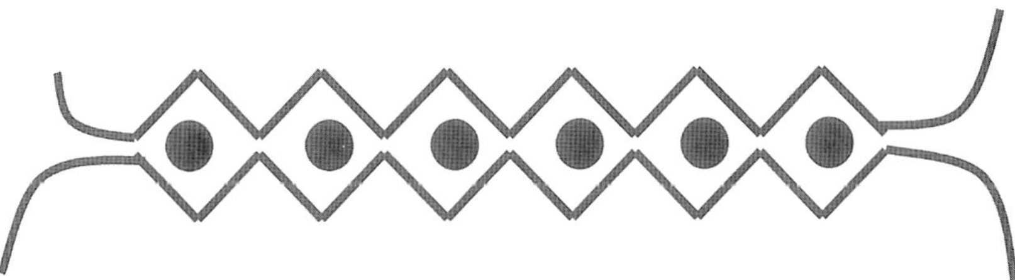

**Fig. 6.8** Schematic model illustrating the 'egg-box' model for the gelation of alginates or pectin. In the case of alginate calcium ions (•) bind guluronic acid blocks. For pectin the calcium ions cross-link blocks of galacturonic acid. Junction zones may involve formation of dimers or higher-order aggregates of the chain blocks.

of an additional type of junction zone.[41] Careful selective extraction of pectins from plant cell walls using chelating agents such as *trans*-1,2-diaminocyclo-hexane-*N,N,N',N'*-tetraacetic acid CDTA result in high methoxy (HM) pectins that gel on addition of calcium.[42] In these materials the blocks must have been created within the plant in order to control association of the pectins. This suggests that commercial pectin extracts may be inhomogeneous, and this is supported by the growing interest in the isolation of 'calcium selective pectins' from commercial extracts. Whereas nuclear magnetic resonance (NMR) methods have proved very useful in characterizing the block distribution in alginates,[43] the characterization of ester distribution within pectins is less well developed. It is possible that pectin populations are far more heterogeneous than alginates. Although there are developments in chemical methods[44] for measuring ester distributions these will only yield average properties of the population. There is a need for methods that can map heterogeneity of individual polymers, and identify and help extract subpopulations with different structures.

Most research has been focused on identifying the nature of the important block structures that lead to association and on characterizing the junction zones that form. Less is known about the long-range structure of the gel network. Two extreme types of structures can be envisaged. Firstly the gels could approximate to classic rubber-like structures. The point cross-links are expanded into blocks and the junction zones are linked via the remaining sections of the polymer chains (Fig. 6.9a). The other alternative is that the blocks make the chains sticky and they then associate latterly forming fibrous structures that can grow, branch and fill the sample volume (Fig. 6.9b). These structures are quite different but there is no clear evidence what sort of long-range structure is actually formed, although electron microscopy and atomic force microscopy of pectin networks favours a fibrous structure.[45,46] Perhaps better methods of imaging gel networks within intact gels, or imaging of gel precursors may provide solutions to this problem in the future.

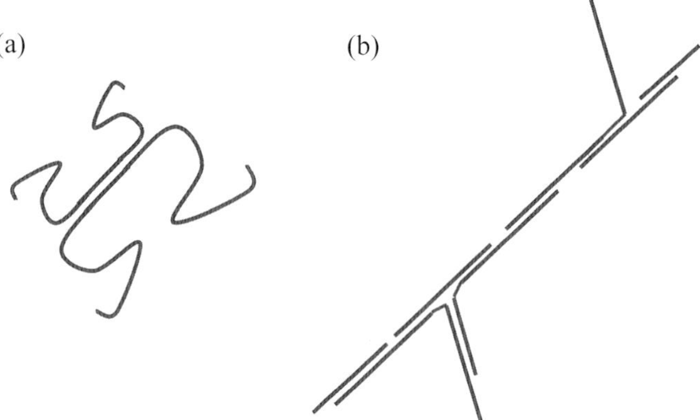



**Fig. 6.9**  Examples of possible alternative long-range structures that may occur within gels. (a) The ordered junction zones are linked by 'disordered' polymer chains. (b) The chains assemble into branched fibrous structures with the junction zones providing the 'glue' to hold the chains within the fibres.

### 6.2.2.3 Higher-order helical aggregates

The ribbon-like structures responsible for the gelation of LM pectin, alginate, glucomannans and galactomannans are examples of two-fold helices.[12] Although the chemical repeat unit in the backbone is a monosaccharide the physical repeat unit is a disaccharide. A number of gel-forming polysaccharides, including xanthan, gellan, furcellaran, agar and the carrageenans, form higher-order helical structures and helix formation is important for gelation.[1,12] The nature of the sugar units, the configuration of the sugar ring and the type of glycosidic linkage determines the ordered secondary conformation of the polysaccharide.[12]

Information on the helical structures of these polysaccharides has been obtained from X-ray diffraction data collected for oriented fibres. The fibres are prepared by stretching lengths cut from thin films of the polysaccharides. The X-ray data provide information on the nature of the helical structure of the polysaccharide, and also information on the association of these helices within the fibre or film. These associated structures are considered as models for the junction zones within hydrated gels. The dehydration step involved in film formation may induce changes in the helical structure or the mode of association. Thus extensive spectroscopic and physical chemical studies have been made in order to confirm that helix formation does occur in solution, and to investigate the nature of helix association involved in network formation and gelation. The studies on these polysaccharides are exhaustive and the literature is extensive. It is not possible to detail the research on each polysaccharide in this review, and the intention is to illustrate the basic principles. There are several review articles that discuss these polymers in more detail and provide a route to the original literature.[1,8–11,16,18,47,48]

Because the helical structures of the polymers depend on the specific chemical structure it is useful to indicate what is currently known about these structures. The best characterized structure is gellan.[1,47–53] X-ray patterns show that gellan forms three-fold helices of pitch 2.82 nm. The axial rise per chemical repeat unit is half the extended length of the chemical repeat unit suggestive of a double-helical structure. Patterns obtained for deacylated gellan are the most crystalline and show that gellan crystallizes into a trigonal unit cell ($a = b = 1.56$ nm and $c$ (fibre axis) $= 2.82$ nm). Molecular modelling of the X-ray data suggests that gellan forms a left-handed three-fold double helix. The packing of the helices in the unit cell has been used as a model for the junction zones of the gel, and additional studies have been made to identify how ester substitution and different cations modify this packing, and hence influence gelation. Early X-ray fibre diffraction studies of ammonium and monovalent ion forms of ι- and κ-carrageenan showed that they formed three-fold helical structures.[54] Modelling of the patterns suggested that both single- and double-helical structures were stereochemically feasible, with some preference being given for the double-helical forms. Better fibre patterns were obtained for divalent cation forms of ι-carrageenan,[55,56] and these patterns, plus data on the potassium salt form,[57] have been used to refine a right-handed three-fold double-helical model. In this model two individual chains of pitch 2.6 nm are parallel and offset by half their pitch, yielding a double helix of pitch 1.3 nm. The patterns for κ-carrageenan are poor, but stereochemically plausible models have been refined against the available data.[58] The best model[58] was a three-fold parallel coaxial double helix with a pitch of 2.5 nm, similar to the ι-carrageenan helix, but with the two chains offset from the half-staggered arrangement by a 28° rotation and an axial translation of 0.1 nm.

Improved molecular transforms[59] of furcellaran and κ-carrageenan were obtained from fibres prepared from mixed films of furcellaran or κ-carrageenan with either galactomannans or glucomannans. These patterns support coaxial double-helical models for furcellaran and κ-carrageenan. However, in the case of κ-carrageenan the data suggest that the two chains are rotated coaxially without any translation along the helix axis. In the case of furcellaran there is no evidence for any translation of the two chains within the double helix.[59] Possibly the half-staggered arrangement in ι-carrageenan is stabilized by local crystallization, and furcellaran and κ-carrageenan would adopt similar structures if they could be induced to crystallize. X-ray data for agarose, the main gelling component of agar, are poor and both possible three-fold double- and single-helical structures have been suggested.[60] X-ray data for xanthan show that it forms a five-fold helical structure of pitch 4.7 nm. Studies[61] on families of polysaccharides based on the xanthan structure, but containing elongated or truncated side chains, have demonstrated that the position of attachment of the side chain on alternate glucose residues is critical for formation of the five-fold helix. Both single- and double-helical structures are stereochemically acceptable and, on the basis of the comparatively poor X-ray data alone, it is not possible to discriminate between these structures.[61]

With the exception of agar these materials are all polyelectrolytes. Dilute solutions of agar are difficult to prepare and study because of the tendency of the polymer to precipitate or gel.[1,17] For the remaining materials there is extensive evidence[1,16,18] to suggest that they adopt the ordered helical structure in solution and that a reversible helix–coil transition occurs on heating and cooling. At sufficiently high dilutions the helix 'melts' and the polymer adopts a coil-like configuration. This arises because repulsive interaction between charges on the helix act to destabilize the helix, and this can be offset by screening the charge. In the absence of added electrolyte, dilution of the polymer solution dilutes the counterions and destabilizes the helix. Increasing the ionic strength increases the setting temperature of the helix. In addition to raising the setting temperature of the helix the addition of salt can give rise to hysteresis in the melting and setting of the helix. This effect only occurs for certain polysaccharides, and for these polymers the types of cations present can be very important. When hysteresis does occur, raising the ionic strength can increase the melting temperature sufficiently to make the transition irreversible on heating. These specific ion effects can be difficult to study because the polysaccharides are often prepared as impure salt forms and certain counterions present may promote hysteresis. These specific ion effects result from binding of the cations on helix formation and association.[1,18] For furcellaran and κ-carrageenan specific binding occurs for $K^+$, $Rb^+$ and $Cs^+$ but not for $Li^+$ or $Na^+$. For purified ι-carrageenan, in the absence of any contaminating κ-material, there does not appear to be any evidence for selective cation binding. For gellan ion binding occurs for all the common monovalent and divalent cations and is only absent for bulky cations such as tetramethylammonium (TMA).[1,16] By careful choice of cations it is possible to study the separate effects of helix formation and ion binding, and to identify their contributions to gelation.

Where detailed studies have been made the evidence suggests that, in the absence of specific cation binding, helix formation on cooling is accompanied by aggregation of the polysaccharides. In the case of sodium ι-carrageenate[1,18] and TMA gellan,[1,16] at sufficiently high concentrations, gelation will occur. The gels are transparent and have low shear moduli. The structures tend to break down on shearing but have the property to 'heal' or recover on removal of the applied stress. For these gels the melting and setting shows no hysteresis, and is reversible and coincident with the helix–coil transition. In these cases aggregation and gelation appears to arise solely through double-helix formation. There are two extreme forms of gel structure that could be envisaged. The earliest models for carrageenan gelation[54] suggested that on cooling, helix formation was initiated leading to multiple links between adjacent chains, and the formation of a rubber-like structure with double-helical junction zones replacing the point cross-links (Fig. 6.9a). The junction zones would be connected by regions of the chains that had not been able to form helical structures. Topological considerations suggest that the multiple links on a chain would need to form in sequence, and this might restrict the level of helix formation in such a complex multiply linked network. However, experimental estimates of the level of helix formation in carrageenan gels[1] suggest that a very high percentage of the chains adopt the helical structure. An alternative is that helix nucleation occurs on contact between chains and the helical structure propagates along the chains. Mismatching of the chains on contact would lead to loose ends, which could form helices on contact with other chains, resulting in elongated filamentous structures that can branch, or even form networks (Fig. 6.9b). Support for the formation of such filaments comes from electron microscopy of ι-carrageenan networks,[62] light scattering studies[63] and AFM images of the aggregation of TMA gellan[1,64] (Fig. 6.10a).

In the presence of specific cations, furcellaran, κ-carrageenan or gellan undergo further aggregation and, at sufficiently high concentrations, form stiffer gels.[1,16,18] Increasing the ionic strength, through the addition of salts containing specific cations, increases the melting and setting temperatures, broadens the hysteresis between setting and melting and, at sufficiently high salt levels, leads to gels that set on cooling but are thermally irreversible. Increasing salt levels leads to increased turbidity and often increased syneresis. Where tested the additional aggregation step appears to be reversible on addition or removal of specific cations. There is strong evidence for the binding of specific cations within the junction zones of the gels and the models for these junction zones are based on the modelling of X-ray fibre diffraction data. The junction zones are pictured as 'crystallites' of aggregated helices bound by associated cations. Although the models for the junction zones are well defined the long-range structure within the gels is less well known. Once again there are basically

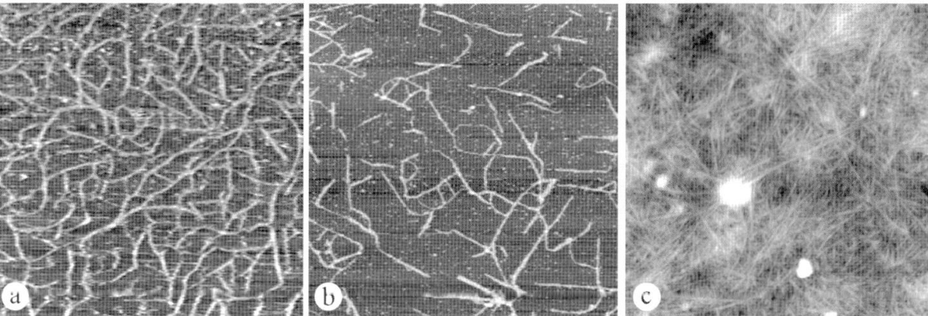

**Fig. 6.10**  Atomic force microscopy images of gellan gel precursors and gellan gels. (a) Branched fibrous aggregates (gel precursors) formed by TMA gellan in the absence of gel-promoting cations. Image size $800 \times 800$ nm. (b) Branched fibrous aggregates formed by gellan in the presence of potassium cations. Note the variable height and thickness of the fibrous gel precursors. Image size $800 \times 800$ nm. (c) Image of the top surface of a hydrated 1.5% acid-set gellan gel showing the fibrous network structure of the gel. Image size $2 \times 2$ μm.

two extreme models (Fig. 6.9). At one extreme the ordered junction zones are considered to be linked by regions of the chains that do not participate in double-helix formation.[65–67] The other extreme model is based on electron microscopy[68,69] and AFM studies of gellan gels and gel precursors,[64] and electron microscopy[70] and AFM studies of carrageenan gelation.[1,71] Cation binding is considered to bind the filamentous aggregates generated via double-helix formation into thicker branched fibrous structures. In the case of gellan gels these fibrous structures have been observed for gel precursors (Fig. 6.10b) and directly within hydrated gellan gels (Fig. 6.10c). In this fibrous model of the gels the elasticity will be dependent on the degree of branching and the extent of aggregation of the helices: higher levels of aggregation will lead to increasingly stiff fibres. Increased aggregation at higher salt levels will give rise to higher levels of demixing and a more turbid structure that is more prone to syneresis. At the higher salt levels the increased melting temperature and the extent of aggregation will make it difficult to completely disperse the polymers in the sol state and will lead to heterogeneous gel structures. This type of effect will be considered later under fluid gels.

The behaviour of the analogous material, agar, is difficult to study because the material readily aggregates or gels forming turbid structures.[1,17] The gels show high levels of thermal hysteresis and are prone to syneresis. It seems reasonable to assume that on cooling the uncharged helices are more insoluble than those formed by the charged carrageenans and the agar helices readily associate forming more highly aggregated networks similar to those of gellan or κ-carrageenan.

Aggregation of helices to form networks is also thought to explain the gelation of HM pectins.[1,25] This type of pectin gel is formed in the presence of sugar at low pH. The mixtures are prepared at high temperature and gel on cooling forming transparent thermally irreversible gels. The nature of the sugar is not significant and these cosolutes are considered to act by reducing the solubility of the pectin. Low pH will reduce the charge on the polysaccharides promoting helix formation, helix association and hence reduced solubility of the polymers. Certainly less acid is required as the DE of the pectin increases, and fully esterified pectins will gel in the absence of added acid. Circular dichroism studies favour helix formation on gelation.[72] Modelling of X-ray fibre diffraction data has been taken to imply junction zones consisting of aggregated three-fold helices with methyl groups occupying channels between the chains.[73,74] The fact that there are only weak equatorial reflections in the X-ray patterns, and that the gels are transparent, suggests that the level of helix aggregation is low. Electron microscopy[45] and atomic force microscopy[46] have supported the idea of fibrous network structures for the gels.

### 6.2.3 What are fluid gels?

Fluid gels are examples of weak gels. The polymers form networks that are elastic when deformed. Increasing the level of deformation does not lead to fracture and irreversible breakdown of the gel structure. Rather, at sufficiently high deformations, the gels flow. If the applied stress is removed then the materials remain deformed but recover their elastic response at low deformation. The gel structure must break to allow flow but can reform or 'heal' on standing. Such structures can be used to suspend particles in emulsions or dispersions in order to prolong shelf life. The materials flow on pouring, spreading or stirring but regain their consistency when the applied stress is removed. The main polysaccharide used to generate these types of structures is xanthan gum. Xanthan is usually called a thickening or stabilizing agent, but association and gelation is the basis of this behaviour.

As with gellan and the carrageenans, extensive studies[1] have shown that xanthan forms a helical structure in solution. From the X-ray data alone it is not possible to determine whether xanthan forms a single or double helix.[61] Although there is still some controversy the consensus of extensive physical chemical studies is that xanthan forms a double-helical structure. The chemical repeat unit is charged and dilution of aqueous solutions dilutes the counterions, lowers the helix–coil transition temperature and dilutes out the helical structure. Addition of salt raises the ionic strength, screens the charge on the polysaccharide, and raises the helix–coil transition temperature. The transition is thermoreversible and seldom shows hysteresis or marked turbidity. However, there is clear evidence that helix formation is accompanied by aggregation.[75] Aggregation is very important for determining the functional properties of xanthan. Under continuous shear aqueous xanthan samples show reversible 'shear-thinning' behaviour at low polysaccharide concentrations. There is strong evidence to suggest that this unusual thixotropic behaviour is due to aggregation of xanthan and the aggregates are called microgels. Removal of aggregates reduces the shear-thinning behaviour resulting in a viscosity profile characteristic of a 'stiff' polysaccharide in solution. The degree of aggregation of xanthan depends on the isolation of the polymer and the methods used to disperse it. Xanthan is produced by bacterial fermentation. It is normally isolated by isopropanol precipitation from the broth and then dried. In the early US Department of Agriculture (USDA) literature on the preparation of xanthan it was noted[76] that salt should be added during precipitation to ensure solubility of the product. Isolation of xanthan polysaccharides from the broth suggests that in the absence of added salt the dried material can be insoluble. With addition of sufficient salt the product is soluble but the thixotropy is poor. At intermediate and increasing salt levels the solubility improves but thixotropy of the dispersed product is worse. AFM images suggest that in the soluble (solution) form the xanthan is present as single molecules (Fig. 6.11a) and that the aggregates (Fig. 6.11b) are larger gel-

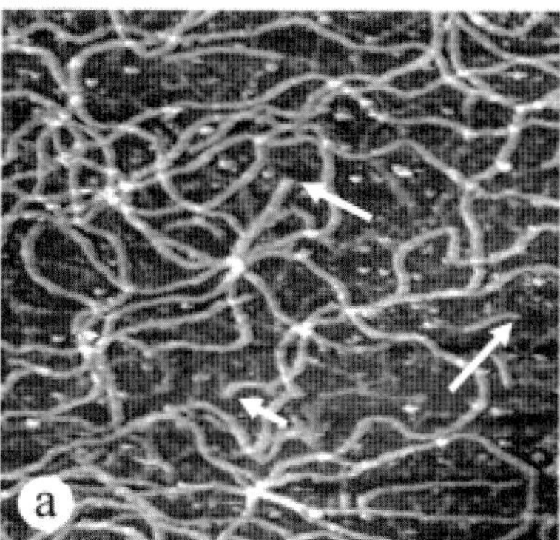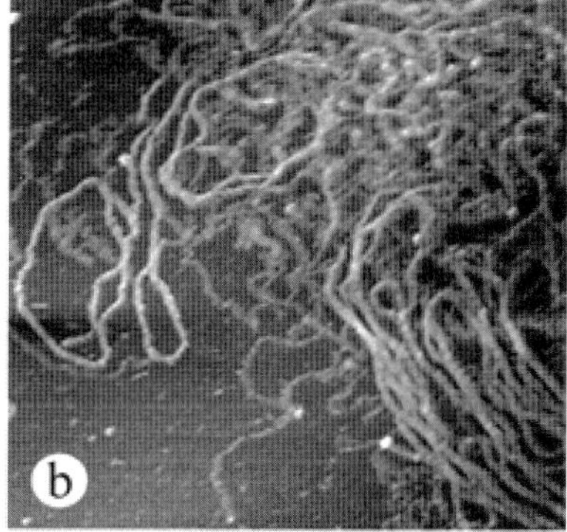

**Fig. 6.11**  Atomic force microscopy images of xanthan gum samples prepared under different conditions. (a) 'Molecular spaghetti': an entangled network of xanthan gum polysaccharide chains. Image size 1.2 × 1.2 µm. Some chain ends are indicated by arrows. (b) Xanthan microgel. Image size 1.4 × 1.4 µm.

like structures (microgels). How are these microgels formed? The most likely explanation is that when salt is added to the broth the xanthan is forced into the ordered helical form and then precipitated and dried as single molecules. In the absence of salt, or at intermediate salt levels, the xanthan will be only partially ordered. Precipitation will increase the xanthan concentration and the concentration of associated counterions, raising the helix–coil transition temperature, driving helix formation and intermolecular association. Addition of salt does not lead to thermal hysteresis suggesting that helix formation would lead to the type of network structure suggested for ι-carrageenan or TMA gellan. The precipitation, filtration and drying steps would fragment this weak gel resulting in dried microgels. Insolubility is considered to be due to the difficulties in hydrating and breaking down (dissolving) these microgels. Thixotropy would be attributed to a dispersion of microgels that can self-heal into a more complete network by reformation of double-helical links broken during extraction from the broth. Microgels would contain helical regions and small regions of denatured xanthan chains. This would account for the surprising observation[77] that cellulases, which cannot degrade xanthan in the helical form, can break down microgel structures.

Thus, whereas ι-carrageenan or TMA gellan can form weak gels under quiescent conditions and will be homogeneous networks, xanthan microgels are formed under shear and will associate to form heterogeneous networks of particles weakly associated with each other. There are other observations that support this view of the xanthan network. A general feature of gellan gelation is that as the ionic strength is increased then both the elastic modulus and the fracture strength rise, pass through a maximum, and then decrease to zero at higher ionic strength.[78] Increased ionic strength leads to an increase in turbidity and an increase in water loss from the gels under compression. At low ionic strength the gels behave as elastic solids but, at high ionic strengths, they show unusual properties. On successive applications of stress the gels progressively stiffen: they shear-harden.[79] The maximum in the modulus has been attributed to a transition from homogeneous networks at low ionic strength to an inhomogeneous network at high ionic strength. It is proposed that at high ionic strengths the aggregates cannot be properly melted and the residual structures present at high temperatures act as nuclei for gelation on cooling, leading to weaknesses where the growing structures merge.[79] Microgel-like structures that mimic the behaviour of xanthan can be generated by stirring or shearing samples such as agar, carrageenans or gellan whilst they are gelling on cooling.[80]

### 6.2.4 Polysaccharide mixtures

Understanding the gelation of polysaccharide mixtures is important in trying to explain the behaviour of commercial polysaccharide extracts, polysaccharide blends, complex polysaccharides such as starch and the role of polysaccharides in food materials. The large number of possible combinations of polysaccharides makes it impossible to review all of the reported studies on polysaccharide mixtures. Fortunately it is possible to divide such mixtures into a few simple classes of materials with characteristic properties. This can most easily be done by first considering binary mixtures of polysaccharides.

If two polysaccharides are mixed together and the binary mixture gelled then it is possible to define a finite number of idealized structures that might be formed for the resultant gels.[81] Although fairly simplistic these models do form the basis for establishing certain basic relationships between the mixed gel and its components. Four types of gel structure can be described (Fig. 6.12). These have been called swollen networks (Fig. 6.12a), interpenetrating networks (Fig. 6.12b), phase-separated networks (Fig. 6.12c) and coupled networks (Fig. 6.12d).

## *6.2.5 Phase-separated networks*

Under equilibrium conditions in dilute solution even chemically similar polysaccharides will phase separate. Examples are amylose and amylopectin,[82] pectins with differing DE values,[83] or pectins and hemicelluloses.[83] However, for fairly concentrated sols, where the mixtures may be entangled, phase separation will be slowed. Rapid gelation will entrap polymers hindering separation and forming fairly uniform networks. Controlling the rates of phase separation and gelation offers a route to manipulating structure and texture of polysaccharide blends.[84,85] If partial phase separation occurs then the mixed gels will appear as inclusions or particles of one phase filling a matrix of the second phase. Gelation of the matrix and/or the 'filler' particles results in the formation of a composite gel.

### *6.2.5.1 Starch*

A common example of a phase-separated polysaccharide network is starch.[1,86] There are a number of reviews that discuss the structure, gelatinization and gelation of starch.[1,13,87,88] Starch is the major storage polysaccharide in plants.[1,13] It consists of water-insoluble spheroidal granules whose shape, size and size distribution vary depending on the plant source.[88] It is clear from optical and X-ray studies that the granules are ordered and partially crystalline.

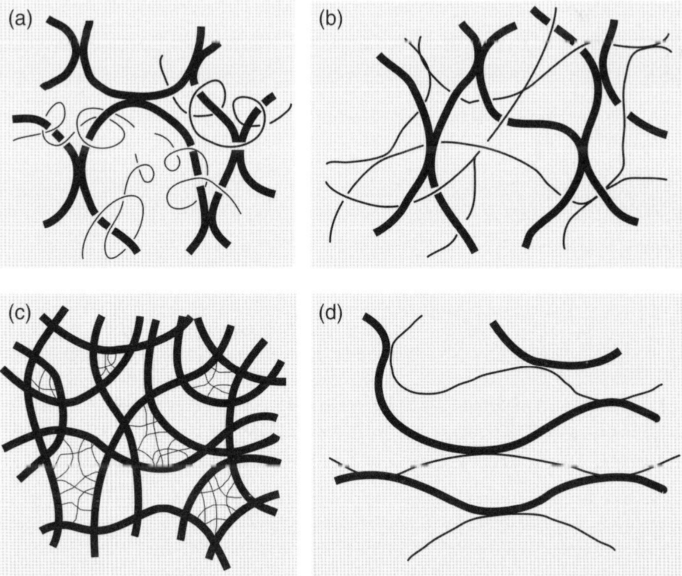

**Fig. 6.12** Schematic models showing idealized structures for two component polymer networks. (a) Swollen networks, (b) interpenetrating networks, (c) phase-separated networks and (d) coupled networks.

Granules can be dissolved in dimethylsulfoxide and fractionated into the two chemically distinct polysaccharides amylose (~20%) and amylopectin (~80%). Starch variants lacking amylose are crystalline; there are no reported starches completely lacking amylopectin, and amylose can be leached from granules without destroying granule crystallinity. These studies suggest that the amylopectin component determines the crystalline structure of the granule, although in high-amylose mutants some amylose may also be present in a crystalline form.[89] The branches of the amylopectin form crystalline lamellae, which are contained within annular bands (growth rings), with the molecular axes oriented radially within the granule.[88] In normal starches these crystallites are embedded within a matrix of amorphous amylose.[88] Starch gels are prepared by dispersing granules in water and then heating and cooling the dispersion. Water is initially believed to be taken up by the amylose within the granule. On heating above a characteristic temperature the crystallites can melt allowing the granules to irreversibly swell and take up water. Amylose is released from the granule leading to swollen granules interpenetrated by a fluid amylose matrix. If the starch concentration is high enough then, when cooled, the amylose gels, resulting in an amylose network interpenetrating the swollen granules (Fig. 6.13). The granules are considered as filler particles reinforcing the amylose network.[90] This is an example of a phase-separated gel because the amylopectin is contained within the remnants of the swollen granule.

The properties of starch gels[90] can be understood in terms of the behaviour of the swollen granules and isolated amylose gels.[91] The gelation of amylose is complex and still contentious.[1] At high temperatures amylose behaves as a flexible polymer. For amylose concentrations above a critical concentration, which can be less than $c^*$, cooling the sol leads to gelation. Normally on cooling the mixture becomes turbid, slowly develops a network structure, and eventually becomes partially crystalline. The development of the turbidity and the network structure, and the extent of crystallization depend on the amylose concentration, but the rate of crystallization is independent of amylose concentration.[1,91] It

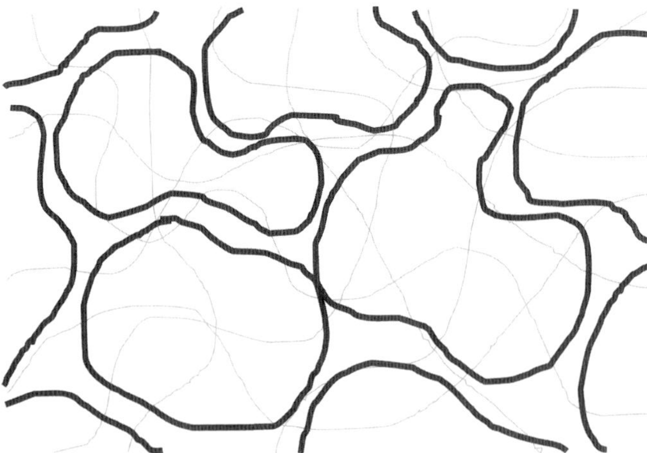

**Fig. 6.13**  Schematic model for composite starch gels, showing swollen gelatinized starch granules filling an interpenetrating amylose gel network.

has been proposed[91] that on cooling the amylose phase separates from solution, causing the samples to become turbid and generating an amorphous network. Crystallization is then considered to occur locally within the amorphous network, generating permanent cross-links or junction zones. Amylose networks are fibrous structures but the detailed molecular nature of the gel network is unclear. Electron microscopy supports the idea of a partially crystalline network structure for the gel.[86,92] A key question is when does helix formation occur? Does helix formation on cooling trigger phase separation, or does phase separation and/or crystallization induce helix formation? Certainly within the crystals the amylose is present in its characteristic double-helical structure. The melting temperature of amylose crystals depends on chain length,[93] and crystals in the amylose network do not melt below 100°C, making the amylose gels thermally irreversible.

If swollen granules are leached free of amylose and cooled then they harden and eventually recrystallize.[90] However, this process is much slower than the crystallization observed for the amylose. This is believed to be a partial re-establishment of the structure of the granule. Because the branch lengths of the amylopectin are short these crystals can be melted at temperatures close to the gelatinization temperature of the starch. This transition is thermoreversible. Isolated amylopectin will also gel on cooling forming opaque thermoreversible gels.[94] The gels take a long time to form and cross-linking is related to crystallization, suggesting that network formation involves cross-linking via crystallization of short branches of the amylopectin molecules.

Thus the formation of starch gels involves gelation and crystallization of the solubilized amylose to form an opaque, thermally irreversible network filled with soft swollen granules. With time the amylopectin within the granules crystallizes hardening the granules and reinforcing the amylose network. This transition contributes to the hardening observed on the staling of starch-based foods. As this transition is thermally reversible the 'staling' can be reversed on heating (Fig. 6.14). If the sample is heavily sheared on preparation then the result

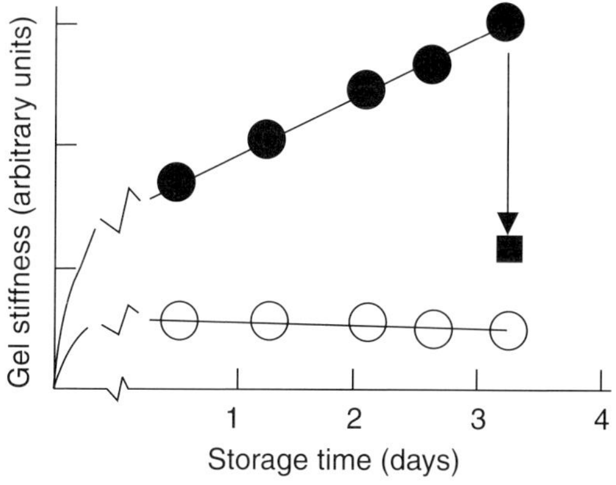


**Fig. 6.14**   Comparison of the gelation of pea starch (●) and pea amylose (o) samples. The modulus of the 4% (w/w) amylose gel (o) develops rapidly and saturates. This structure is irreversible to heating below 100°C. A 20% (w/w) starch gel will contain approximately 4% amylose. For the starch gel there is a rapid build-up of modulus due to the gelation of the released amylose, but the modulus of the starch gel is enhanced by the filling action of the swollen starch granules. As the granules harden, due to recrystallization of the amylopectin, this filling effect is enhanced. This effect is thermoreversible and can be eliminated on heating to 95°C and then recooling to room temperature (■).

is an approximation to a mixture of amylose and amylopectin. The amylopectin is believed to inhibit amylose association, possibly due to amylose–amylopectin interactions.

Blends of starch with other hydrocolloids are often claimed to produce novel synergistic behaviour. The most straightforward explanation for these effects is that on gelatinization the swollen granules exclude the added hydrocolloid, increasing its local concentration and enhancing its viscosity. However, this is an interesting area worthy of further research.

### 6.2.5.2 Semi-refined carrageenans

Semi-refined carrageenans will also form composite networks. The milder extraction procedures result in a carrageenan preparation that contains fragments of algal cell wall cellulose (Fig. 6.15). The presence of the cellulose modifies the behaviour of the carrageenan gel, the most notable example being the induced turbidity of the semi-refined samples.

## 6.2.6 Swollen networks

These types of structure are most likely to arise from mixtures of a gelling and a non-gelling polysaccharide, or mixtures of two gelling polysaccharides under conditions where only one of the polymers is induced to gel. The non-gelling polymer is considered to reside within and swell the gelled network (Fig. 6.12a). If the rate of demixing of incompatible polysaccharides is slow compared with the rate of gelation then the non-gelling polymer will be fairly uniformly distributed within the gel network.

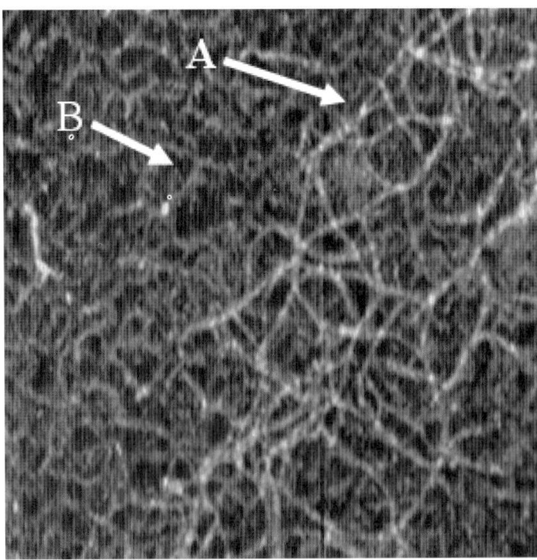


**Fig. 6.15**  Atomic force microscopy image of a semi-refined carrageenan sample. Image size 700 × 700 nm. The cellulose network (A) is thicker and stiffer than the ι-carrageenan network (B) and this introduces contrast into the image, allowing the two networks to be identified. The cellulosic fragments are contained within the carrageenan network.

## *6.2.7 Interpenetrating networks*

Interpenetrating networks consist of two independent networks that each span the entire sample volume and interpenetrate throughout each other (Fig. 6.12b). True interpenetrating networks at the molecular level were thought to be unlikely to form because of the tendency of polysaccharides to phase-separate from each other. However, if charged and uncharged polysaccharides are mixed then phase separation can be inhibited.[95] If the two polymers were to phase-separate then counterions associated with the charged polysaccharide would become localized within regions of the gel. The consequent large change in entropy of mixing opposes separation of the polymers. This principle has formed the basis for the development of a number of interpenetrating networks.[96–99] There is some suggestion that screening of the charged polymer promotes separation and can be used to control and modify the structure within the gel.

## *6.2.8 Coupled networks*

These are formed by mixtures of polysaccharides under conditions where the individual components alone will not gel, but the mixtures do gel. Although the mechanisms for gelation in most of these systems are still controversial there is considerable evidence in all of these cases that intermolecular binding between the two polysaccharides contributes to formation of a permanent network. There are a number of review articles that consider in detail the literature on these systems.[1,8–11,16] The intention here is to summarize the situation and indicate areas where the knowledge and understanding are lacking.

If intermolecular binding between different polysaccharides is to occur then there needs to be some stereochemical similarity between the two polysaccharide structures. The conditions under which gelation is induced need to favour formation of mixed (hetero-) junction zones rather than individual (homo-) junction zones. In the simple schematic diagram shown in Fig. 6.12d the only linkages are heterojunction zones. In real gels there may be a mixture of both hetero- and homo-junction zones.

There are only a few polysaccharide systems that are generally considered to form coupled gels:[1,8–11] pectin-alginate, xanthan-galactomannan, xanthan-glucomannan mixtures and mixtures of certain algal polysaccharides (agar, carrageenans and furcellaran) with either galactomannans or glucomannans. There are also a number of variants of the xanthan structure that will form coupled gels with galactomannans or glucomannans.[1,10,11,100–103] The study of these systems has helped to develop the models for xanthan mixed gels that are used in foods.

### *6.2.8.1 Pectin-alginate gels*

Pectin-alginate mixtures will gel under conditions for which the components will not gel singly: gels are formed in the absence of calcium or high concentrations of sugar, at low pH.[1,8,10] The stiffness of the gels has been found to depend on the M/G ratio of the alginate, the DE of the pectin, the concentration of any added sugar, the pH and the ratio of alginate to pectin. The stiffest gels are formed for 1:1 mixtures of high-G alginates and high-DE pectins. Sugar is not essential for gelation but can be used to modify the setting properties and melting points of the gels. Calcium present prior to gelation inhibits gelation but calcium added after gelation strengthens the gels. Gelation can be induced by cooling mixtures from high temperatures under appropriate conditions or, at room temperature, by lowering the pH. Gelation appears to begin when the pH drops below 3.4.[104] On the basis of the near mirror-image symmetry between polygalacturonic acid blocks and polyguluronic acid blocks, and the observation of marked changes in circular dichroism on gelation, it has been proposed[105] that the junction zones are pseudo-egg-box structures formed between alginate G blocks and methyl esterified galacturonic acid blocks. The addition of calcium may promote further alginate G block self-association, thus enhancing the stiffness of the gels. Gelation is favoured by high-DE pectins and hence essentially mixtures of highly charged and weakly charged polysaccharides. This would inhibit phase separation of the polymers. At pH < 3.4 both the polymers are uncharged and the mixture should start to phase-separate, but gelation would arrest this effect, explaining the transparency of the gels. Alginate G–G block association can occur at low pH and hence, certainly in the presence of added calcium, it is likely that these gels contain mixtures of hetero- and homo-junction zones.

### 6.2.8.2 Xanthan-glucomannan gels

As described earlier xanthan forms only weak gels and native glucomannans such as konjac mannan do not gel. However, mixtures of the polysaccharides form thermoreversible gels on heating and cooling.[1,8–11] The gels are transparent and there is no thermal hysteresis between setting and melting. The strongest evidence for intermolecular binding between the different polysaccharides comes from X-ray diffraction studies.[81,102,106,107] X-ray diffraction patterns obtained from oriented fibres, prepared from mixed xanthan-konjac mannan gels, show new patterns (Fig. 6.16) that are not a sum of the X-ray patterns of the component polysaccharides, as would be expected for a simple mixture of the two polymers.[1,11,106] Similar data have been obtained for mixed gels formed between konjac mannan and xanthan-like polysaccharides.[102,103] In all cases the natural ordered conformation of the xanthan (or xanthan-like) polymer is a five-fold helix, that of konjac mannan is a two-fold helix, and the

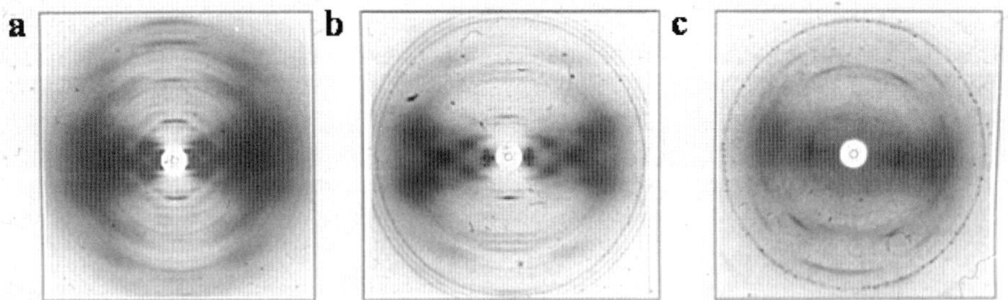

**Fig. 6.16**   X-ray diffraction patterns obtained for oriented fibres prepared from (a) xanthan, (b) a 50:50 mixture of xanthan and konjac mannan and (c) konjac mannan. The fibre axes are vertical. The fibre patterns are indicative of (a) a five-fold helix, (b) a six-fold helix and (c) a two-fold helix. The unique patterns shown in (b) confirm intermolecular binding between the two polysaccharides, and provide a basis for building models of the new structure. Details of the sample preparation and discussion of the results is reported elsewhere.[107]

mixtures show evidence for a new six-fold helical structure. The intriguing question is how do these new junction zones form?

The key to understanding gelation lies in the thermoreversibility of the gels. Experimental data show that heating the mixtures above the helix–coil transition temperature of xanthan, and then cooling the mixtures enhances gelation. Xanthan and the xanthan-like polysaccharides consist of a cellulosic backbone solubilized by charged side chains attached to alternate glucose residues.[61] The family of polysaccharides differ in the length and composition of the side chain, and the esterification of the chemical repeat unit. All show a helix–coil transition and, above this transition temperature, the cellulosic backbone is exposed. The side-chain substitution pattern means that the cellulosic backbones are ribbon-like structures with an unsubstituted and a substituted face. It has already been mentioned that the backbone of glucomannans strongly resembles cellulose. Although acetylation of the glucomannan inhibits large-scale association and crystallization it is possible for regions of the glucomannan to associate with the bare cellulosic face of xanthan, and for the complex to twist into a six-fold helix, with the charged xanthan side chains decorating the outer surface of the helix. Such a model structure has been demonstrated through analysis of the X-ray data for fibres prepared from acetan (a xanthan-like polymer)-konjac mannan gels[103] (Fig. 6.17).

Clearly gelation is favoured at low ionic strength and on heating and cooling the mixtures. The absence of hysteresis in setting and melting behaviour suggests that double-helix formation alone is responsible for polymer association. Addition of salt after gelation has occurred should stabilize the mixed helix and stabilize the gels. Factors that stabilize the xanthan helix should inhibit gelation. Thus mixtures of xanthan and the glucomannan at room temperature, or mixtures at high temperature at high ionic strength, might not be expected to gel. However, gelation can occur and there is evidence that, in these types of systems, the gel structure slowly evolves as a function of time.[101] In such cases it is possible that the glucomannan may be able to find and bind to regions of the xanthan chain that are not associated as xanthan helices, or that the formation of hetero-junction is favoured, and drives a xanthan helix-to-coil transition allowing mixed junction zones to form.

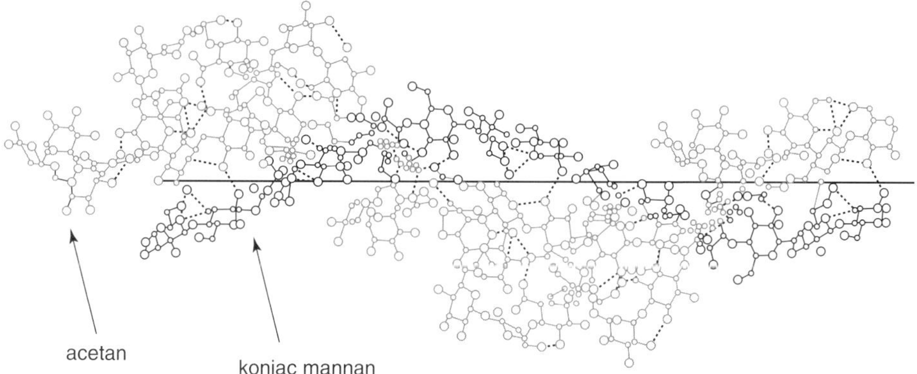



**Fig. 6.17**    Model of the six-fold left-handed parallel mixed double helix formed by the binding of acetan and konjac mannan.[103]

The fact that xanthan is charged and the glucomannans are uncharged may explain why little demixing occurs, and why the gels are transparent and show little evidence of syneresis. Uniform mixing of the polymers would favour intermolecular contacts between the different polysaccharides and formation of mixed junction zones. Addition of salt would be expected to increase the possibility of demixing and favour enhanced turbidity and syneresis of the gels. Most physical chemical studies on xanthan and mixed gels are normally performed using well-prepared solutions of xanthan where mixed junction zone formation would be favoured. In real applications it is likely that the xanthan is present as a dispersion of weakly associated microgels. Under these conditions the glucomannans may interpenetrate the microgels and cross-link them converting the weak gels to strong gels. Thus in these systems there may be xanthan–xanthan and xanthan–glucomannan cross-links, and the overall distribution of xanthan within the network would be difficult to distinguish from that of a dispersion of xanthan alone.

### 6.2.8.3 Xanthan-galactomannan gels

Mixtures of xanthan with galactomannans such as carob or tara gum form thermoreversible gels under conditions for which the individual components alone do not gel.[1,8–11,22] The gels are transparent, and show little evidence of syneresis or hysteresis in their setting and melting behaviour. As with xanthan-glucomannan gels, factors that favour denaturation of the xanthan helix favour gelation, and gelation is favoured with those galactomannans most prone to show self-association. Thus mixed gel formation is favoured for xanthan-tara and xanthan-carob combinations, but it is difficult to gel mixtures of xanthan and guar gum. It is not simply the M/G ratio that is important. Clearly it is the distribution of unsubstituted mannan blocks that is important because different galactomannans with the same M/G ratios show different propensities to gel, and again the galactomannans most likely to self-associate are most likely to form mixed gels.

In general it is considered that gelation involves the formation of hetero-junction zones between denatured xanthan and 'bare' mannan blocks of the galactomannans. In these cases X-ray diffraction patterns obtained for oriented fibres prepared from mixed gels show patterns that are not simply sums of the patterns for the individual components.[81,106] The X-ray data are discussed in more detail elsewhere[1,11] but the interpretation of the new patterns is more difficult than is the case for xanthan-glucomannan gels. The patterns for the mixed gels are similar to those seen for the galactomannans alone, but show systematic absences of certain reflections. It is proposed that galactomannans can cocrystallize with denatured xanthan chain segments, and that the longer side chains on the xanthan molecules disrupt certain planes in the crystal lattice. This disruption is believed to account for the missing reflections. This model for the mixed junction zones is stereochemically feasible but the idea of cocrystallization is difficult to test at the present time. Once again factors that stabilize the xanthan helix should inhibit gelation. Thus mixtures of xanthan and galactomannans at room temperature, or mixtures at high temperature at high ionic strength, might not be expected to gel. Gelation can occur under these restrictions and, in such cases, it is possible that the galactomannan may be able to find and bind to regions of the xanthan chain that are not associated as xanthan helices, or that the formation of hetero-junction zones is favoured, and drives a xanthan helix-to-coil transition allowing mixed junction zones to form.

Xanthan is charged and the galactomannans are uncharged and this would explain why little demixing occurs and the gels are transparent, and show little evidence of syneresis. Again uniform mixing of the polymers would favour intermolecular contacts between the different polysaccharides and formation of mixed junction zones. Addition of salt would be expected to increase the possibility of demixing and favour enhanced turbidity and syneresis of the gels. As argued for xanthan-glucomannan gels, in real applications it is likely that the xanthan is present as a dispersion of weakly associated microgels. Under these conditions the galactomannans may interpenetrate the microgels and cross-link them, converting the weak gels to strong gels. Thus in these systems there may be xanthan–xanthan and xanthan–galactomannan cross-links, and the overall distribution of xanthan within the network would be difficult to distinguish from that of a dispersion of xanthan alone.

### 6.2.8.4 Algal polysaccharide glucomannan or galactomannan mixed gels

These types of gels are formed by mixtures of the gel-forming algal polysaccharides such as agarose, furcellaran or κ-carrageenan with plant galactomannans or glucomannans.[1,8–11,22] Because the algal polysaccharide alone will gel the synergism seen for these mixtures manifests itself through enhanced rheological properties at a given polysaccharide concentration, or gelation at concentrations below that at which the algal polysaccharide alone will gel. Gelation with galactomannans is favoured for those galactomannans that show most evidence for self-association. The gels formed are thermoreversible and transparent, but do show thermal hysteresis in their setting and melting behaviour.

For these systems the evidence for intermolecular binding is perhaps more circumstantial.[1,9–11] The most studied systems are κ-carrageenan mixed gels. The kinks in the κ-carrageenan chains can be selectively cleaved to produce small segmented carrageenan fragments that undergo helix–coil transitions, but do not gel at normal concentrations. Mixtures of these segments with galactomannans do gel, and this was attributed to carrageenan–galactomannan binding. Evidence for immobilization of galactomannans on gelation comes from NMR studies of gelation, although this could arise from self-association of the galactomannans. ESR and DSC studies have been interpreted in terms of intermolecular binding between the two different polysaccharides. Such studies have shown that factors that inhibit self-association of κ-carrageenan and its gelation inhibit mixed gel formation. Thus the potassium salt of κ-carrageenan will form mixed gels with galactomannans or glucomannans but the sodium salt form does not form mixed gels. These observations may explain why κ-carrageenan will form mixed gels but ι-carrageenan, which does not form helical bundles, will not form mixed gels. This is supported by light scattering studies on the aggregation of low molecular weight κ-carrageenan and ι-carrageenans in the presence of galactomannans. Mixed gel formation follows the level of insolubility or tendency to aggregate for the algal polysaccharides: agarose > furcellaran > κ-carrageenan >> ι-carrageenan. There is direct evidence that mixed gel formation reduces the aggregation of the algal polysaccharide and the hysteresis of the gelation and melting process.

There is no direct evidence for specific intermolecular binding from X-ray diffraction studies. X-ray patterns of oriented fibres prepared from mixed gels always yielded molecular transforms of the helical structures of the algal polysaccharides.[59,81,108,109] This is consistent with the idea that the galactomannans or glucomannans inhibit aggregation and crystallization of the algal polysaccharides. Interestingly, annealing of oriented κ-carrageenan-carob fibres resulted in crystallization of carob within the fibres.[81] Crystallization of carob resulted in diffraction rings superimposed on the aligned patterns indicating that the galactomannan did not cocrystallize with the aligned fibres. Although no evidence for specific intermolecular binding was observed it was suggested[81] that random adsorption of segments of the galactomannans or glucomannans to the surfaces of aggregates of the algal polysaccharide helices may occur. This type of structure is emerging as the model for association in these mixed gels and is really a refinement of the earliest model for association suggested by Dea and coworkers.[110] Support for this model has come from gel permeation studies of galactomannans on columns packed with aggregated agarose,[111] confirming related studies[112] suggesting binding to carrageenan aggregates.

Electron microscopy of mixed gels failed to reveal the location of the galactomannans in galactomannan-carrageenan networks, but the images were considered consistent with additional galactomannan cross-linking of a carrageenan network.[113] Galactomannans can be fluorescently labelled and the labels do not inhibit gelation with carrageenans. Mapping of the fluorescent label to locate the galactomannans, and the use of sulfur or potassium as markers to map κ-carrageenan, failed to reveal any phase separation of the polymers within mixed gels.[81] Although there are isolated reports of phase separation in these gels this seems highly unlikely to occur in such mixtures of a charged and neutral polymer, particularly if cross-linking does occur between the polymers.

The idea of galactomannan or glucomannan adsorption to algal polysaccharide aggregates means that gelation is favoured by factors, such as high ionic strength, that promote cation-mediated association of the algal polysaccharide helices. Here increased ionic strength favours mixed gel formation. This is in contrast with the behaviour of xanthan mixed gels. However, if nonspecific adsorption to aggregates is important it is difficult to understand why similar mixed gels are not formed with gellan.

# 6.3 Protein gels

## *6.3.1 What are proteins?*

The basic building blocks of proteins are amino acids. Proteins are high molecular weight biopolymers. Unlike polysaccharides proteins seldom adopt a single type of secondary structure. Rather, different sequences of amino acids fold into particular structural arrangements, of which the two most common forms are α-helices and β-sheet structures. Estimates of the secondary structures of proteins can be made by analysis of circular dichroism spectra. The presence of these ordered structures within the protein generates a characteristic size and shape that is consolidated via a range of interchain interactions including electrostatic interactions, hydrophobic interactions, hydrogen bonds and sometimes covalent linkages, such as disulfide bonds. Food proteins come in a variety of sizes and shapes,[2,114] ranging from small globular structures seen for the egg albumins, serum albumins and the typical milk proteins such as β-lactoglobulin, α-lactalbumin or the caseins, through the disc-like structures of the plant storage proteins, to rod-like (myosin and actin) or fibrous structures (gelatin) of the meat proteins.

Crystal structures are available for most of the common globular proteins, and there is extensive information available on the three-dimensional (3-D) structures of myosin, actin and gelatin. For the globular proteins the crystal structures are believed to represent the native structures of the proteins, particularly for the blood, milk and egg albumins. Many seed globulins are naturally present as crystalline arrays of proteins within the plant tissue. The muscle proteins actin and myosin can be isolated from muscle without substantially altering the ordered structure of the molecules. The caseins are a family of proteins ($\alpha_{s1}$-, $\alpha_{s2}$-, $\beta$- and $\kappa$-casein) present in milk as an aggregated structure known as the casein micelle. The micelles are spherical colloidal structures typically >100 nm in diameter. Sodium caseinates are extracts from milk used to stabilize emulsions, which are mixtures of the different proteins. Gelatin is unusual in that it is extracted by denaturation of native collagen structures. The ordered structure of gelatin is generally considered to be formed as the extract tries to reassemble back into the collagen structure.[3–5]

## 6.3.2 How do proteins form networks?

Protein gelation has been reviewed by several authors[2–7] and the following account is a summary of the main features of protein gelation. The types of network structures and gels formed by proteins will depend on the type of protein and the conditions used to prepare the gel. It is possible to break down the behaviour into three broad classes: globular proteins, fibrous proteins and the casein micelle.

### 6.3.2.1 Globular proteins

For purified proteins the initial step in the preparation of gels involves preparing a sol or solution of the protein. This involves dissolving the protein in aqueous solution usually at low temperatures. The actual state of the sol will depend on the type of protein and the nature of the solvent. In the simplest situation, such as for serum albumins, the sol will be a simple solution of individual proteins, although strong charge–charge interactions between proteins can lead to higher-order structure formation in 'solution'. In this chapter we will mainly be interested in heat-induced gelation of proteins at neutral pH.

In general, although solvation of the protein will lead to some degree of expansion and mobility of the less-ordered regions of the structure, it is generally assumed that the solution structure approximates to the crystal structure of the protein. This is likely to be the situation for the serum albumins. For $\beta$-lactoglobulin the monomer associates in solution to form a dimer, and further association to higher-order oligomers may occur. Such equilibria will be sensitive to pH and ionic strength. The seed globulins are even more complex structures. The proteins are built up from smaller subunits and there is a complex association-disassociation equilibrium that will depend on pH and ionic strength. Variation of pH can lead to structural modification of proteins. Examples include the low pH-induced 'molten globule' transition of $\alpha$-lactalbumin or the expansion of serum albumins.

Heat-set protein gels are formed by heating the sol at a predetermined rate to a defined temperature, holding at that temperature for a finite period of time, and then cooling to room temperature. The form and properties of the gel will depend on the thermal history of the sample, but protein association normally begins when the proteins are heated above a characteristic temperature. Depending on the initial pH and ionic strength it is possible to obtain either transparent or turbid gels. At low ionic strength turbidity increases, around the isoelectric point of the protein, where the net charge is zero. Similarly, at neutral pH, increasing the ionic strength reduces charge–charge interactions and enhances turbidity. So why do the proteins start to associate above a characteristic temperature, and what determines the turbidity of the gel?

DSC studies show an endothermic transition on heating. The proteins adsorb energy and this is generally considered to give rise to some degree of protein unfolding. Originally it was believed that the secondary structure of the protein was completely destroyed on heating, but it has now been clearly established that most proteins remain globular, and retain the majority of their secondary and tertiary structure. Although there is only a small change in secondary structure on heating this is sufficient to allow the proteins to aggregate. Spectroscopic studies suggest that the main structural changes are in β-sheet structure with aggregation involving the formation of intermolecular β-sheet. This is easiest to observe for proteins, such as the serum albumins, that contain little native β-sheet structure, but has also been demonstrated for proteins such as β-lactoglobulin or the seed storage protein glycinin, both of which contain substantial levels of β-sheet in their native states. Aggregation may involve other structural changes such as hydrophobic bonding or covalent linkages such as disulfide bonding. These may supplement β-sheet formation and can convert reversible aggregates into permanently associated structures.

Thus heating makes the proteins 'sticky' and facilitates association. The form of association depends on the protein and the experimental conditions. However, there are two types of aggregation process. Firstly, when there are strong repulsive interactions between proteins then the tendency is to form fibrous aggregates (Fig. 6.18) and the resultant gels tend to be transparent. So why do the proteins form fibres? For disc-like structures such as the seed

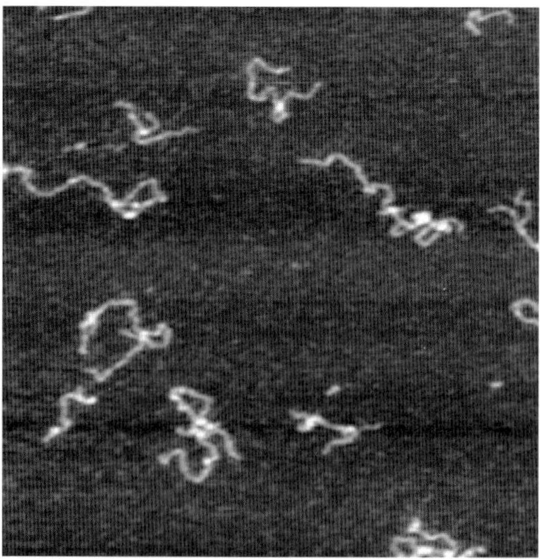

**Fig. 6.18**  Atomic force microscopy image (scan size 2 × 2 μm) showing the types of fibrous aggregates formed by the milk protein β-lactoglobulin when heated at neutral pH.

globulins it might be expected that the faces of the disc are most likely to associate and stick together, forming aggregates resembling stacks of coins.[115] For essentially spherical proteins it is more likely that the proteins can stick at different sites on their surfaces, but that they prefer to form long chains because this will lead to the maximum separation of charge in the formed aggregate. Fibre formation can be demonstrated and characterized by a range of indirect methods including flow and electric birefringence or light scattering, and through direct observation of the aggregates[2,6,115–118] with electron or atomic force microscopy (Fig. 6.18). These methods can be used to characterize the detailed stiffness and shape of the aggregates. At sufficiently high levels of aggregation the fibres will become entangled producing viscoelastic behaviour at high protein concentrations. However, the heating of most proteins at concentrations above 5% gives rise to permanent gels.[2] Microscopic studies of gels suggest that the networks are interconnected fibrous structures.[2,6] Light-scattering studies, such as those on β-lactoglobulin under conditions where the repulsive interactions between the proteins are partially screened, shows formation of three-dimensional branched aggregates.[119,120] What does not seem to be clear is whether, at the higher concentrations at which gelation occurs, linear aggregates are still formed, which then cross-link to form a network, or whether at these concentrations the proteins form branched, rather than linear aggregates, which grow into the final gel network.

The second type of gel structures are formed when the repulsive interactions are screened at high ionic strength, or when the pH is close to the isoelectric point of the protein. In this case globular aggregates are formed and the gels tend to be turbid. This is considered to be a demixing or coagulation of the proteins leading to the growth of large globular aggregates, which then associate to form colloid-like gel structures. In addition to being turbid such gels will fracture more easily at the links between coagulates, and tend to syneresis and to release water on compression.

### 6.3.2.2 Fibrous proteins

Broadly speaking the gelation of rod-like proteins such as myosin is similar in behaviour to that of the globular proteins, but the shape and size of the basic unit will alter the nature of the final network. However, the behaviour of the fibrous protein gelatin is very different.[3–5] The gelation of gelatin resembles that of thermoreversible polysaccharide gelling agents such as gellan or the carrageenans. Gelatin is perhaps the classic example of a gelling agent and has been extensively studied for many decades. Gelatins are essentially denatured and degenerated collagens, and gelation may loosely be considered as a frustrated attempt to re-establish the ordered collagen structure. The structure and properties of the gels will depend on the source and method of extraction. Detailed structural information and the effects of preparative procedures on function are described in several reviews.[3–5,121]

The important feature of the collagen structure relevant to the gelation of gelatin is the fact that collagen consists of three so-called α-chains that form a triple helix stabilized by hydrogen bonding. In bone or hide the collagen rod-like structures are further assembled into three-dimensional or two-dimensional structures with characteristic staggering or spacing of the collagen molecules. In Europe most gelatins are extracted from bone and cattle hide by a liming process. Extraction isolates the α-chains, degrades the material altering the molecular weight profile, and the alkaline conditions can modify certain amino acids changing the pI of the product. Commercial gelatins are heterogeneous materials and not all of the material contributes to network formation.

Isolated α-chains will gel, and there is evidence that one type of α-chain ($\alpha_2$-chains) is less susceptible to gelation, and is preferentially found in the non-gelling fraction of commercial gelatins. At high temperatures the α-chains are considered to be essentially denatured. On cooling the observed changes in optical rotation have been taken to indicate a coil–helix transition. The suggestion is that junction zones are formed by the nucleation and growth of triple helices. At low concentrations the kinetics of gelation favours a first-order reaction whereas at higher concentrations the reactions are third-order. This seems to indicate intra-helix formation at low concentrations and interhelix formation at higher concentrations. The initial stage of association on cooling is followed by a longer, slower maturing, or ageing of the network. Physical chemical studies support the idea of triple-helical junction zones but yield little information on the long-range structure within the gel. If each chain is to be linked to several others through formation of triple-helical junction zones, separated by lengths of disordered chains (Fig. 6.9a), then there would appear to be topological problems in assembling such a structure, unless each junction can form in sequence along each chain. Similar constraints may also apply to the transient models of junction zones unzipping and reforming during annealing of the gel structures. An alternative is the development of the type of structure seen with the gelation of gellan, in which assembly of α-chains into triple helices leads to the growth of branched aggregated filaments (Fig. 6.9b). Here the effects of quenching or annealing could be attributed to competition between nucleation and growth of helical segments. Certainly electron microscopy of gelatin gels seems to favour this sort of fine network structure.[122,123] Interestingly, Ledward in his review on the gelation of gelatin[3] cites electron microscopy studies that suggest that ageing of these gels is accompanied by aggregation of triple helices into larger fibres, with some evidence for staggering of the helices into banded structures resembling native collagen. Recent atomic force microscopy studies of gelatin aggregation[124] also favour this type of association (Fig. 6.19). The fine fibrous open networks account for the transparency of the gels.

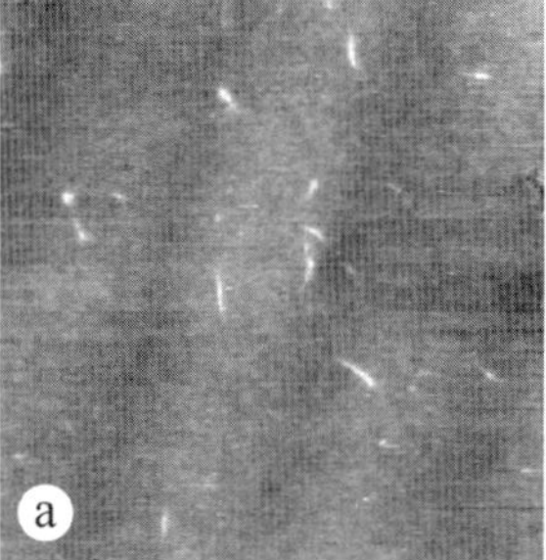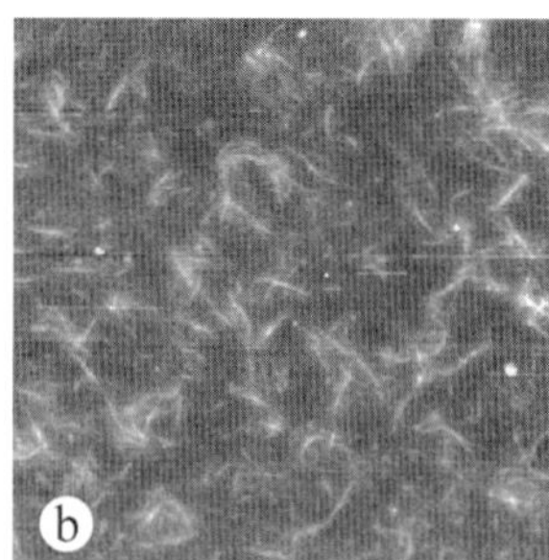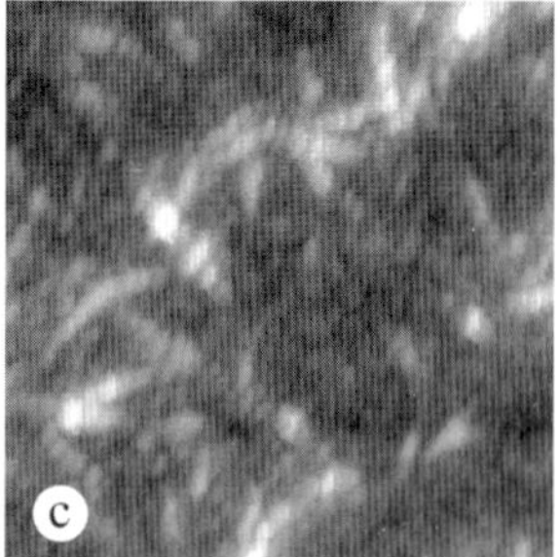



**Fig. 6.19** Atomic force microscopy images showing the time-dependent development of a gelatin network. Scan sizes are (a, b) 4 × 4 μm and (c) 600 × 600 nm. Detailed experimental conditions are described elsewhere.[124] The images show the early stage of gelation (a), which consists of a fine network formed from thin fibres, believed to be triple helices. A few brighter, thicker aggregates of these fibres can be seen. The modulus of the gel increases slowly with time and this is seen (b) to involve 'growth' of these aggregated junction zones within the fine network leading to a coarser network structure. Higher-resolution images (c) of these aggregated junction zones show banding suggestive of the reformation of a 'collagen-like' structure.

### 6.3.2.3 Casein gels

The caseins are a group of phosphorylated proteins ($\alpha_{S1}$-, $\alpha_{S2}$-, $\beta$- and $\kappa$-caseins) that are found in milk in an aggregated micelle structure:[125] the casein micelle. Sodium caseinate is a mixture of casein proteins obtained by precipitation of casein micelles from milk at pH 4.6. The $\alpha_{S1}$-, $\alpha_{S2}$- and $\beta$-caseins tend to precipitate in the presence of small quantities of calcium, whereas $\kappa$-casein does not and, in mixtures with the other caseins, $\kappa$-casein appears to inhibit their precipitation. An emerging model,[7] which describes the association of the caseins, treats the molecules as block copolymers containing hydrophobic and hydrophilic regions (Fig. 6.20). The $\beta$-caseins are considered to contain a single hydrophobic block and a single hydrophilic block allowing them to assemble into detergent micelle-like structures (Fig. 6.20a) resembling the block type junction zones already described for certain polysaccharide gels. The $\alpha_{S1}$-caseins are considered as tri-block copolymers with a central hydrophobic block sandwiched between hydrophilic blocks, allowing them to assemble into polymeric chains (Fig. 6.20b). Analysis of the amino acid sequences of the $\alpha_{S2}$-caseins has been taken to suggest a chimeric structure containing features characteristic of both $\alpha_{S1}$- and $\beta$-caseins. The association of the caseins into casein micelles is envisaged to involve the association of hydrophobic blocks, allowing the $\beta$-casein to cross-link $\alpha_{S1}$- and $\alpha_{S2}$-casein polymeric chains into a network structure for the globular casein micelle. The $\kappa$-caseins act to poison polymerization and populate the surface of the micelle. At the surface the $\kappa$-caseins sterically stabilize the micelles against aggregation. These associated micelles are considered to differ from natural micelles in milk in that they lack the small nanoclusters of crystalline calcium phosphate. It has been proposed[126] that the hydrophobic blocks on the caseins can adsorb to these nanoparticles trapping them within native micelles. The enzymatic degradation of the brush border structures provided by the $\kappa$-caseins, as in the action of rennet, destabilizes the micelles permitting micelle association and gelation. This type of network can be thought of as a colloidal or particulate gel, but equally can be considered as a coagulated structure.

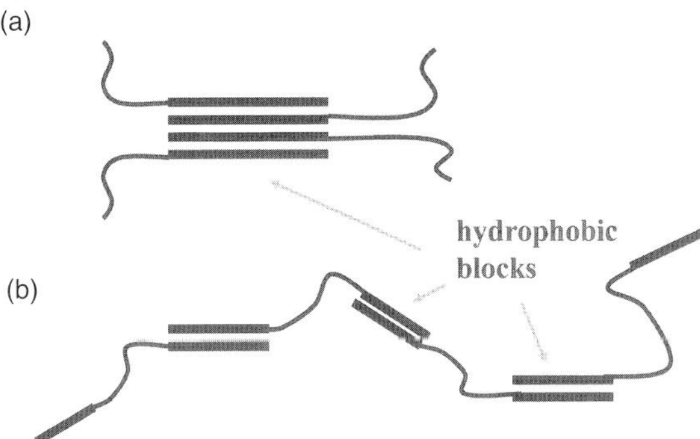


**Fig. 6.20**   Schematic model of the suggested block copolymer structures of (a) the $\beta$-caseins and (b) the $\alpha_{S1}$-caseins as described by Horne.[7] The structure of the $\beta$-caseins allows them to form micellar aggregates whereas that of the $\alpha_{S1}$-caseins favours linear aggregate structures.

### 6.3.3 Protein mixtures

Most of the proteins used as gelling agents in the food industry will be mixtures of proteins rather than pure proteins. Examples are caseinates, whey and soya isolates. There is a vast literature on the gelation of such complex mixtures and it would not be possible to adequately summarize such work in this short chapter. In order to interpret the behaviour of these systems it is necessary to relate the behaviour of the crude protein isolates to that of mixtures of purified proteins. As we have seen for mixed polysaccharide systems simple mixing of two proteins can lead to several classes of mixed gel: swollen, coupled, interpenetrating and phase-separated networks (Fig. 6.12). In the case of proteins there is also the possibility of forming a simple mixed network in which both self- and non-self-interactions contribute to the network. It seems likely that binary mixtures of globular proteins will form either mixed or phase-separated networks. Given the relatively nonspecific nature of the association of globular proteins the most likely type of structure to be formed is a simple mixed system in which all proteins freely interact with each other to form a network. The nature of such networks should easily be revealed by labelling and microscopy. Different proteins can be distinguished by antibody labelling and transmission electron microscopy.[2] Where studies have been made for simple mixtures of proteins, early results[127,128] were taken to favour synergistic interactions, but more recent results[129] seem to favour a homogeneous mixed network structure. Provided both proteins within the mixture are heated to sufficiently high temperatures to generate 'sticky' patches on their surfaces then association and gelation prevents significant phase separation. For such mixtures the major features dictating gelation and the properties of the gels will mirror the behaviour of single protein systems. However, if different proteins respond to the heating regime differently, becoming sticky over different time frames, then preferential self-association could occur, leading to coupled or interpenetrating networks, and this possibility makes this an area worth more detailed study.

### 6.3.4 Interfacial protein networks

Proteins accumulate at air–water or oil–water interfaces in order to reduce the surface or interfacial tension.[130] Thus the surface of a protein solution will differ in structure from the bulk, and this difference can be enhanced by drying of the surface layer. Such surface structures involve the concentration of the protein, but are the structures formed protein glasses or gels? To answer this question we need to consider what happens as proteins adsorb at surfaces.

The adsorption of proteins at an air–water interface is accompanied by a time-dependent decrease in surface tension or the equivalent increase in surface pressure. The surface pressure rises to a pseudo-equilibrium value and then slowly increases with time. The long-term slow increase in surface pressure is attributed to a maturation or ageing of the protein structure. Once a protein structure is formed the proteins lose their ability to diffuse at the interface or to exchange with proteins in the bulk. As the protein structure develops the surface rheology changes and the protein structures are generally viscoelastic. There is considerable experimental evidence to suggest that protein adsorption leads to partial unfolding of the protein structure. As in the bulk this change in structure is considered to lead to 'sticky' proteins that associate to form networks. The surface structures formed may be further consolidated by additional types of linkage such as disulfide bonding. For interfaces it is easier to study the formed structures than the early stages of association and network precursors. Interfacial protein structures can be formed in a Langmuir trough and transferred to suitable substrates for imaging by Langmuir–Blodgett techniques.[131,132] For protein structures formed by the adsorption of protein to the interface from the bulk, rather than films formed by spreading protein directly at the air–water interface, the methods have been modified to allow visualization of the interfacial film and to eliminate passive adsorption of bulk protein onto the substrate.[133]

The monolayer interfacial protein films can be imaged in the atomic force microscope (AFM) and the type of structure formed depends on the nature of the protein. There are parallels between the types of network structures formed by proteins at interfaces and those formed in the bulk, although the proteins in the interfacial networks are far more highly concentrated. Equally there are clear differences seen in the structures formed by a fibrous protein such as gelatin or globular proteins.

### 6.3.4.1 Interfacial gelatin networks

The AFM images of the interfacial structures formed by gelatin reinforce the idea of two-dimensional gel networks rather than the formation of protein glasses. The results obtained are consistent with the bulk model of gelation, triggered by triple-helix formation and the formation of fine fibrous networks, followed by further lateral association into thicker fibres or bundles of helices[124] (Fig. 6.19). Occasionally these bundles displayed a 'collagen-like' periodicity consistent with the idea of the attempted self-assembly of collagen structures.[124] The studies emphasized the importance of the thermal history of the gelation process: gelatin samples that had been quenched did not form large fibrous bundles, in contrast to slowly cooled solutions, and the two extremes resulted in protein structures with very different interfacial rheology.

### 6.3.4.2 Globular protein networks

The AFM images of the interfacial structures[131] formed by globular proteins are similar for most globular proteins studied and resemble highly packed arrays of globular particles (Fig. 6.21). At first sight, based on the images alone, it would seem to be difficult to distinguish between glassy or gel structures. However, the AFM images are three-dimensional images and it is useful to study the height variation across the structure. Such closer inspection of the images suggests the presence of 'holes' in the structure, arrowed in Fig. 6.21, and a wide variation in the height of the individual proteins. On compression of the films the holes disappear and the height of the film becomes uniform. This would seem to suggest that different proteins are unfolded to different extents and flatten to different degrees on the interface. Different degrees of unfolding would expose different structural features and are likely to lead to different extents of interaction between different neighbouring proteins. Certainly if the extent of unfolding and interaction depended on the space available when the proteins adsorbed to the interface this might account for the heterogeneity of the structure and the presence of holes. The last proteins to adsorb would have little space to unfold and interact with their neighbours. Any such proteins not bound to the network structure would

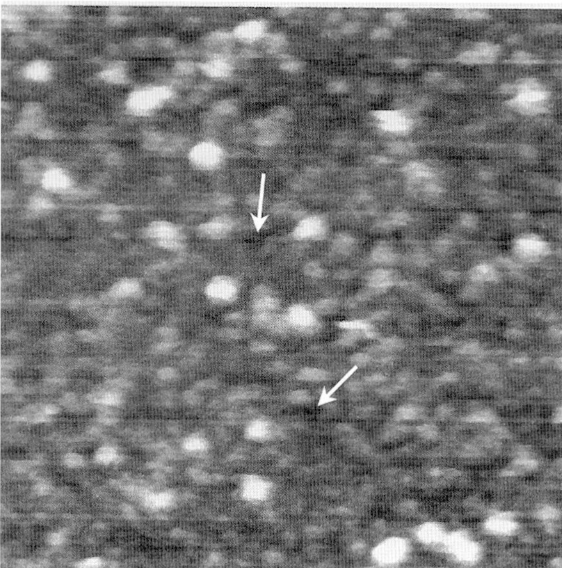

**Fig. 6.21**   Atomic force microscopy image of β-lactoglobulin interfacial film. The film has been transferred from an air–water interface by the Langmuir–Blodgett technique, and has been imaged under butanol. Note the uneven height variation across the image. The dark regions (arrowed) represent holes where weakly attached, largely 'native-like' proteins have dropped out of the network. The variation in height for the remaining proteins suggests different levels of unfolding, and presumably different levels of attachment within the network.

tend to be displaced on sampling and on the washing of the structures prior to imaging. Thus the images, and certainly other interfacial characteristics, favour a gel network. The heterogeneity of the protein network implies that the structural changes on adsorption will be different for different proteins within the network. Current methods used to study changes in protein structure during adsorption tend to spatially average the network structure. There is clearly a need to develop methods for probing structural changes in individual proteins within such networks.

### 6.3.5 Interfacial protein networks in foods

At present the discussions have considered protein networks formed at the air–water interfaces present in protein solutions. Such structures may be of importance in the characterization of the sols, particularly if the protein films dry out at the surface. The interfacial structures formed may influence the formation of bulk structures and may even nucleate aggregation in the early stages of bulk gelation. The commonly used trick of coating sols or gels with oils to eliminate drying effects may actually be counterproductive, because unfolding and association at oil–water interfaces may be even more extensive and lead to stiffer network structures.

Of more interest in foods are the interfacial protein structures formed at air–water or oil–water interfaces during the formation of food foams and emulsions.[130] In most food systems the proteins of interest will be globular proteins although, in most cases, there will be mixtures of several proteins present at the interfaces.[134] Even the commercial proteins used as emulsifiers or foam stabilizers will not be purified proteins but mixtures or protein isolates. In addition there will be other components such as lipids or added surfactants that will compete for occupancy of the interface.[134] In fact the competition between these small molecules and the proteins is a major source of instability in the preparation and stabilization of food foams and emulsions. Both surfactants and proteins alone can stabilize the interfaces. Many of us will be familiar with producing surfactant-stabilized interfaces through blowing soap bubbles. Such structures tend to be short-lived and most of us can recall the disappointment as the thin liquid film drained and the bubble burst. By contrast protein-stabilized structures are more long-lasting: familiar examples are meringues and the head on a glass of beer. When both surfactants and proteins are present at the interface the surfactant, if available in sufficient quantity, will win control of the interface and destabilize the protein structure. That is why we remove the egg yolk (the lipoproteins) when we make meringues, and why we can clean the residues of the beer foam from dirty beer glasses with soap solutions. So how and why does this competition occur and can we control it?

Before we consider the behaviour of complex mixtures of several proteins and mixtures of surfactants it is worth considering a simpler protein-surfactant mixture. Surfactants are more surface active than proteins and, in a straight conflict for occupation of the interface, surfactants will displace individual proteins. Because proteins combine to form gel-like networks at the interfaces they cannot be individually displaced and the surfactants must disrupt the network before they can expel proteins into the bulk. Fortunately for the surfactants these protein networks have weaknesses. Some proteins are only weakly integrated into, or not attached at all, to the network (Fig. 6.21). These proteins can be displaced creating holes into which the surfactant can adsorb. Once the surfactant gains a foothold on the interface it expands and the nucleated surfactant domains grow in area. As the area on the interface occupied by surfactant increases the area occupied by protein has to decrease. Because the proteins are linked together no proteins become detached and escape into the bulk. Rather, individual proteins, and then the network itself, fold forming a thicker protein layer extending further into the bulk medium. Eventually the surfactant domains are bounded by narrow protein filaments and the weakest protein–protein linkages fail, breaking the network. At this point the interface becomes an ocean of surfactant containing islands of protein that can be broken up and expelled into the bulk. This process of displacement[133–135] (Fig. 6.22),

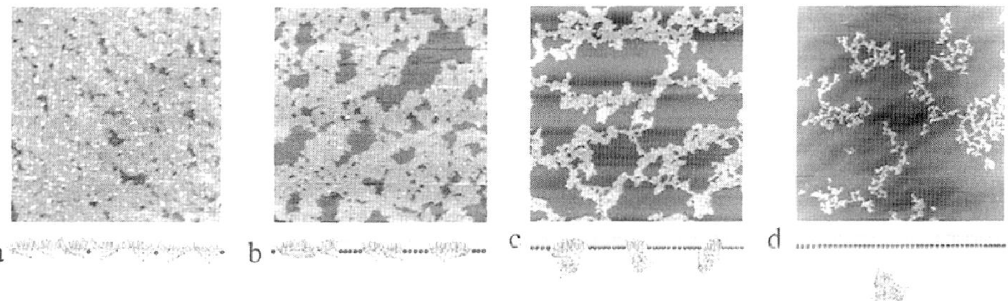

**Fig. 6.22**   Atomic force microscopic images illustrating the 'orogenic' displacement of the protein β-lactoglobulin from an air–water interface by the nonionic surfactant Tween 20. The images show the growth of surfactant domains (black) within a protein network (grey). Images sizes are (a) $1 \times 1$ μm, (b) $3.2 \times 3.2$ μm, (c) $6 \times 6$ μm and (d) $10 \times 10$ μm. The cartoons beneath micrographs illustrate the characteristic features of the displacement process.

which involves folding and failure of the protein networks, has been called an 'orogenic displacement mechanism'.

The orogenic displacement model has been found to apply for all globular proteins studied to date, for air–water, oil–water and solid–water interfaces and for water-soluble, oil-soluble, charged and uncharged surfactants.[133–136] The model is generic because if the proteins form interfacial networks then these networks need to be broken to release proteins. Differences can be seen for different proteins that reflect their usefulness as foam or emulsion stabilizers. Thus for β-casein, which is comparatively poor at stabilizing air–water interfaces, the displacement with nonionic surfactants, such as Tween 20, involves the growth of circular domains, indicating uniform compression and folding of the weak protein structure with subsequent failure at comparatively low surface pressures.[133] By contrast β-casein is more effective at stabilizing oil–water interfaces, and in this case domain growth is harder, with the growing domains having a ragged boundary, indicative of a heterogeneous structure, with stress redistribution and folding at the weaker regions of the protein network.[135] The growth of large irregular domains is characteristic of the displacement of most proteins at air–water or oil–water interfaces by nonionic surfactants. Although these surfactants can and do bind to the proteins this interaction does not seem to alter the surface pressure at which the protein network fails.[136] For ionic surfactants there are slight differences in behaviour. Strong repulsive interactions between charged surfactants favour dispersion of the surfactants uniformly across the interface and, in this case, invasion of the interface with surfactant favours nucleation of domains with little domain growth.[136] Screening of the charged surfactants facilitates domain growth resulting in displacement patterns similar to those of uncharged surfactants. Protein–surfactant binding can, at least for certain proteins and surfactants, alter the stability of the protein network enhancing its ability to resist displacement.[136]

In most foods there will be more than one type of protein present at the interface. What sort of mixed protein networks are formed? It is possible to label proteins and to image the structures formed at high resolution using different types of microscopy.[137,138] The resolution is sufficient to establish that no significant phase separation appears to occur for mixtures of relevant food proteins. It appears that the unfolding and association on adsorption precludes diffusion and separation of the proteins, and the resultant structures are homogeneous networks. Studies on the competitive displacement of simple mixtures of proteins suggest that the behaviour can be understood in terms of the displacement of the component proteins. The final failure of the mixed network appears to be dominated by the component protein that alone best resists displacement.

The fact that the models can be extended to describe complex mixed systems means that they can be applied to describe the behaviour of commercial materials such as protein isolates.[139] Indeed, in such cases the deviations in behaviour from that of simple mixtures can be used to identify factors induced during isolation, processing or storage of isolates that decrease or enhance their behaviour. Particular components in mixtures can be more effective than other components, and an understanding of this allows the composition of the materials to be monitored and controlled to enhance or stabilize its functionality.

The generic nature of the competitive displacement process means that it has wide applications in food products produced by the brewing, baking and dairy industries. The molecular understanding allows clear strategies for enhancing stability. These broadly involve improving the cross-linking of the protein network to resist displacement or restricting the ability of surfactants, or surfactant-like molecules, to reach and invade the interface.[140–143]

There are many opportunities for manipulating the types of interfacial structures present in foods through processing or controlling the composition of the product. However, in the future there may be additional opportunities for rationally designing new types of interfacial structures. Interfacial networks are examples of two-dimensional nanostructures. This is an example of an area where nanotechnology could be used to assemble novel structures with novel functionality. This would involve the creation of multilayered structures in which individual molecular layers are added layer by layer. Such 'fuzzy nanoassemblies' are well studied in the synthetic polymer area[144] and are beginning to be investigated for food biopolymers and for controlling the properties of emulsions.[145] There are opportunities for controlling the interfacial structure to moderate and control droplet interactions in order to prevent creaming, or coalescence, or to promote controlled association of droplets. Interfacial structures could be manipulated to facilitate binding of droplets to bulk polymeric structures, or to regulate encapsulation or release of molecules contained within the droplets.

## 6.4 Polysaccharide-protein gels

The literature on food-related structures containing polysaccharide-protein mixed gels is vast. Even supposedly simple foods such as ice cream, cheese, yoghurt, custard, cakes or bread are highly complex multicomponent structures where different components gel or influence gelation at different stages of processing in order to produce the final structure and texture of the food product. For many foods there will be both interfacial networks and bulk networks, and possible interactions between these structures. In the present chapter the intention is not to try to unravel the behaviour of particular food products, but rather to lay foundations upon which such descriptions can be developed. The assumption is that labelling and various kinds of microscopy can be used to identify what and where different components are present in these complex structures. The belief is that these complex structures can be assigned to particular classes of structures in which the structure and properties can be described in terms of the behaviour of the pure components.

For simple binary mixtures the four main classes of structure have already been described: swollen, coupled, interpenetrating and phase-separated networks (Fig. 6.12).

Swollen networks are most likely to be encountered in mixtures of neutral polysaccharides with proteins. The mixture of a neutral and charged biopolymer should inhibit phase separation[95] and result in the soluble polysaccharide being uniformly distributed within the protein gel network. In the case of mixtures of proteins with galactomannans or glucomannans it might be possible to generate interpenetrating networks by selectively and separately gelling the two components. Perhaps the main examples of mixtures of neutral polysaccharides and proteins are starch-protein mixtures. Whether the proteins are gluten proteins, egg, milk or plant storage proteins the final structure present in foods is a phase-separated structure. The problem here is in intimately mixing the two components. The starch polysaccharides are locked up within the granule structure and only released upon heating and gelatinization. In the case of baked products the gluten networks are formed before gelatinization of the starch and will hinder mixing of the components. Similarly with heat-set globular proteins the heating stage will promote protein aggregation before or during gelatinization of the starch. The swelling of the granules will remove water from the protein phase, concentrating the protein and promoting protein gelation. On cooling, gelation of the proteins and separate gelation of the released amylose will hinder mixing of the two biopolymers preserving the phase-separated structure. This is an unusual example of where gelation competes with mixing of the polymers preserving an unexpected phase-separated structure.

Given the discussions on coupled networks for polysaccharide-polysaccharide mixtures, and the need for compatible structures, it might be expected that coupled protein-polysaccharide gels would be very rare. However, proteins and polysaccharides will interact and form complexes or condensates, and this can lead to gelation.[146] Controlled deposition and binding of monolayers of proteins and polysaccharides provides a route towards the construction or design of interfacial structures. Despite this there are few reports of protein-polysaccharide coupled gels.[8,10,146] Perhaps the best-known example is the synergism between κ-casein and κ-carrageenan.[147,148] This is believed to be an electrostatic interaction that occurs above the isoelectric point of the protein. Micellar aggregates of κ-casein can be disrupted by κ-carrageenan to produce linear aggregates of κ-casein and κ-carrageenan.[149,150] In the case of milk gels, κ-carrageenan is believed to cross-link micelles by binding to the κ-casein on the surface of the micelle. Above the isoelectric point of the protein both the protein and the polysaccharide are negatively charged. Although the protein has a net negative charge it is believed that the carrageenan can interact with patches of positive charge on the surface of the protein. There are reports of interactions between polysaccharides and proteins that lead to gelation,[8,10,146,151] but this is a largely unexplored area, particularly with regard to its importance for food gels.

The vast majority of polysaccharide-protein mixtures will form phase-separated gels. This is a large and expanding area of research, which is covered in a number of review articles[1,8,84,95,152–156] together with a large number of articles on particular mixed systems. These materials can be regarded as composite structures and can be formed between polysaccharides and either gelatin or various globular proteins. Strictly speaking these are partially phase-separated structures in which phase separation has been arrested by gelation of one or both of the components. This is in contrast to the starch-protein mixtures where gelation inhibits mixing rather than demixing of the components.

Thus the gels can be considered as composite structures[1,8,153,156] consisting of filler particles with a modulus $G_f$ contained within a matrix of modulus $G_m$. For such structures where the moduli of the filler and the matrix are fairly similar the modulus of the composite structure will lie between upper and lower bounds defined for analysis of polymer blends (Fig. 6.23). Ideally at low and high fractions of individual components the composite should approximate to the behaviour of the lower (Fig. 6.23b) or upper (Fig. 6.23a) boundaries. At intermediate fractions there will be a fairly sharp transition (Fig. 6.23c) between the two modes of behaviour. Attempts to model the behaviour of the composites in terms of the behaviour of the individual pure phases has shown that during gelation water redistribution occurs and the effective polymer concentrations within the two phases vary. For protein-polysaccharide mixtures the two phases are relatively easy to visualize and such studies have revealed that each phase is often not pure and contains multiple inclusions of the other phase. However, the network structures formed within each phase are characteristic of the structures formed by the pure phases. Recent direct AFM studies on hydrated mixed protein-polysaccharide

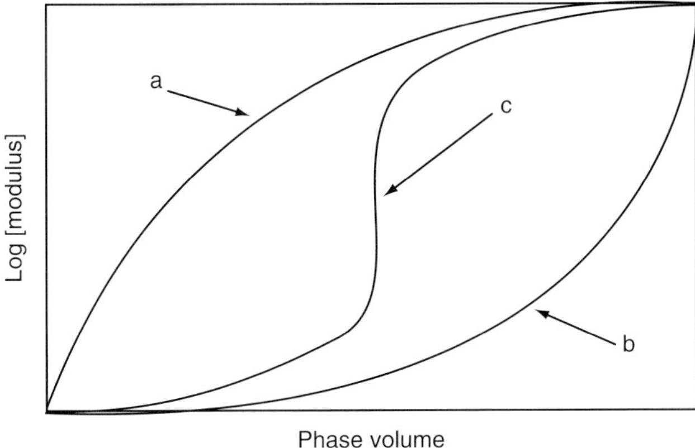



**Fig. 6.23**  Schematic diagram showing the upper, isostrain (a) and lower, isostress (b) boundaries for the modulus of a phase-separated binary gel system. The line (c) indicates the expected transition in behaviour corresponding to the inversion of the nature of the major supporting phase.

mixed gels has allowed the network structure of the two phases to be visualized and compared with those of the protein or polysaccharide gels.[157]

## 6.5 Conclusions

In this chapter an attempt has been made to review the basic ideas underlying the gelation of food biopolymers. Where possible reviews have been referenced that provide access to the background literature, and the main physical principles have been identified and summarized. It is hoped that the information provided will allow a better understanding of the complex structures found in food systems. The view presented is a personal one and, where possible, I have tried to indicate areas where I believe there is scope for future research and understanding. Emerging areas are the study of the gelation of mixed proteins, and the investigation of the gelation of high solids biopolymer systems and biopolymer mixtures. The few studies[158] of the latter area indicate that there is much to study and understand for these systems.

## 6.6 References

1  Morris, V.J. (1998) Gelation of polysaccharides. In: Hill, S.E., Ledward, D.A. & Mitchell, J.R. (eds) *Functional Properties of Food Macromolecules*, 2nd edn, pp. 143–226. Aspen, Gaithersburg, MD.

2  Clark, A.H. (1998) Gelation of globular proteins. In: Hill, S.E., Ledward, D.A. & Mitchell, J.R. (eds) *Functional Properties of Food Macromolecules*, 2nd edn, pp. 77–142. Aspen, Gaithersburg, MD.

3  Ledward, D.A. (1986) Gelation of gelatin. In: Mitchell, J.R. & Ledward, D.A. (eds) *Functional Properties of Food Macromolecules*, 1st edn, pp. 171–201. Elsevier Applied Science, London & New York.

4  Stainsby, G. (1985) Gelatin gels. In: *Collagen as a Food: Advances in Meat Research*, Vol. 4, pp. 209–222. van Nostrand, Reinhold, London.
5  Ross-Murphy, S.B. (1997) Structure and rheology of gelatin gels. *Imaging Sci. J.* **45**, 205–209.
6  Clark, A.H., Kavanagh, G.M. & Ross-Murphy, S.B. (2001) Globular protein gelation – theory and experiment. *Food Hydrocolloids* **15**, 383–400.
7  Horne, D.S. (2002) Casein structure, self-assembly and gelation. *Curr. Opin. Colloid Interface Sci.* **7**, 456–461.
8  Morris, E.R. (1990) Mixed polymer gels. In: Harris, P. (ed.) *Food Gels,* pp. 291–359. Elsevier Applied Science, London & New York.
9  Williams, P.A. & Phillips, G.O. (1995) Interactions in mixed polysaccharide systems. In: Stephen, A.M. (ed.) *Food Polysaccharides and Their Applications*, pp. 463–500. Marcel Dekker, New York.
10  Morris, E.R. (1995) Polysaccharide synergism – more questions then answers. In: Harding, S.E., Hill, S.E. & Mitchell, J.R. (eds) *Biopolymer Mixtures*, pp. 247–288. Nottingham University Press, Nottingham, UK.
11  Morris, V.J. (1995) Synergistic interactions with galactomannans and glucomannans. In: Harding, S.E., Hill, S.E. & Mitchell, J.R. (eds) *Biopolymer Mixtures*, pp. 289–314. Nottingham University Press, Nottingham, UK.
12  Rees, D.A. (1969) Structure, composition and mechanism in the formation of polysaccharide gels and networks. *Adv. Carbohydr. Chem. Biochem.* **24**, 267–313.
13  Zobel, H.F. & Stephen, A.M. (1995) Starch: structure, analysis, and application. In: Stephen, A.M. (ed.) *Food Polysaccharides and Their Applications,* pp. 19–66. Marcel Dekker, New York.
14  Gunning, A.P., Giardina, T.P., Faulds, C.B. *et al.* (2003) Surfactant mediated solubilisation of amylose and visualisation by atomic force microscopy. *Carbohydr. Polym.* **51**, 177–182.
15  Coffey, D.G., Bell, D.A. & Henderson, A. (1995) Cellulose and cellulose derivatives. In: Stephen, A.M. (ed.) *Food Polysaccharides and Their Applications,* pp. 123–153. Marcel Dekker, New York.
16  Morris, V.J. (1995) Bacterial polysaccharides. In: Stephen, A.M. (ed.) *Food Polysaccharides and Their Applications,* pp. 341–375. Marcel Dekker, New York.
17  Stanley, N.F. (1995) Agars. In: Stephen, A.M. (ed.) *Food Polysaccharides and Their Applications,* pp. 187–204. Marcel Dekker, New York.
18  Piculell, L. (1995) Gelling carrageenans. In: Stephen, A.M. (ed.) *Food Polysaccharides and Their Applications,* pp. 205–244. Marcel Dekker, New York.
19  Bellion, C., Hamer, G.K. & Yaphe, W. (1982) The degradation of *Eucheuma-spinosum* and *Euceuma-cottoninii* carrageenans by iota-carrageenases and kappa-carrageenases from marine bacteria. *Can. J. Microbiol.* **28**, 874–880.
20  Rochas, C. & Heyraud. A. (1981) Acid and enzymic hydrolysis of kappa-carrageenan. *Polym. Bull.* **5**, 81–86.
21  Nishinari, K., Williams, P.A. & Phillips, G.O. (1992) Review of the physicochemical characteristics and properties of konjac mannan. *Food Hydrocolloids* **6**, 199–222.
22  Dea, I.C.M. & Morrison, A. (1975) Chemistry and interactions of seed galactomannans. *Adv. Carbohydr. Chem. Biochem.* **31**, 241–312.
23  Grant Reid, J.S. & Edwards, M.E. (1995) Galactomannans and other cell wall storage polysaccharides in seeds. In: Stephen, A.M. (ed.) *Food Polysaccharides and Their Applications,* pp. 155–186. Marcel Dekker, New York.
24  Moe, S.T., Draget, K.I., Skjåk-Bræk & Smidsrød, O. (1995) Alginates. In: Stephen, A.M. (ed.) *Food Polysaccharides and Their Applications,* pp. 245–286. Marcel Dekker, New York.
25  Voragen, A.G.J., Pilnik, W., Thibault, J-F. *et al.* (1995) Pectins. In: Stephen, A.M. (ed.) *Food Polysaccharides and Their Applications,* pp. 287–339. Marcel Dekker, New York.
26  O'Neill, M.A. & York, W.S. (2003) The composition and structure of plant primary cell walls. In: Rose, J.K.C. (ed.) *The Plant Cell Wall. Annual Plant Reviews,* Vol. 8, pp.1–54. Blackwell Publishing, Oxford, UK.

27  Micard, V. & Thibault, J-F. (1999) Oxidative gelation of sugar-beet pectins: use of laccases and hydration properties of the cross-linked pectins. *Carbohydr. Polym.* **39**, 265–273.

28  Thibault, J-F. (1988) Characterization and oxidative crosslinking of sugar-beet pectins extracted from cossettes and pulps under different conditions. *Carbohydr. Polym.* **8**, 209–223.

29  Ishii, T. (1997) Structure and function of feruloylated polysaccharides. *Plant Sci.* **127**, 111–127.

30  Williams, M.A.K., Foster, T.J., Martin, D.R. *et al.* (2000) A molecular description of the gelation mechanism of konjac mannan. *Biomacromolecules* **1**, 440–450.

31  Millane, R.P. & Hendrixson, T.L. (1994) Crystal structures of mannan and glucomannans. *Carbohydr. Polym.* **25**, 245–251.

32  Robinson, G., Ross-Murphy, S.B. & Morris, E.R. (1982) Viscosity molecular weight relationships, intrinsic chain flexibility, and dynamic solution properties of guar galactomannan. *Carbohydr. Res.* **107**, 17–32.

33  Kohn, R. (1975) Ion binding on polyuronates – alginate and pectin. *Pure Appl. Chem.* **42**, 371–397.

34  Atkins, E.D.T.A., Mackie, W., Parker, K.D. & Smolko, E.E. (1971) Crystalline structures of poly-D-mannuronic acid and poly-L-guluronic acid. *Polym. Lett.* **9**, 311–316.

35  Atkins, E.D.T.A., Nieduszynski, I.A., Mackie, W. *et al.* (1973) Structural components of alginic acid II. The crystalline structure of poly-$\alpha$-L-guluronic acid. Results of x-ray diffraction and polarised infrared studies. *Biopolymers* **12**, 1879–1887.

36  Atkins, E.D.T.A., Hopper, E.D.A. & Isaac, D.H. (1973) Polysaccharide conformation: Effect of side-group geometry on four diequatorially (1→4) linked polysaccharides. *Carbohydr. Res.* **27**, 29–37.

37  Grant, G.T., Morris, E.R., Rees, D.A. *et al.* (1973) Biological interactions between polysaccharides and divalent cations: The egg-box model. *FEBS Lett.* **32**, 195–198..

38  Gidley, M.J., Morris, E.R., Murray, E.J. *et al.* (1979) Spectroscopic and stoichiometric characterisation of the Ca-mediated association of pectate chains in gels and in the solid state. *J. Chem. Soc. Chem. Comm.* 990–992.

39  Davies, M.A.F., Gidley, M.J., Morris, E.R. *et al.* (1980) Intermolecular association in pectin solutions. *Int. J. Biol. Macromol.* **2**, 330–332.

40  Gidley, M.J., Morris, E.R., Murray, E.J. *et al.* (1980) Evidence for two mechanisms of interchain association in calcium pectate gels. *Int. J. Biol. Macromol* **2**, 333–334.

41  Racape, E., Thibault, J-F., Reitsma, J.C.E. & Pilnik, W. (1989) Properties of amidated pectins 2. Poly-electrolyte behavior and calcium-binding of amidated pectins and amidated pectic acids. *Biopolymers* **28**, 1435–1448.

42  MacDougall, A.J., Needs, P.W., Rigby, N.M. & Ring, S.G. (1996) Calcium gelation of pectic polysaccharides isolated from unripe tomato fruit. *Carbohydr. Res.* **293**, 235–249.

43  Grasdalen, H. (1983) High-field, H1 NMR spectroscopy of alginate: sequential structure and linkage conformations. *Carbohydr. Res.* **118**, 255–260.

44  Needs, P.W., Rigby, N.M., Ring, S.G. & MacDougall, A.J. (2001) Specific degradation of pectins via a carbidiimide-mediated Lossen rearrangement of methyl-esterified galacturonic acid residues. *Carbohydr. Res.* **333**, 47–58.

45  Löfgren, C., Walkenström, P. & Hermansson, A.M. (2002) Microstructure and rheological behaviour of pure and mixed pectin gels. *Biomacromolecules* **3**, 1144–1153.

46  Fishman, M.L., Cooke, P.H. & Coffin, D.R. (2004) Nanostructure of native pectin sugar acid gels visualized by atomic force microscopy. *Biomacromolecules* **5**, 334–341.

47  Chandrasekaran, R. (1999) X-ray and modelling studies on the structure-function correlations of polysaccharides. *Macromol. Symp.* **120**, 17–29.

48  Chandrasekaran, R. (1997) Molecular architecture of polysaccharide helices in oriented fibers. *Adv. Carbohydr. Chem. Biochem.* **52**, 311–439.

49  Chandrasekaran, R. & Radha, A. (1995) Molecular architectures and functional properties of gellan gum and related polysaccharides. *Trends Food Sci. Technol.* **6**, 143–148.

50  Chandrasekaran, R., Radha, A. & Thailambal, V.G. (1992) Roles of potassium ions, acetyl and L-glycerate groups in native gellan double helix – an x-ray study. *Carbohydr. Res.* **224**, 1–17.

51  Millane, R.P., Arnott, S. & Atkins, E.D.T. (1988) The crystal structure of gellan. *Carbohydr. Res.* **175**, 1–15.

52  Chandrasekaran, R. & Thailambal, V.G. (1990) The influence of calcium ions, acetate and L-glycerate groups on the gellan double helix. *Carbohydr. Polym.* **12**, 431–442.

53  Chandrasekaran, R., Puigjaner, L.C., Joyce, K.L. & Arnott, S. (1988) Cation interactions in gellan – an x-ray study of the potassium salt. *Carbohydr. Res.* **181**, 23–40.

54  Anderson, N.S., Campbell, J.W., Harding, M.M. *et al.* (1969) X-ray diffraction studies of polysaccharide sulphates: Double helix models for κ- and ι-carrageenans. *J. Mol. Biol.* **45**, 85–99.

55  Arnott, S., Scott, W.F., Rees, D.A. & McNab, G.G.A. (1974) ι-Carrageenan, molecular structure and packing of polysaccharide double helices in oriented fibres of divalent calcium salts. *J. Mol. Biol.* 253–267.

56  Janaswamy, S. & Chandrasekaran, R. (2002) Effect of calcium ions on the organisation of iota-carrageenan helices: An x-ray investigation. *Carbohydr. Res.* **337**, 523–535.

57  Janaswamy, S. & Chandrasekaran, R. (2001) Three dimensional structure of the sodium salt of iota-carrageenan. *Carbohydr. Res.* **335**, 181–194.

58  Millane, R.P., Chandrasekaran, R., Arnott, S. & Dea, I.C.M. (1988) The molecular structure of kappa-carrageenan and comparison with iota-carrageenan. *Carbohydr. Res.* **182**, 1–17.

59  Cairns, P., Atkins, E.D.T., Miles, M.J. & Morris, V.J. (1991) Molecular transforms of kappa carrageenan and furcellaran from mixed gel structures. *Int. J. Biol. Macromol.* **13**, 65–68.

60  Arnott, S., Fulmer, A., Scott, W.E. *et al.* (1974) The agarose double helix and its function in agarose gel formation. *J. Mol. Biol.* **90**, 269–284.

61  Millane, R.P. (1992) Molecular and crystal structures of polysaccharides with cellulosic backbones. In: Chandrasekaran, R. (ed.) *Frontiers of Carbohydrate Research* 2, pp. 168–190. Elsevier Applied Science, New York.

62  Brigham, J.E., Gidley, M.J., Hoffman, R.A. & Smith, C.G. (1994) Microscopic imaging of network strands in agar, carrageenan, locust bean gum and kappa-carrageenan locust bean gum gels. *Food Hydrocolloids* **8**, 331–344.

63  Gunning, A.P. & Morris, V.J. (1990) Light scattering studies of tetra-methyl ammonium gellan. *Int. J. Biol. Macromol.* **12**, 338–341.

64  Gunning, A.P., Kirby, A.R., Ridout, M.J. *et al.* (1996) Investigation of gellan networks and gels by atomic force microscopy. *Macromolecules* **29**, 6791–6796.

65  Morris, E.R., Rees, D.A. & Robinson, G. (1980) Cation-specific aggregation of carrageenan helices: Domain model of polymer gel structure. *J. Mol. Biol.* **138**, 349–362.

66  Robinson, G., Morris, E.R. & Rees, D.A. (1980) Role of double helices in carrageenan gelation: the domain model. *J. Chem. Soc. Chem. Comm.* 152–153.

67  Robinson, G., Manning, C.E. & Morris, E.R. (1991) Conformation and physical properties of the bacterial polysaccharides gellan, welan, and rhamsan. In: Dickinson, E. (ed.) *Polymers, Gels and Colloids*, pp. 2–33. Royal Society of Chemistry Special Publication number 82. RSC, Cambridge, UK.

68  Atkin, N., Abeysekara, R.M., Kronestedt-Robards, E.C. & Robards, A.W. (2000) Direct visualization of changes in deacetylated Na⁺ gellan polymer morphology during the sol-gel transition. *Biopolymers* **54**, 195–210.

69  Ikeda, S., Nitta, Y., Temsiripong, T. *et al.* (2004) Atomic force microscopy studies on cation-induced network formation of gellan. *Food Hydrocolloids* **18**, 727–735.

70  Lundin, L. & Hermansson, A.M. (1997) Rheology and microstructure of Ca- and Na-κ-carrageenan and locust bean gum gels. *Carbohydr. Polym.* **34**, 365–375.

71  Ikeda, S., Morris, V.J. & Nishinari, K. (2001) Microstructure of aggregates and non-aggregated κ-carrageenan helices visualized by atomic force microscopy. *Biomacromolecules* **2**, 1331–1337.

72  Morris, E.R., Gidley, M.J., Murray, E.J. *et al.* (1980) Characterisation of pectin gelation under conditions of low water activity, by circular dichroism, competitive inhibition and mechanical properties. *Int. J. Biol. Macromol.* **2**, 327–330.

73  Walkinshaw, M.D. & Arnott, S. (1981) Conformation and interactions of pectins 1. X-ray diffraction analysis of sodium pectate in neutral and acidified forms. *J. Mol. Biol.* **153**, 1055–1073.

74 Walkinshaw, M.D. & Arnott, S. (1981) Conformation and interactions of pectins 2. Models for junction zones in pectinic acid and calcium pectate gels. *J. Mol. Biol.* **153**, 1075–1085.

75 Ross-Murphy, S.B., Morris, V.J. & Morris, E.R. (1983) Molecular viscoelasticity of xanthan polysaccharide. *Faraday Symp. Chem. S.* **18**, 115–129.

76 Jeanes, A., Rogovin, P., Cadmus, M.C. *et al.* (1974) Polysaccharide (xanthan) of *Xanthomonas campestris* NRRL-B1459. Procedures for culture maintenance and polysaccharide production, purification, and analysis. USDA Report ARS-NC-51, 1–14.

77 Rinaudo, M., Milas, M. & Kohler, N. (1983) Enzymatic clarification process for improving the injectivity and filterability of xanthan gums. US Patent 4,416,990.

78 Sanderson, G. & Clark, R.C. (1983) Gellan gum. *Food Technol.* **37**, 63–70.

79 Morris, V.J., Tsiami, A. & Brownsey, G.J. (1995) Work hardening effects in gellan gels. *J. Carbohydr. Chem.* **14**, 667–675.

80 Norton, I.T., Jarvis, D.A. & Foster, T.J. (1999) A molecular model for the formation and properties of fluid gels. *Int. J. Biol. Macromol.* **26**, 255–261.

81 Cairns, P., Miles, M.J., Morris, V.J. & Brownsey, G.J. (1987) X-ray fibre diffraction studies of synergistic binary polysaccharide gels. *Carbohydr. Res.* **160**, 411–423.

82 Kalichevsky, M.T. & Ring, S.G. (1987) Incompatibility of amylase and amylopectin in aqueous solution. *Carbohydr. Res.* **162**, 323–328.

83 MacDougall, A.J., Rigby, N.M. & Ring, S.G. (1997) Phase separation of plant cell wall polysaccharides and its implications for cell wall assembly. *Plant Physiol.* **114**, 353–362.

84 Brown, C.R.T., Foster, T.J, Norton, I.T & Underdown, J. (1995) Influence of shear on the microstructure of mixed biopolymer systems. In: Harding, S.E., Hill, S.E. & Mitchell, J.R. (eds) *Biopolymer Mixtures*, pp. 65–83. Nottingham University Press, Nottingham, UK.

85 Norton, I.T. & Frith, W.J. (2001) Microstructure design in mixed biopolymer composites. *Food Hydrocolloids* **15**, 543–553.

86 Hermansson, A.M., Kidman, S. & Svegmark, K. (1995) Starch – A phase-separated biopolymer system. In: Harding, S.E., Hill, S.E. & Mitchell, J.R. (eds) *Biopolymer Mixtures*, pp. 225–245. Nottingham University Press, Nottingham, UK.

87 Parker, R. & Ring, S.G. (2001) Aspects of the physical chemistry of starch. *J. Cereal Chem.* **34**, 1–17.

88 Gallant, D.J., Bouchet, B. & Baldwin, PM. (1997) Microscopy of starch: evidence of a new level of granule organization. *Carbohydr. Polym.* **32**, 177–191.

89 Ridout, M.J., Parker, M.L., Hedley, C.L. *et al.* (2003) Atomic force microscopy of pea starch granules: Granule architecture of wild-type parent, *r*, and *rb* single mutants, and the *rrb* double mutant. *Carbohydr. Res.* **338**, 2135–2147.

90 Miles, M.J., Morris, V.J., Orford, P.D. & Ring, S.G. (1985) The roles of amylose and amylopectin in the gelation and retrogradation of starch. *Carbohydr. Res.* **135**, 271–281.

91 Miles, M.J., Morris, V.J. & Ring, S.G. (1985) Gelation of amylose. *Carbohydr. Res.* **135**, 257–269.

92 Leloup, V.M., Colonna, P., Ring, S.G. *et al.* (1992) Microstructure of amylose gels. *Carbohydr. Polym.* **18**, 189–197.

93 Moates, G.K., Noel, T.R., Parker, R. & Ring, S.G. (1997) The effect of chain length and solvent interactions on the dissolution of the B-type crystalline polymorph of amylose in water. *Carbohydr. Res.* **298**, 327–333.

94 Ring, S.G., Colonna, P., I'Anson, K. *et al.* (1987) Gelation and crystallisation of amylopectin. *Carbohydr. Res.* **162**, 277–293.

95 Piculell, L., Bergfeldt, K. & Nilsson, S. (1995) Factors determining phase behaviour of multicomponent polymer systems. In: Harding, S.E., Hill, S.E. & Mitchell, J.R. (eds) *Biopolymer Mixtures*, pp. 13–35. Nottingham University Press, Nottingham, UK.

96 Amici, E., Clark, A.H., Normand, V. & Johnson, N.B. (2002) Interpenetrating network formation in agarose-kappa-Carrageenan gel composites. *Biomacromolecules* **3**, 466–474.

97 Amici, E., Clark, A.H., Normand, V. & Johnson, N.B. (2001) Interpenetrating network formation in agarose-sodium gellan gel composites. *Carbohydr. Polym.* **46**, 383–391.

98  Amici, E., Clark, A.H., Normand, V. & Johnson, N.B. (2000) Interpenetrating network formation in gellan-agarose gel composites. *Biomacromolecules* **1**, 721–729.

99  Clark, A.H., Eyre, S.C.E., Ferdinando, D.P. & Lagarrigue, S. (1999) Interpenetrating network formation in gellan-maltodextrin gel composites. *Macromolecules* **32**, 7897–7906.

100  Ojinnaka, C., Brownsey, G.J., Morris, E.R. & Morris, V.J. (1998) Effect of deacetylation on the synergistic interaction between acetan and locust bean gum or konjac mannan. *Carbohydr. Res.* **305**, 101–108.

101  Ridout, M.J., Brownsey, G.J. & Morris, V.J. (1998) Synergistic interactions of acetan with carob or konjac mannan. *Macromolecules* **31**, 2539–2544.

102  Ridout, M.J., Cairns, P., Brownsey, G.J. & Morris, V.J. (1998) Evidence for intermolecular binding between deacetylated acetan and the glucomannan konjac mannan. *Carbohydr. Res.* **309**, 375–379.

103  Chandrasekaran, R., Lanaswamy, S. & Morris, V.J. (2003) Acetan:glucomannan interactions – a molecular modeling study. *Carbohydr. Res.* **338**, 2689–2898.

104  Morris, V.J. & Chilvers, G.R. (1984) Cold-setting alginate-pectin gels. *J. Sci. Food Agr.* **35**, 1370–1376.

105  Thom, D., Dea, I.C.M., Morris, E.R. & Powell, D.A. (1982) Interchain associations of alginate and pectins. *Prog. Food Nutr. Sci.* **6**, 97–108.

106  Cairns, P., Miles, M.J. & Morris, V.J. (1986) Intermolecular binding of xanthan and carob gum. *Nature* **322**, 89–90.

107  Brownsey, G.J., Cairns, P., Miles, M.J. & Morris, V.J. (1988) Evidence for intermolecular binding between xanthan and the glucomannan (Konjac mannan). *Carbohydr. Res.* **176**, 329–334.

108  Cairns, P., Miles, M.J. & Morris, V.J. (1986) X-ray fibre diffraction studies of kappa carrageenan-tara gum mixed gels. *Int. J. Biol. Macromol.* **8**, 124–127.

109  Miles, M.J., Morris, V.J. & Carrol, V. (1984) Carob gum-kappa carrageenan mixed gels – Mechanical properties and X-ray fibre diffraction studies. *Macromolecules* **17**, 2443–2445.

110  Dea, I.C.M., McKinnon, A.A. & Rees, D.A. (1972) Tertiary and quaternary structures in aqueous polysaccharide systems which model cell wall cohesion: Reversible changes in conformation and association of agarose, carrageenan and galactomannans. *J. Mol. Biol.* **68**, 153–172.

111  Viebke, C. & Piculell, L. (1996) Adsorption of galactomannans onto agarose. *Carbohydr. Polym.* **29**, 1–5.

112  Parker, A., Lelimousin, D., Miniou, C. & Boulenguer, P. (1995) Binding of galactomannans to kappa-carrageenan after cold mixing. *Carbohydr. Res.* **272**, 91–96.

113  Lundin, L. & Hermansson, A.M. (1995) Influence of locust bean gum on the rheological behaviour and microstructure of K-kappa-carrageenan. *Carbohydr. Polym.* **28**, 91–99.

114  Harding, S.E. (1998) Dilute solution viscosity of food biopolymers. In: Hill, S.E., Ledward, D.A. & Mitchell, J.R. (eds) *Functional Properties of Food Macromolecules*, 2nd edn, pp. 1–49. Aspen, Gaithersburg, MD.

115  Mills, E.N.C., Huang, L., Gunning, A.P. & Morris, V.J. (2001) Formation of thermally-induced aggregates of the soya globulin β-conglycinin. *Biochim. Biophys. Acta* **1547**, 339–350.

116  Gosal, W.S., Clark, A.H., Pudney, P.D.A. & Ross-Murphy, S.B. (2002) Novel amyloid fibrillar networks derived from a globular protein: beta-lactoglobulin. *Langmuir* **18**, 7174–7181.

117  Ikeda, S. & Morris, V.J. (2002) Fine-stranded and particulate aggregates of heat-denatured whey proteins visualised by atomic force microscopy. *Biomacromolecules* **3**, 382–389.

118  Arnaudov, L.N., de Vries, R., Ippel, H. & van Mierlo, C.P.M. (2003) Multiple steps during the formation of β-lactoglobulin fibrils. *Biomacromolecules* **4**, 1614–1622.

119  Durand, D., Gimel, J.C. & Nicolai, T. (2002) Aggregation, gelation and phase separation of heat denatured globular proteins. *Physica A* **304**, 253–265.

120  Gimel, J.C., Durand, D. & Nicolai, T. (1994) Structure and distribution of aggregates formed after heat-induced denaturation of globular proteins. *Macromolecules* **27**, 583–589.

121  Ward, A.G. & Courts, A. (1977) *The Science and Technology of Gelatin*. Academic Press, London.

122  Lewis, D.F. (1981) The use of microscopy to explain the behaviour of foodstuffs – A review of work carried out at the Leatherhead Food Research Association. *Scan. Electron Micros.* Part 3, 391–404.

123  Tohyama, K. & Miller, W.G. (1981) Network structure in gels of rod-like polypeptides. *Nature* **289**, 813–814.

124  Mackie, A.R., Gunning, A.P., Ridout, M.J. & Morris, V.J. (1998) Gelation of gelatin – Observations at the air/water interface and in the bulk. *Biopolymers* **46**, 245–252.

125  Swaisgood, H.E. (1992) Chemistry of the caseins. In: Fox, P.F. (ed.) *Advanced Dairy Chemistry 1. Proteins,* pp. 63–110. Elsevier, New York.

126  Holt, C. (1998) Casein structure and casein-calcium phosphate interactions. In: *Proceedings of the 25th International Dairy Congress, Aarhus, Denmark,* pp. 200–208. Danish National Committee of International Dairy Federation, Copenhagen.

127  Gezimati, J. & Creamer, L.K. (1997) Heat-induced interactions and gelation of mixtures of β-lactoglobulin and α-lactalbumin. *J. Agr. Food Chem.* **45**, 1130–1136.

128  Bauer, R., Rischel, C., Hansen, S. & Ogendal, L. (1999) Heat-induced gelation of whey protein at high pH studied by combined UV spectroscopy and refractive index measurement after size exclusion chromatography and by *in-situ* dynamic light scattering. *Int. J. Food Sci. Technol.* **34**, 557–563.

129  Kavanagh, G.M., Clark, A.H., Gosal, W.S. & Ross-Murphy, S.B. (2000) Heat-induced gelation of β-lactoglobulin/α-lactalbumin blends at pH 3 and pH 7. *Macromolecules* **33**, 7029–7037.

130  Dickinson, E., Murray, B.S. & Stainsby, G. (1998) Protein adsorption at the air-water and oil-water interface. In: Dickinson, E. & Stainsby, G. (eds) *Advances in Food Emulsions and Foams,* pp. 123–162. Elsevier Applied Science, London.

131  Gunning, A.P., Wilde, P.J., Clark, D.C. *et al.* (1996) Atomic force microscopy of interfacial protein films. *J. Colloid Interf. Sci.* **183**, 600–602.

132  Morris, V.J., Kirby, A.R. & Gunning, A.P. (1999) Interfacial systems. In: *Atomic Force Microscopy for Biologists,* pp.160–208. Imperial College Press, London.

133  Mackie, A.R., Gunning, A.P., Wilde, P.J. & Morris, V.J. (1999) The orogenic displacement of protein from the air/water interface by competitive adsorption. *J. Colloid Interf. Sci.* **210**, 157–166.

134  Wilde, P.J. (2000) Interfaces: their role in foam and emulsion behaviour. *Curr. Opin. Colloid Interface Sci.* **5**, 176–181.

135  Mackie, A.R., Gunning, A.P., Wilde, P.J. & Morris, V.J. (2000) Orogenic displacement of protein from the oil-water interface. *Langmuir* **16**, 2242–2247.

136  Gunning, P.A., Mackie, A.R., Gunning, A.P. *et al.* (2004) The effect of surfactant type on surfactant-protein interactions at the air-water interface. *Biomacromolecules* **5**, 984–991.

137  Gunning, A.P., Mackie, A.R., Kirby, A.R. & Morris, V.J. (2001) Scanning near-field optical microscopy of phase separated regions in a mixed interfacial protein (BSA) surfactant (Tween 20) film. *Langmuir* **17**, 2013–2018.

138  Mackie, A.R., Gunning, A.P., Ridout, M.J. *et al.* (2001) Orogenic displacement in mixed β-lactoglobulin/β-casein films at the air/water interface. *Langmuir* **17**, 6593–6598.

139  Woodward, N.C., Wilde, P.J., Mackie, A.R. *et al.* (2004) Effect of processing on the displacement of whey proteins: Applying the orogenic model to a real system. *J. Agr. Food Chem.* **52**, 1287–1292.

140  Cooper, D.J., Husband, F.A., Mills, E.N.C. & Wilde, P.J. (2002) Role of beer lipid-binding proteins in preventing lipid destabilization of foam. *J. Agr. Food Chem.* **50**, 7645–7650.

141  Sarker, D.K. & Wilde, P.J. (1999) Restoration of protein foam stability through electrostatic propylene glycol alginate-mediated protein-protein interactions. *Colloid Surface B* **15**, 203–213.

142  Sarker, D.K., Wilde, P.J. & Clark, D.C. (1998) Enhancement of protein foam stability by formation of wheat arabinoxylan-protein crosslinks. *Cereal Chem.* **75**, 493–499.

143  Sarker, D.K., Wilde, P.J. & Clark, D.C. (1995) Control of surfactant-induced destabilization of foams through polyphenol-mediated protein-protein interactions. *J. Agr. Food Chem.* **43**, 295–300.

144  Decher, G. & Schlenoff, J.B. (2003) *Multilayer Thin Films – Sequential Assembly of Nanocomposite Materials*. Wiley-VCH Verlag, Weinheim.

145  Ogawa, S., Decker, E.A. & McClements, D.J. (2004) Production and characterization of O/W emulsions containing droplets stabilized by lecithin-chitosan-pectin mutilayered membranes. *J. Agr. Food Chem.* **52**, 3595–3600.

146  MacDougal, A.J., Brett, G.M., Morris V.J. *et al.* (2001) The effect of peptide-pectin interactions on the gelation of a plant cell wall pectin. *Carbohydr. Res.* **335**, 115–126.

147  Lin, C.F. (1977) Interaction of sulfated polysaccharides with proteins. In: Graham, H.D. (ed.) *Food Colloids*, pp. 320–346. Avi Publishing Co., Westport, CT.

148  Hansen, P.M.T. (1982) Hydrocolloid-protein interactions: Relationship to stabilization of fluid milk products. A Review. In: Phillips, G.O., Wedlock, D.J. & Williams, P.A. (eds) *Gums & Stabilizers for the Food Industry,* pp. 127–138. Pergamon Press, Oxford.

149  Snoeren, Th.H.M. (1976) Kappa carrageenan. A study on its physicochemical properties, sol-gel transition and interaction with milk proteins. PhD thesis, Nederlands Instituut voor Zuivelonderzoak, Ede, Holland.

150  Snoeren, Th.H.M., Both, P. & Schmidt, D.G. (1976) An electron microscopic study of carrageenan and its interaction with κ-casein. *Neth. Milk Dairy J.* **30**, 132–141.

151  Marudova, M., MacDougall, A.J. & Ring, S.G. (2004) Physicochemical studies of pectin/poly-L-lysine gelation. *Carbohydr. Res.* **339**, 209–216.

152  Clark, A.H. (1995) Kinetics of demixing. In: Harding, S.E., Hill, S.E. & Mitchell, J.R. (eds) *Biopolymer Mixtures*, pp. 37–64. Nottingham University Press, Nottingham, UK.

153  Ross-Murphy, S.B. (1995) Small deformation rheological behaviour of biopolymer mixtures. Harding, S.E., Hill, S.E. & Mitchell, J.R. (eds) *Biopolymer Mixtures*, pp. 85–98. Nottingham University Press, Nottingham, UK.

154  Abeysekera, R.M. & Robards, A.W. (1995) Microscopy as an analytical tool in the study of phase separation of starch-gelatin binary mixtures. Harding, S.E., Hill, S.E. & Mitchell, J.R. (eds) *Biopolymer Mixtures*, pp. 143–160. Nottingham University Press, Nottingham, UK.

155  Wolf, B., Scirocco, R., Frith, W.J. & Norton, I.T. (2000) Shear-induced anisotropic microstructure in phase-separated biopolymer mixtures. *Food Hydrocolloids* **14**, 217–225.

156  Normand, V., Pudney, P.D.A., Aymard, P. & Norton, I.T. (2000) Weighted-average isostrain and isostress model to describe the kinetic evolution of the mechanical properties of a composite gel: Application to the system gelatin: maltodextrin. *J. Appl. Polym. Sci.* **77**, 1465–1477.

157  Roesch, R., Cox, S., Compton, S. *et al.* (2004) Kappa-carrageenan and beta-lactoglobulin interactions visualized by atomic force microscopy. *Food Hydrocolloids* **18**, 429–439.

158  Kasapis, S. (1995) Phase separation in hydrocolloid gels. In: Harding, S.E., Hill, S.E. & Mitchell, J.R. (1995) *Biopolymer Mixtures*, pp. 193–224. Nottingham University Press, Nottingham, UK.

# Chapter 7
# Wheat-Flour Dough Rheology

*Robert S. Anderssen*

## 7.1 Introduction

Independently of whether it is achieved with an industrial mixer, a small-scale commercial device, a recording mixer or hand kneading, the formation of a wheat-flour dough is an evolving deformation and flow process. Wheat-flour dough rheology is therefore a dynamic modelling consideration, not the static state of peak dough development (PDD) as is tacitly assumed in many studies and publications. Though, from a cereal science and industrial perspective, the investigation of the rheology of this particular static state is quite important, even crucial and inciteful on occasions, it avoids the real issue of the molecular dynamics occurring within the dough during its formation. From the perspective of plant breeding, which is the source for the pragmatic ideas, questions and challenges that motivate cereal science research, the molecular dynamics is the key fundamental rheological issue. The accumulation of the elastic potential energy of the dough during its formation occurs at the molecular level. All other rheological matters depend on how this occurs. Consequently, a molecular model of how the elastic potential energy is accumulated, as a dough is formed, is the foundation stone of wheat-flour dough rheology. It leads naturally to the mathematical modelling of the mixing of a wheat-flour dough as an open-loop hysteresis phenomenon.

However, before a model for the accumulation of the elastic potential energy can be proposed, there are two matters that require detailed discussion. Firstly, the basic information about the developmental rheology of a wheat-flour dough must be reviewed. Such information has been recovered in different ways. Various forms of indirect measurements include walk-in-refrigerator experiments, temperature measurements, and high-resolution monitoring of the stress–strain behaviour of a dough during mixing on a Mixograph. More traditional cereal science measurements involve high-performance liquid chromatography (HPLC) analysis of molecular structure, summary measurements of the molecular weight distributions (MWDs) of the polymeric protein components, such as the unextractable polymeric proteins (UPP), the extension testing of dough mixed to PDD, and near infrared monitoring of protein and moisture.

Secondly, there is a need to explain and examine the molecular connection between what happens in a dough and what the indirect measurements reveal about its molecular structure. In one way or another, the indirect measurements are monitoring the changing MWDs of the gluten components and their changing relationship to the other components within a dough. Because the current measurement protocols allow one to collect summary information about the MWDs, such as UPP, the MWDs have become a key link concept between the indirect measurements and the molecular dynamics within the dough. Consequently, in the subsequent discussion, this central role played by the MWDs will be a key issue in the deliberations.

The importance of the analysis and interpretation of these various indirect measurements (in the recovery of information about the rheology of a wheat-flour dough) is stressed in this chapter by making the indirect measurements the starting point for the identification of the currently available key pieces of the wheat-flour dough rheology jigsaw puzzle. The motivation, in part, is to highlight the real nature of the science that underpins the study of wheat-flour dough rheology, and its central importance to all aspects of cereal science research.

Initially, however, it is necessary to draw a clear distinction between:

- the cereal science/technology perspective of wheat-flour dough rheology, and
- the formulation, solution and analysis of mathematical models that link various rheological measurements to the molecular dynamics that they monitor.

The former is often performed as an activity independent of the latter. The latter, however, is the framework on which the former must be built to give it quantification and rigour. It is the latter that is the major focus of this chapter, but this cannot be achieved without an appropriate review of the former.

### 7.1.1 The two independent aspects of cereal science and technology: molecular biorheology and process biorheology

The primary goal that underpins the cereal science of wheat-flour dough is the elucidation of an understanding of the genetic and processing basis of end-product quality. Rheology plays a central, though sometimes passive and not so obvious, role in this endeavour in two essentially different, though interrelated, ways.

#### 7.1.1.1 Genetics as the key to plant breeding: molecular biorheology

From the *modus operandi* perspective of the plant breeder, end-product quality, first and foremost, depends on the genetics of the wheat grown and processed to produce the desired product. In essence, if the genetics is inappropriate for the end product to be produced, no manipulation of the environmental conditions during the growing and harvesting, or of the processing steps during the manufacture, can recover the situation. At one level the exemplification is trite. Without some appropriate chemical intervention, soft wheats alone do not make good breads, hard wheats alone are inappropriate for (sweet) biscuit and cake making, and durum wheat must be used for high-quality pasta. At the cutting edge of plant breeding, the exemplification is deep. To breed bread wheats for marginal agricultural conditions in which only a low protein content is achieved, a gliadin-glutenin composition suitable for producing a good bread must be present in a high proportion in the limited protein.[1] From a plant breeding perspective, the genetics of a wheat controls the expression and organization of the various molecules that build the starches, gluten polymers and other components that make a wheat kernel. To define end-product quality for the plant breeder, molecular information must be recovered from the measurements performed to assess the factors that contribute to and determine quality. In one form or another, such measurements are rheological. Even electrophoresis measurements have a rheological aspect as they involve the reptation of a molecule down a column in response to a voltage gradient.[2,3] Consequently, the underlying science involves the recovery of molecular information from rheological measurements performed on biomaterials. An appropriate name is 'molecular biorheology'.

### 7.1.1.2 Process rheology as the key to efficiently maximizing end-product quality: process biorheology

From the *modus operandi* perspective of the different manufacturing steps involved after the growing of the wheat (harvesting, receivals, storage, milling, mixing, baking), the industrial processing of wheat-flour dough products, as well as maximizing end-product quality with respect to a given wheat genotype, must be designed, monitored and performed efficiently. For example, a good baker blends inexpensive low-protein flour with expensive high-protein flour and exploits knowledge about their mixing to produce a superior bread (maximize quality) and, hence, profit. All stages of the processing of a wheat to produce an end product involve flow and deformation in one form or another. Because wheat is a granular material, even storage involves flow and deformation considerations. Some of the measurements at receivals, such as falling number, are clearly rheological. The underlying science is an example of 'process (bio-)rheology' where the emphasis is on the rheology of the specific process under examination as distinct from experimental and theoretical rheology.

These two rheological endeavours, molecular biorheology and process biorheology, though complementary and supplementary to each other, are, from a scientific and rheological perspective, distinct activities. For the former, any rheological measurement that yields molecular insight is acceptable, whereas only measurements that accurately simulate and model the flow and deformation of the specific process being examined are acceptable for the latter. Though the focus of this chapter is mainly molecular biorheology, including the recovery of information from various forms of indirect measurements, like the MWD components of the proteins such as UPP, and the modelling of the accumulation of elastic potential of the dough during its mixing, the relevance of process biorheology will be examined when and where appropriate or as the need arises.

## 7.1.2 The pervasive nature of wheat-flour dough rheology in cereal science and technology

At this stage, it is crucial to stress that, because of the number and complexity of the steps involved in going from the genetics of a plant, through the processing (harvest, receival, storage, milling, mixing, baking), to end-product quality, the required understanding is sought in different ways using various forms of indirect measurements. Such indirect measurements range across a broad spectrum of choices and are exploited in various ways to recover different types of information, which relate back, in one way or another, to the rheology of dough formation and processing. It is this connection back to the rheology that

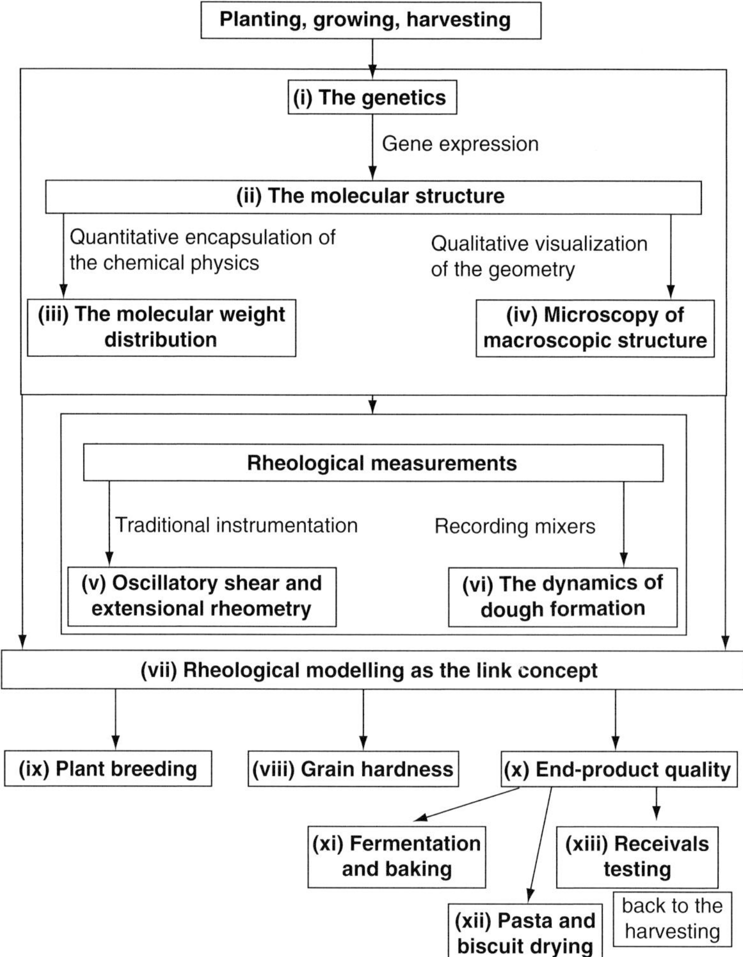



**Fig. 7.1**  A schematic summary of the logic and information flow in the pathway from genetics to end-product quality via rheology.

is a key central issue in this chapter. Figure 7.1 contains a schematic summary of the type of logic and information flow involved in the pathway from genetics to end-product quality via rheology.

In order to set the scene for the subsequent discussion, as well as give a general, albeit brief, introduction to the background science of wheat-flour dough rheology, it is first necessary to review the key (cereal science) experimental protocols that are utilized to collect relevant information about the genetics and the processing of wheat-flour dough. The aim is a summary of how these different protocols contribute alone and to each other to build the current molecular and processing understanding of wheat-flour dough rheology. In the subsequent discussion, the protocols start with a brief survey of some of the measurement modalities that yield information about how the genetics are connected to the rheology through various indirect measurements of key components of the molecular weight distribution (MWD) of the proteins such as the total and the unextractable polymeric proteins (TPP and UPP). Then, following a path that traces an increasing macroscopic summary of how the measurements see the rheological status and the processing of a wheat-flour dough, the protocols work through to an analysis of measurements that monitor information about end-product quality. All of these protocols, through the changing MWD of the various polymeric components, have a connection back to the accumulation of elastic potential energy by the dough during its mixing.

## (i) The genetics

The genetics controls the expression of the proteins that will eventually determine the nature and extent of the accumulation of elastic potential energy. Fundamental genetics is a collage crafted from information about the structure of the wheat genome, the expression of the proteins that build and organize the various components within a wheat, and the biochemical pathways that control the building and organizing in response to genetic memory and environmental inputs. Insight is obtained in various ways including microarray experiments that allow specific genes to be identified that appear to be connected to the phenotype of interest,[4,5] the construction of genotype-phenotype maps that formalize the nature of genotype-phenotype epistatis,[6-9] and bulk segregant analysis that allows biological averaging to be performed to enhance the process of gene identification. At a genetic level, a key phenotype is the expression of the gliadins and glutenins in bread wheats. In recent cereal science research,[10-12] various measures of the MW of such molecules have played a crucial role in developing the current quite sophisticated understanding of the molecular dynamics of wheat-flour dough formation.

The genetics is determined by the gene expression activity controlled by the wheat genome. The next level above the associated molecular dynamics is the molecular structure that the expressed proteins build.

## (ii) The molecular structure

The cross-linking that occurs within the dough during its mixing depends heavily on the chemical structure of the gluten molecules and their molecular weight distributions. The size, shape and number of the molecular segments in various components of wheat-flour doughs, such as the gliadins and the glutenins,[13] can be assessed using a number of (commercial) measurement protocols including high-performance liquid chromatography (HPLC), size exclusion (SE-)HPLC, reverse phase (RP-)HPLC, SDS-PAGE, lactate PAGE, etc. Such research gives insight about the molecular components that are important in the formation of a dough, and thereby are crucial in building a model of the molecular biorheology of dough formation. For example, such work has identified the key role played by the different glutenin subunits in building the polymer networks within a dough and in controlling its qualitative rheology as assessed by extension, Farinograph, Mixograph and Alveograph tests.[14-16] From a plant breeding perspective, such information must be coupled with the three-dimensional organization of the macromolecular components (gluten polymers, starch granules, etc.) within a wheat as it grows, in order to formulate a genetic characterization of a wheat's developmental biology. For example, the enhancement of the proportional presence of particular glutenin subunit components requires knowledge about the time of formation of such components.

In order to characterize the molecular structure in a way that can be used in mathematical models, there is a need to quantify it in terms of well-established chemical and physical concepts such as molecular weight distributions (MWDs), lengths of molecular chains, the proportional presence of the various chemical components, etc.

(iii) The link concept role of the molecular weight distribution (MWD)

In many applications, the molecular characterization of synthetic polymers and biomaterials is taken to be the MWDs of the molecular chains and structures of their various components. However, such MWDs can only be measured indirectly. For synthetic polymers, although HPLC and PAGE-type methods are often applied, it is more common to perform rheological measurements on the polymer, and then to solve an appropriate mixing rule, which relates the rheological measurements to the MWD of the polymer,[17,18] to recover some appropriate estimate (functional) of the MWD. Though the related mathematics is nontrivial and challenging,[19,20] this is often the preferred approach, because the rheological measurements are easy to perform and the molecular structure, and the associated MWD, of the polymer is quite simple, often taking the form of a single monodisperse peak. The reverse is essentially the situation for biomaterials, because of the complex structure of the molecular components of which they are formed. In most cereal science studies (e.g. refs 12 and 21), this difficulty is resolved by using not the MWDs but some appropriate summary such as the total and the unextractable polymeric protein content[22] (TPP and UPP) (Fig. 7.2). Even though TPP and UPP are highly smoothed indirect measurements of the different MWD components, they have proved very successful in developing the current understanding of the genetic and molecular basis of wheat-flour dough rheology and its impact on quality. For example, a Web of Science search on the seminal Gupta *et al.* paper,[22] which gives considerably more than 100 hits, identifies the importance and relevance of the utility of TPP and UPP, and related measures, in assessing the MWDs of the gluten components in wheat-flour doughs.

Information about the molecular structure can also be obtained from various microscopy studies – transmission electron microscopy (TEM), scanning electron microscopy (SEM), etc. They visualize and monitor the various ways in which molecular structures are organized to form high-level macroscopic structures, such as the starches and gluten polymers, as well as the organization of such macroscopic components within a wheat-flour dough. Such visualization is complementary and supplementary to the type of information discussed in (iii) because the concepts discussed there measure the presence and size of the various components independent of their geometric interrelationships. Clearly, such geometric details are essential to understanding the dynamics of the processes occurring.

(iv) The microscopic picture

Stereological assessments, based on various microscopic modalities, have been used to determine the macromolecular structure of wheat kernels[23] as well as wheat-flour and pasta doughs.[24,25] The three-dimensional reconstructions of dough structure that are obtained from such stereological data have been utilized for various purposes.[26] For example, it is clear from a study of the micrographs of extruded and rolled pasta dough that moisture layers are formed around the starch granules that give, because of the 70–75% of starch in durum semolina and wheat flour, the connected porosity structure that facilitates the drying of pasta and biscuits.[27] Microscopic studies have allowed the formation of the gluten network to be

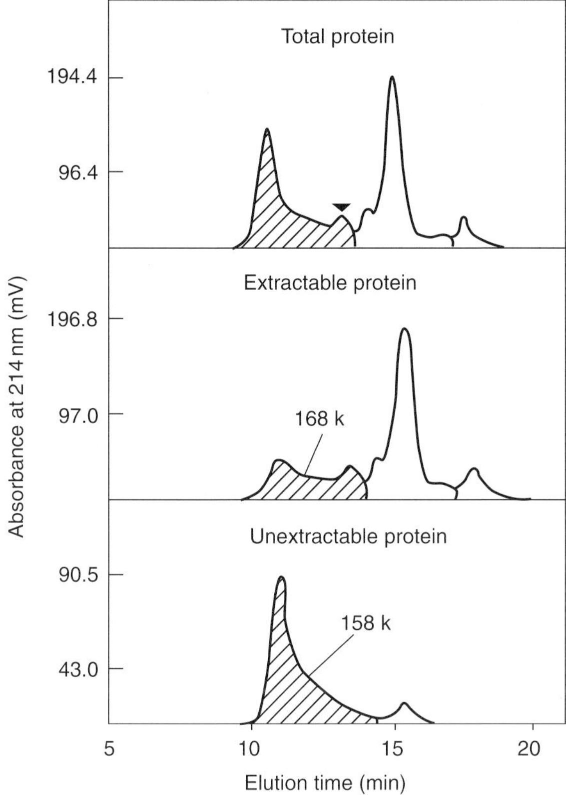


**Fig. 7.2**   The successive panels from top to bottom show for SE-HPLC separations of the total polymeric protein (TPP), the extractable polymeric protein and the unextractable polymeric protein (UPP). From Gupta *et al.* (1993),[22] with permission of Elsevier.

tracked in the development of a wheat-flour dough.[26] However, before a scientific basis for modelling the macroscopic processes can be formulated, the type of measurements, in terms of how they are performed and the type of information they 'see', need to be identified. As new microscopic technologies are developed, such as immunofluorescence microscopy, enhanced insight about microstructure within a dough will result. In addition, image analysis techniques are being utilized to assess the microscope images.[28]

To bring the details discussed in (i) to (iv) into a wheat-flour dough perspective, they must be connected, through appropriate measurements, to the dynamics of the flow and deformation that occur in the mixing of a wheat-flour dough. Traditional rheological measurements enter as one approach to accommodating this need as, in one way or another, they are the basis for characterizing the current stress–strain status of a material.

(v) The basic rheological measurements
The formulation of rheological models for wheat-flour dough formation must be based on what can be measured rheologically. Oscillatory shear and extensional measurements of the rheology of a dough at some predetermined state, such as peak dough development (PDD), are a common mode of analysis. Such data are the starting point for the formulation of

constitutive relationships for the viscoelasticity of a dough.[29] They also play a key role as the link concept between the genetics of a dough and its baking performance. For example, HPLC results are regularly compared with extensional measurements to hypothesize about the genetics of and genetic control within wheat,[13] while baking performance is related back to appropriate rheological measurements,[16] such as the classic $R_{max}$ (maximum of the force resisting extension) and $Ext_{rupture}$ (length of the extension at the time of rupture). The storage modulus $G'$ and the loss modulus $G''$, derived from oscillatory shear measurements under the assumption that the dough is behaving like a linear viscoelastic material, have not been as successful in assessing the process biorheology of a wheat-flour dough as extensional measurements.[30–32] This is a direct consequence of the fact that the rheology of the proofing and the baking of a dough is extensional, because it relates to the rheology of the sheets within the dough that contain and constrain the pockets of $CO_2$. However, the formation of a wheat-flour dough is a dynamic process that first involves the accumulation of elastic potential energy until peak dough development is attained, after which it degrades. In order to formulate models of the accumulation, there is a need for measurements that see the evolving molecular dynamics of dough formation in terms of its mixing.

However, it is not simply a matter of performing arbitrary rheological measurements to see the relevant molecular dynamics. In general, traditional rheological measurements are unable to identify the nature of the stress–strain dynamics occurring during the mixing and development of a dough, as, among other things, the development involves the accumulation of elastic potential energy. In order to obtain such insight, one must turn to the information measured by recording mixers.

(vi) Measurement of the developmental rheology of wheat-flour dough formation
Farinogram and mixogram characterizations and comparisons of the qualitative/quantitative rheology of different varieties of wheats and the wheat-flour doughs that they make,[16,33] allows one to recover information about the evolving molecular dynamics in the formation of a wheat-flour dough. It was the invention of recording mixers, such as the Valiograph (Jeno Hankoczy in Hungary in 1912), Farinograph and Mixograph, that placed cereal science on a rigorous footing back in the 1930s. The qualitative structure of such characterizations contained the key information about the basic qualitative rheological phases involved in wheat-flour dough formation: hydration, dough development, maximum bandwidth, peak dough development, and breakdown. In many ways, there is a clearer understanding of the situation in the earlier literature,[34] and the papers by Voisey *et al.* referenced in that paper[34]), than one finds in some more recent publications. For example, explanations based on mixing time place the wrong emphasis on the relevance of the rheological information in farinograms and mixograms. The correct characterization is the strain (revolutions) applied to the dough by the mixing mechanism. In part, this has occurred because of an overemphasis scientifically on the importance of a dough at PDD, because of its industrial significance. PDD is the key indicator from an industrial assessment perspective, because it is more or less the point to which dough is mixed before it is baked. Consequently, it is a key process biorheology matter, but only one of many key factors in the mixing from a molecular biorheology perspective. Farinograms and mixograms can be viewed as the indirect measurements that allow the elastic potential energy to be modelled.

On the one hand, the above deliberations have established the crucial importance of rheological measurements. On the other hand, it is not simply a matter of collecting together the various types of observations, measurements and images outlined above and then assuming that there is nothing more to do. From a modelling perspective, one is only at the 'conceptualization' (brainstorming) stage prior to model formulation. The next step is to obtain the insight required, so that either:

- appropriate predictions can be made about end-product quality on the basis of genetic assumptions and considerations and how the processing biorheology will be performed (the 'forward experimental and modelling process', which is the starting point for the accumulation of the scientific information on which 'mathematical' models are formulated, solved and interpreted), or
- informative insight can be inferred about the type of genetics and processing biorheology that is likely to achieve a specific end-product result (the 'inverse process' for which an 'explicit' model is required in order to exploit the information collected in the 'forward process' for focusing plant breeding).

In one way or another, even though some may only be informative correlations, 'mathematical' models must be formulated. The obvious strategy, which is gradually becoming a key aspect of cereal science, is to perform the model formulation in terms of the rheological observations so that the rheology performs a nontrivial linking between the genetics plus processing biorheology and end-product quality. Because of rheology's quantitative and rigorous mathematical structures, based on stress–strain encapsulations, such a linking allows the genetics and the end-product quality to be related via the rheology, and, thereby, gives both a much more rigorous foundation on which to perform the 'forward' and 'inverse' activities.

(vii) Linking the genetics to end-product quality via rheology

It is the rheological measurements of (v) and (vi) that play the role of the link concepts[35] that connect the genetics to end-product quality.[13] In order to relate the measurements of (iv) and (v) to end-product quality, much of current cereal chemistry research focuses on correlating the information obtained from various experiments with the genetics, the flour, the milling or the mixing with baking performance.[15] On some occasions, the link is only to some rheological experiment, such as the monitoring of the changing glutenin subunits in the wheat. On other occasions, the link is only between rheological measurements and the actual baking, such as correlations of $R_{max}$ and $Ext_{rupture}$ with loaf volume, and the effect of salt on dough rheology,[36] in order to maximize the utility of such correlations as link concepts between the genetics and end-product quality. In fact, the experiments are often specifically planned, often using statistical experimental design protocols, to monitor specific correlations. Sometimes, it represents a short-term expedient rather than an advance in scientific measurement. The measurements of small-scale testing must be correlated with traditional recording mixers, rheological measurements like those in (v) and (vi), and industrial mixers. A recent study of the linking in terms of large-deformation properties of wheat-flour doughs was published by Sliwinski and co-workers.[37,38]

However, the rheology of wheat-flour dough development is not the only rheology that is important in relating the genetics plus processing to end-product quality. Because grain hardness is such a strongly controlled genetic property of wheat and has a huge impact on how a particular wheat should be milled to produce flour of a specified quality, it plays an equally significant role.

(viii) The importance and relevance of the grain hardness phenotype: the grain hardness and milling connection
Grain hardness, because it is controlled genetically, and because it is a basic criterion for the classification of wheat with respect to end products, is important from both a molecular and process biorheology perspective. At the processing level, it has a major impact on milling. At a molecular level, it allows one to differentiate between the macromolecular structures within wheat kernels that determine the rheological nature of hardness. Rheological measurements of the wheat before it is milled to produce wheat flour are crucial, because grain hardness is clearly genetic.[39] Various measurements of hardness are utilized to exploit the significance of this fact. Some are qualitative, like the hardness index generated by the software on an SKCS 4100 instrument, while others are quantitative, like the rheological phases in the crush-response profiles generated by the SKCS 4100 instrument.[40] In addition, grain hardness is a reliable predictor of protein content, and has a nontrivial connection to dough rheology. Furthermore, because hardness correlates strongly with milling, mixing and baking performance, one is again back to rheology. In fact, this represents a compelling reason why the future quantification of cereal science must turn to rheology for its foundation. This applies not only to the more obvious ramifications of cereal science rheology, such as milling and mixing, but also to the construction of genotype-phenotype maps and the identification of quantitative trait loci (QTL).

In order to appreciate fully the significance and relevance of rheology in the pathway from genetics to end-product quality via processing, it is important to understand how the different types of rheological results influence and are exploited for practical purposes. In the following items, (ix)–(xiii), a number of representative examples are discussed in order to highlight the significance of the role played by the molecular and processing biorheology. For example, the plant breeder requires insight about what is occurring at the DNA and molecular level in terms of how they influence the subsequent rheology, as well as using rheology to suggest genetics plus processing possibilities to achieve a specified end-product quality (i.e. solving the underlying 'inverse problem').

(ix) The plant breeding perspective
In the cereal science that supports plant breeding, one is looking for precursors that are good indirect predictors of the presence of molecules that determine some specified and desired end-product quality. Such precursors range over a number of possibilities including HPLC techniques and indirect measurements of rheological properties. The identification of such precursors is a fundamental motivation behind the experimentation discussed in (vi) and the reason why rheological link concepts are so important. For example, a goal in the development of NIR (Near InfraRed) assessment protocols is to have good calibration-and-prediction predictors of desired rheological properties that allow one to perform rapid screening as a replacement for more complex rheological tests that would require more material or involve expensive and time-consuming measurements. For the early stages of plant breeding, the goal is to test a broad spectrum of possibilities rapidly. The development of small-scale

testing protocols[14] has given the plant breeder the ability to perform rheological screening at a much earlier stage than previously.

For the manufacturer of wheat-flour dough products, the process biorheology must be focused accordingly. As explained below and mentioned earlier, the process biorheology is a much more focused and constrained activity than molecular biorheology.

(x) Assessing the process biorheology that determines end-product quality
The impact of (vi) and (vii) forces a need, even in a scientific context, to identify the rheology associated with the processing to produce end products of high quality. As explained and illustrated in some detail by Dobraszczyk and Morgenstern,[41] within the industrial context, if one wishes to understand the rheology of food processing, then it is absolutely

"…necessary to define the set of deformation conditions which the food sees in practice and perform (rheological) tests under similar conditions."

In part, it illustrates the crucial need, in rheological modelling, to conceptualize and respect, in terms of subsequent measurements and their interpretation, the nature of the rheological flows and deformations occurring during each stage of the processing. In a way, this is the role and purpose of process biorheology – as discussed above in (ii). In addition, this generates the need to understand the relationship between rheometers and what they actually measure. This has already been reviewed briefly in (v) and (vi).

Molecular biorheology still involves some very challenging issues. As illustrated below, the essential challenge is collection of appropriate data that will allow appropriate rheological models to be formulated.

(xi) The fermentation and baking perspective: bubble rheology
The nature of the bubble expansion in the fermentation of a dough and the baking of a bread is directly related to:

- the elastic potential energy stored in the dough, and
- the form of the cross-linking of the molecules within the dough sheets that form within the walls of the bubbles.

The analysis of the fermentation and baking bubble rheology is at the cutting edge of quantitative cereal science. Dobraszczyk and Morgenstern[41] have examined the importance and utility of strain hardening of a dough at PDD as an appropriate measure of bubble rheology. Strain hardening is a good illustration of this need to understand what a rheological measurement tells one. It is a good indicator for fermentation and baking, but not for molecular dynamics. For molecular dynamics, any measurement is appropriate if it tracks the stress–strain development of a dough and sees the development of the open-loop hysteretic structure. For fermentation and baking, as stressed above, the only rheological measurements that are appropriate are the ones that see the processing circumstances under investigation. One must limit attention to measurements, such as strain hardening, that contain the relevant information about the matter under consideration and support the associated decision-making.

The baking and drying of food doughs are excellent illustrations of how one first needs a thorough understanding of the molecular biorheology background before one is in a position to put the associated process biorheology on a rigorous footing that is of direct assistance to industry. For pasta and biscuit drying, this involves a study of the molecular biorheology of porosity.

## (xii) The drying of pasta and biscuits

The drying of pasta and biscuits is another process that plays a major role in determining end-product quality. Here, however, the physical, chemical and rheological processes involved are quite different from bread making and require a quite different approach. Though the mixing and extrusion are important in developing the dough for the subsequent drying, it is a process biorheology matter outside the scope of the current chapter. More details about how the drying depends on the porosity of the water channels around the starch granules can be found in McNabb and Anderssen.[27]

Molecular and process biorheology play key roles even before the milling. Examples include the filling that occurs in grain development, as the wheat plant grows, and the international protocols used for assessing the status and properties of wheat that the farmer has just harvested.

## (xiii) The rheology of receival station testing

Even the assessment of the price to be paid for a wheat at a receival station can involve rheological considerations. When weather damage may have occurred, the falling number test[42] is used to assess the $\alpha$-amylase content of the wheat. In reality[43] the test measures the thermal stability of a gel formed from a mixture of ground wheat and water that is heated and mixed in a predetermined manner. However, this is not a matter that is pursued in this chapter, but its importance from a wheat-flour dough rheology perspective cannot be overlooked.

Disease control, enhanced crop productivity, design protocols for industrial mixers and related matters are key secondary considerations within the framework of the above-mentioned activities. For example, insight about the design of industrial mixers comes as a corollary from the above types of investigations and deliberations.

Though end-product quality depends on each of the steps from plant breeding through to the manufacture of the product, a guaranteed high quality for the end product can only be achieved if the starting raw material (the wheat) has appropriate credentials. It is this fact, more than any other, that drives the importance of genetics in plant breeding and cereal science. In fact, high-quality raw materials can often cover deficiencies in the milling, mixing and baking. On the other hand, as mentioned above, there is a move to improve the process biorheology of the milling, mixing and baking so that the initial quality of the raw material does not have to be so high to cover current deficiencies in industrial mixing equipment. In addition, such improvements would allow more marginal crops to be used to make good end products. But this can only be achieved through an improved understanding of the related wheat-flour dough rheology.

Always, one comes back to rheology in one form or another as the milling, mixing and baking each involve a different type of specific rheology that must be understood to place such activities on a rigorous scientific basis.

In the cereal science literature, investigating the genetic basis of end-product quality often appears to have little to do with dough rheology. The matter under investigation, such as the biochemical pathway for $\alpha$-amylase synthesis in the aleurone layer, is highly focused and specific and disconnected from the bigger picture. The fact that the milling, mixing and baking of a bread (pasta, biscuit) involves a number of independent rheological considerations tends to be overlooked. For example, in a baking test, the emphasis is often on the choice of the double-haploid line. The fact that a wheat must be milled, a dough must be formed and baking is involved tends to be taken for granted. Standard procedures are followed rigorously essentially to remove such matters from consideration. The need for multiphase statistical experimental designs, that take milling, mixing and baking effects into account, is argued to be circumvented by following the standard procedure. Nevertheless, rheology enters the deliberations. It is tacitly assumed that a dough is a dough no matter how it is formed, as long as the standard procedure is followed. The robustness of dough formation is the reason why this is not an unreasonable starting point in many current cereal science endeavours. However, it will and must change.

### 7.1.3 The rheology perspective: the recovery of information from indirect measurements

In the end, as in all areas of science, knowledge is gained by formulating mathematical models of how the various measurement protocols 'see' the underlying molecular and macromolecular structure of a wheat-flour dough and using these models to recover relevant molecular information. This has sometimes been achieved by judiciously designed experiments, but this is not the basis for the development of a quantitative cereal science. For the modelling of the accumulation of the elastic potential energy, one must turn to the modelling of viscoelasticity and open-loop hysteretic structures. As an illustration of the type of mathematics involved, such concepts will be examined in some detail in Section 7.6.

Each different model requires a separate analysis and this fact is also discussed in some detail. The theory and technology for the recovery of information from indirect measurements is comprehensive, sophisticated and huge. It ranges across various forms of regularization and stabilization, the use of joint inversion protocols to combine the recovery from different indirect measurement assessments of the same phenomenon, and the use of calibration-and-prediction when it is not possible to formulate explicit mathematical models that relate the indirect measurements to the particular information to be recovered about the phenomenon of interest.

Understanding the rheology of a wheat-flour dough not only depends on performing one or more of the above cereal science/technology activities, but also on exploiting, in some hierarchical manner, the information obtained from an analysis of the resulting indirect measurement modalities. This involves the concatenation of the types of information outlined in (i) to (xiii) to perform phenomenological modelling of the mechanisms of wheat-flour dough rheology. Some representative examples include:

- The loop-and-train model of Belton for gluten elasticity of a dough.[44]
- Macropolymer models of the microscopic structure of a dough.[45]
- Hysteretic modelling of the mixing of wheat-flour dough and associated rheological interpretations.[33,46]

## 7.2 Background, preliminaries and notation

In order to avoid unnecessary repetition and duplication, as well as to avoid an interruption to the flow of the deliberations, this chapter does not review earlier material, for which excellent recent publications are available. However, for the reader who is unfamiliar with the background literature of cereal science, a brief summary is given in Appendix 1. A summary of notation and terminology is given in Appendix 2.

## 7.3 The phenomenology of wheat-flour dough formation

When water by itself, or with other appropriate ingredients such as salt, is added to wheat flours and mixed, unique materials, wheat-flour doughs, are formed with the following properties, which all have a rheological connection.

(1)  As a result of the interaction of the water with the proteins in the flour, which the mixing accommodates and assists, a connected network is formed, which is viscoelastic. This is the most obvious visual feature and tactile property of a wheat-flour dough.

(2)  At a molecular level, the nature of the viscoelasticity and the associated structure within the dough depends not only on the amount of water added and the nature of the mixing, but also on whether the flour comes from a hard or soft wheat. This becomes the first point where molecular differences play a crucial differentiation role.

(3)  The flour from hard wheats tends to make good bread doughs where the strong viscoelasticity controls the size of the gas bubbles that form during fermentation, whereas the flours from soft wheats, because of the porosity resulting from the water channels formed around the undamaged starch granules, tend to make good biscuit and cake doughs. From a molecular biorheology perspective, the proteins play the key role in bread doughs, whereas the starch granule and protein interaction take centre stage in biscuit or cake doughs.

(4)  The nature of the proteins in hard wheat flours is such that bread doughs have superior gas retention properties, because the rate of diffusion of gases through very thin dough sheets is very small.

(5)  Bread doughs, when cooked, form a solid foam.

(6)  The porosity in biscuit, cake and pasta doughs allow their drying to be performed so that the end product does not check (crack or craze) during or subsequent to manufacture.

Clearly, a detailed rheological analysis of such matters must take explicit account of the molecular differences as well as how these differences contribute molecularly to the observed properties.

The phenomenology of the formation of a dough is succinctly encapsulated in the graph generated on a recording mixer. Because it is a pin-mixer, the associated rheological flow is simplest on a Mixograph. In terms of the mixograms thereby generated (Fig. 7.3), the following five rheological phases, as a function of the increasing number ($N_r$) of revolutions of the mixer, can be identified:

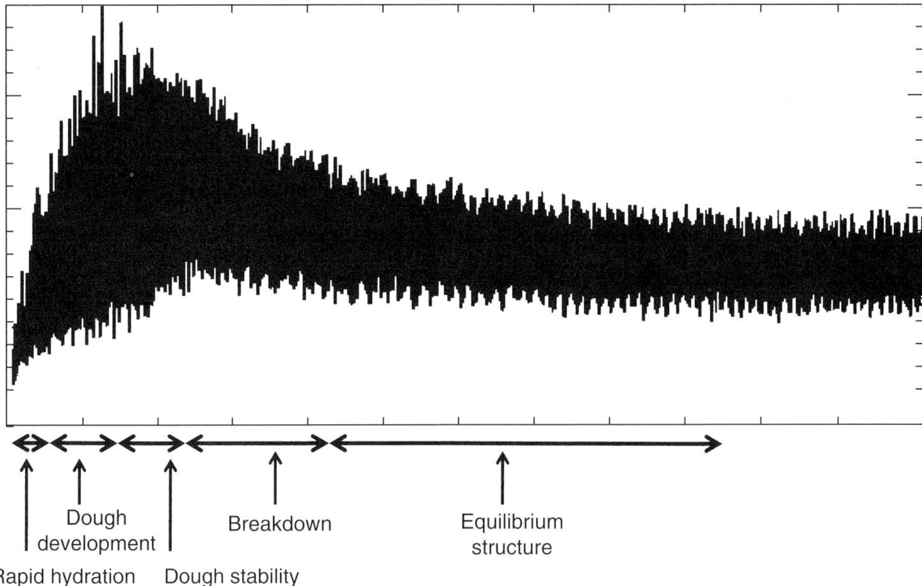



**Fig. 7.3**   A graphical representation of the proposed five mixogram phases: rapid hydration; dough development; dough stability; breakdown; and equilibrium structure. The importance of the width $W(N_r)$ of variation between the maximum and minimum values of $F(N_r)$ is apparent from the mixogram, which clearly shows how it starts to widen out during rapid hydration, reaches a maximum during dough stability and asymptotes to a fixed value during equilibrium structure.

(1)  **Rapid hydration.** The force $F(N_r)$ with which the dough resists the mixing increases rapidly (almost linearly as a function of $N_r$) with the width of the variation $W(N_r)$ between the maximum and minimum values of $F(N_r)$ opening out slowly. To first order, it could be argued that the initial response of the dough to the addition of the water and the mixing is essentially elastic.

(2)  **Dough development.** The change from the rapid hydration phase to that of dough development is definite and specific. There is a clear change to the slope of $F(N_r)$, which is smaller than that during the rapid hydration. The corresponding increase in $F(N_r)$ remains essentially linear, while $W(N_r)$ starts to open out rapidly. Because $W(N_r)$ can be viewed as an indicative measure of the extensional viscosity of a dough, one could identify the start of the dough development phase as the transition point from an elastic-like to a viscoelastic-like response by the dough to the mixing.

(3)  **Dough stability.** The change from the dough development phase to that of dough stability is not so definite. In some situations, dough stability occurs quite rapidly over a small number of revolutions, while in others, as shown in Fig. 7.3, it is a gradual process. From many cereal science and technology aspects, as well as rheologically, it is the key phase, as it involves the following three important subfeatures:

(a)  Maximum bandwidth. This corresponds to the maximum value of $W(N_r)$.

(b)  Peak dough development (PDD). This corresponds to the maximum value of $F(N_r)$.

(c)  Commencement of breakdown. After PDD, the value of $W(N_r)$ commences gradually to decrease while $F(N_r)$ remains essentially constant. It is unclear at this stage as to the importance of this secondary feature. Nevertheless, its potential importance should not be ignored.

(4) **Breakdown.** Though the change from the dough stability phase to that of breakdown is a gradual process, the breakdown phase itself has a clearly definite structure. The force $F(N_r)$ tends to follow a quadratic decrease, as a function of $N_r$, whereas the decrease in $W(N_r)$ is quite gradual.

(5) **Asymptotic equilibrium.** A key observation, the importance of which has essentially been ignored, is that the breakdown phase tends to asymptote slowly to constant values for $F(N_r)$ and $W(N_r)$. However, it is included here because it must contain information about the molecular structure of the wheat that made the flour that made the dough. Clearly, the transition from the breakdown phase to that of the asymptotic equilibrium is quite gradual, but, nevertheless, when the mixing has been recorded for a sufficiently large number of revolutions the asymptotic behaviour is clearly visible.

A more detailed discussion of the significance and relevance of these features is the central theme of the discussion in Sections 7.5 and 7.6.

There is no unique way in which to summarize the information in a mixogram. In earlier work,[15] the graphical structure of a mixogram (Fig. 7.4) was assessed in terms of the following qualitative rheological features: MR – mixing revolutions to peak resistance (pseudostrain; dimensionless); PR – peak resistance of the dough (force); BWPR – bandwidth at peak resistance (force); RBD – resistance characterization of the breakdown (%); BWBD – bandwidth characterization of the breakdown (%); RMBW – revolutions to maximum bandwidth (pseudostrain; dimensionless); MBW – maximum bandwidth (force). Not surprisingly, there is a strong connection between these qualitative measures and the five rheological phases identified above.

Such information has been utilized in various ways as link concepts, as outlined in item (vii) of Section 7.1.2, above. Such qualitative characterizations of the developmental rheology

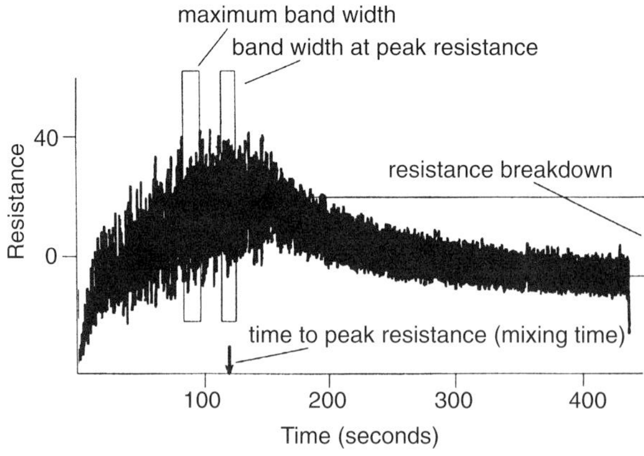



**Fig. 7.4**  The seven phases, MR, PR, BWPR, RBD, BWBD, RMBW and MBW, that are currently used in the qualitative rheological assessment of a mixogram (notation explained in text). From Gras *et al.* (2001),[15] with permission of CSIRO Publishing.

of a wheat-flour dough have proved fundamental in underpinning the current understanding of the cereal science of wheat. The examples are numerous.[15,16] Though there is a need for the formulation of more quantitative and molecular models of dough formation, which the current chapter partially addresses, the above qualitative characterization, and alternative versions of it, will continue to play the major decision-support role in:

- the research and development of wheat-flour dough rheology;
- the associated cereal science and the application of such results to improving plant breeding;
- the processing of wheat-flour dough products from harvest to end product;
- the enhancement of end-product quality.

Such qualitative rheological assessments represent an important initial exploratory phase in assessing plant breeding and cereal science results in a hierarchical manner.

Information about the developmental stress–strain pattern occurring within a dough, as encapsulated in a mixogram, can be recovered in various ways. Some of the possibilities include:

(a) The formulation of a mathematical model for the relative motion between the fixed and moving pins.[47] Such information is basic to determining the strain and strain rates that the relative motion between the moving and fixed pins in a Mixograph applies to a dough.

(b) The decision to define and measure 'mixing time' (which, when specified as minutes and seconds, is only relative to the speed of the mixer) as the 'number of revolutions of the mixer' (since the commencement of the mixing). Among other reasons, the industrial, as well as scientific, popularity of 'mixing time' (even though it is not rheologically meaningful or appropriate) relates to the fact that time is easily measured and applied in an industrial context, whereas 'revolutions' requires the installation, on industrial mixers, of a device that performs the counting. (Interestingly, some of the early industrial mixers included such a counter.)

**Caveat:** In addition, a similar comment applies to papers that discuss the work performed on the dough in terms of the current consumed by the mixer. There is no doubt that such observations are indirect measurements of the overall work being performed on a dough and have some use in an industrial context. However, from both a molecular and a process biorheology perspective, such observations have smoothed out most of the essential information about the way a dough responds to its mixing. The real work on a dough is performed revolution by revolution, and the failure to take this into account can lead to misunderstandings about the nature of dough development during mixing.

(c) Standard software has been developed[48] to determine the seven qualitative rheological parameters defined above. As already mentioned, they encapsulate sufficient information about the developmental rheology of a dough during mixing to have played a key role in progressing cereal science and technology to the position that it holds today.

(d) From a hysteretic perspective, there is need to understand the extent to which the developmental rheology of a dough is rate independent. This reduces to an investigation of the dependence, on mixing speed, of different aspects of dough formation and development as a function of the number of revolutions. It is now known that mixing to peak dough development is essentially rate independent.[46]

(e)  Historically, a mixogram was originally recorded as a pencil trace of a moving paper scroll. As a consequence, in the subsequent development of electronic Mixographs, a data rate of 10 Hz was chosen so that the resulting mixograms had the same visual structure as their paper counterparts. The opportunity of using much higher data rates (e.g. 250 Hz), in order to record the stress–strain pattern of the mixing revolution by revolution, was overlooked. The structure in the resulting high-resolution mixograms is where the explicit stress–strain information about the developmental rheology is contained. Such matters have been examined and discussed.[33,46]

**Caveat:** The information contained in a mixogram, though different from that obtained from other recording mixers, is indicative of the developmental rheology occurring. Interestingly, what was used historically to characterize the qualitative rheological features of a mixogram appear to have a stronger connection with the underlying quantitative rheology than previously appreciated. This is a point that will become apparent at subsequent stages of this chapter. Independently of how it is achieved, all forms of mechanisms that develop dough involve a hysteretic process of one form or another, which successively stimulates, through a process of stress and relaxation, the development of the molecular structure to achieve the fully developed dough state.

**Note:** Similar comments apply to the graphs generated by other recording mixers, such as Farinographs, though the differences in the associated rheological flows occurring in different mixers cannot be ignored.

But, to build models of the rheology of dough development, it is first necessary to review the nature of wheat-flour dough rheology modelling as it relates to the recovery of information from indirect measurements.

## 7.4 Wheat-flour dough rheology modelling from an indirect measurement perspective: a plethora of models

When one reflects on the *modus operandi* of modelling, it is clear that, with respect to the available observational data, 'One formulates models to answer questions!' about the context within which the data have been (or will be) collected. Having identified the particular question to be investigated, the formulation of an appropriate (conceptual, inductive and/or deductive) model reduces to the construction of a (mathematical) relationship between the available data and an appropriate encapsulation of the information that might answer the question. Often the relationship is quite simple because the question being asked can be resolved by comparing the data from one scenario with another.[49] Examples include the use of SDS-PAGE to determine the presence or absence of specific molecular components in one wheat variety compared with another, and the use of HPLC and related measurements to determine, in different wheat varieties, the proportional presence of gliadins and glutenins as well as estimates of TPP and UPP.[13,22] In such comparative assessment situations, the underlying mathematical modelling, because of its elementary nature, is usually taken for granted. As mentioned in Shewry *et al.*[13] (p. 221), alternative interpretations of the data tend to be overlooked.

The need to construct models relates to the fact that, except for special circumstances, one is unable to measure directly the information required to answer the specific question under examination. Interestingly, it is often conceptualized and even articulated that one is measuring the required information when in fact one is assuming (often unwittingly) that the measurements correlate directly with the information, which is only true in a limited sense. For example, it is often tacitly assumed that the bands in an SDS-PAGE experiment measure the molecular weight of the identified polymer, when in fact one is measuring the speed that a particular molecular structure reptates down the column. By assuming that all the bands in the column correspond to different sizes of the same molecular structure reptating down the column, one is able to correlate the bands with molecular weights, especially when polymers of the same molecular structure and known molecular weights are included as controls.

Such situations, where the mathematical modelling is reduced to a simple comparison of the actual measurements, are representative of a popular strategy for analysing and interpreting indirect measurements. Because it aptly summarizes the basic *modus operandi* being applied, it has been conceptualized as[50,51] 'the direct use of indirect measurements'. It essentially reduces the mathematical modelling to comparative assessment and correlations. This is a commendable strategy when the underlying assumptions are robust. In the SDS-PAGE example, the possibility is overlooked that a specific band might correspond to two different types of molecules of different structure, and, hence, molecular weights.

Once the data have been identified that will be utilized to answer the question, it is the question that will drive the modelling. The challenge is the formulation of a model that relates the indirect measurements, as output, to some appropriate information, as input, which will assist in answering the question. Though the same model can often answer a variety of questions, the nature of modelling is such that, as the question changes, it will often be necessary to change the model. Thus, in modelling, there is no single cure-all model, as is so often assumed, but a plethora.

A representative example in wheat-flour dough rheology relates to the questions that can be answered using measurements of the uniaxial extension (Fig. 7.5) of a dough. They include an analysis of the effect of transglutaminase on the rheological properties,[52] genetic and environmental variation for grain quality traits,[53] estimation and utilization of glutenin gene effects in wheat breeding programmes,[54] effects of nitrogen and sulfur fertilization[55] and the prediction of dough properties on the basis of glutenin subunit composition.[10] The formulation of a model depends on the level of sophistication required. In many papers, as explained in Section 7.3, the qualitative rheological parameters of Gras *et al.*[15] were all that were required to resolve the decision-making under consideration. Because various facets (phases) of the extension of a dough can be measured, such as the maximum resistance of a dough to its extension ($R_{max}$) and the amount of extension to rupture ($Ext_{rupture}$), uniaxial extension has played a crucial role as a link (concept) between the genetics of the wheat, from which the wheat flour was derived, and some appropriate end-product quality. This reinforces point (vii) of Section 7.1.2, above.

When the question relates to how well $R_{max}$ and/or $Ext_{rupture}$ can predict the baking performance of a dough mixed to PDD, the associated (mathematical) modelling reduces to

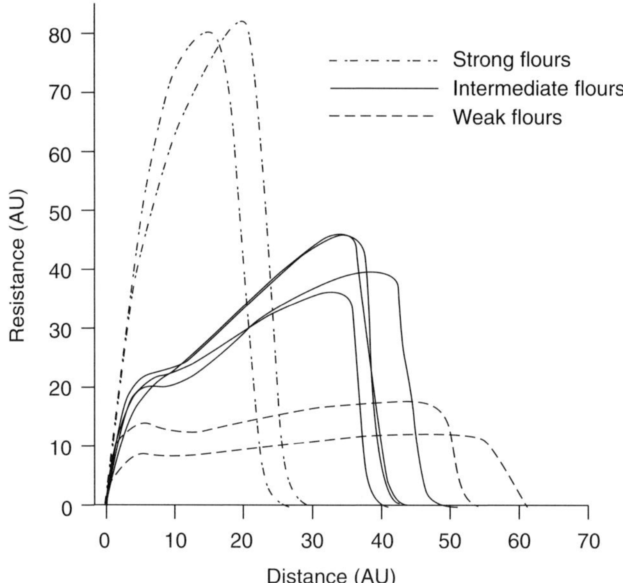



**Fig. 7.5**    The different stress–strain patterns that result when extension tests are performed on the doughs from strong, intermediate and weak flours. From Anderssen *et al.*[16] with permission of Elsevier.

constructing correlations between the chosen uniaxial extension measurements and some end-product measure such as loaf volume or crumb colour. When the question relates to assessing the relationship of uniaxial to biaxial extension occurring in the walls of bubbles in the dough, greater sophistication is required such as a strain hardening assessment.[56] As the sophistication of the question increases, so must the associated mathematical modelling.

**Caveat:** In science generally, not only in cereal science, much modelling is confusing or aimless simply because it is not being performed in response to some specific question. It is the easiest thing in the world to write down some algebra and equations that purport to be a model for some process or to feed (excellent) data into some computer package that generates a multivariate statistical analysis model that has no relevance to the application context because the coefficients in the model have no meaningful connection back to the original experiments. Confusion and aimlessness are certainly the situation when the modelling is performed to justify the collection of the data rather than the utilization of the data to answer some specific question about the science being investigated. It is certainly the situation in cereal science where many of the publications are simply a compendium of experimental results with very little attempt to understand or explain the structure within the actual measurements or to see the connection to the underlying rheology. A good example is the continued use of the hardness index (HI), which in 1963 was an important historical and laudable development,[57] to assess grain hardness instead of more explicit rheological characterizations.[40] Another is the overemphasis, in a uniaxial extension test, on $R_{max}$ and $Ext_{rupture}$ as the only phases in an extensogram that are important in correlating with some end-product quality measure such as loaf volume. As indicated in Anderssen *et al.*,[16] there is information in the other phases in an extensogram that, on occasions, may be more relevant to end-product quality than $R_{max}$ and $Ext_{rupture}$. Consequently, the need is first to identify and then remain focused on the specific question that encapsulates the R&D under investigation, and, furthermore, to formulate any model in a way that explicitly acknowledges the underling cereal science and its connection to the genetics, rheology and end-product quality. Achieving this will help guide the experiments and the collection of the data, as well as meet the much more demanding challenge of formulating a model that truly reflects the matter under consideration and gives insight from and about the relevant information in the associated data.

The question gives the focus, but the data, and the associated modelling, control what can be achieved. As a result, one way to review the various aspects of wheat-flour dough rheology in cereal science is to examine how different classes of data have given different insights into the nature of wheat-flour dough rheology. In order to make such ideas explicit for wheat-flour dough rheology, the next section will give an exemplification of some of the different types of indirect measurement modalities that have been utilized to answer key questions about dough rheology.

The overall goal is the identification of the current knowledge about dough rheology that will allow subsequent deliberations about the accumulation of elastic potential energy to be pursued successfully.

## 7.5 The indirect measurement modalities that directly underpin the rheology of wheat-flour dough formation

The goal of this section is to examine the various key indirect measurement modalities that have yielded crucial information and insights about the molecular and rheological dynamics of the formation of wheat-flour doughs. For each of the modalities examined, the nature of the molecular/rheological information being monitored by the measurements is identified and explained. In particular, the order in which they are discussed below relates directly to the increasing sophistication of the information they monitor. They start with the measurements that identify the relative importance of the water and temperature during the initial formation of a dough, and end with measurements that have a direct connection to the molecular dynamics occurring at the polymer chain level.

### 7.5.1 The walk-in-refrigerator experiments

A number of authors have highlighted[58,59] the fact that, when trying to assess quantitatively the formation of a dough during mixing, the *hydration* and the *energy input* occur simultaneously, and, as a consequence, the relative importance of each is obscured. In order to decouple the chemical hydration of the water with the flour from the mechanical energy input, various forms of walk-in-refrigerator experiments have been performed. In a walk-in-refrigerator (at a temperature of approximately $-8°C$), various mixtures of very small (fine) ice particles and flour are prepared in suitable containers with the ice particles evenly distributed within the flour. The containers are then placed, for a number of hours, in a normal room-temperature environment until a uniform temperature through the mixtures is achieved. The dough thereby formed is the result of the chemical hydration alone of the water with the flour.

The importance of such walk-in-refrigerator experiments relates to their role in assessing the action of water on a wheat flour in the formation of a dough independent of any mixing. An excellent summary of the earlier work can be found in Daniels.[58] Recent investigations[59] have confirmed and placed the earlier results on a quite rigorous footing. In these publications, *undeveloped dough* is defined as wheat flour that has become fully hydrated without being mixed, and *developed dough* as the dough obtained through the mixing of undeveloped dough. It was established, using both creep and oscillatory shear experiments, that, although there are differences between undeveloped and developed dough, the similarities are quite strong; in particular, only a small amount of additional energy was required to transform the undeveloped dough to a developed status similar to a dough obtained using standard mixing. Among other things, it confirmed that when flour is mixed with an appropriate amount of water, the resulting undeveloped dough has rheological properties consistent with the formation of a protein network, and, furthermore, that energy input via some form of elongation-rupture-relaxation process is required for the modification of the protein structure of the undeveloped dough to produce the developed dough.

On the basis of such results, there is general agreement that hydration plays a major initial role in modifying the protein structure in a flour that, along with the assistance of the subsequent mixing, produces a 'normal' dough. As a corollary, normal mixing plays the double role of equally distributing the water in the flour and developing the newly hydrated undeveloped sections within the dough.

Such measurements have clarified the following aspects about the normal developmental rheology of wheat-flour dough formation.

- The interaction of the water with the flour rapidly builds locally connected gluten networks within the dough during its initial formation, independently of the mixing.
- Initially, the major role of the mixing is the equidistribution of the water within the flour, which assists with the formation of the locally connected undeveloped dough networks.
- The subsequent mixing plays the dual role of moving the locally hydratedly connected networks within the dough in an ergodic manner to form larger locally connected networks and to stimulate the opening out of these networks in a manner that builds the final global network that is indicative of a fully developed dough.


### 7.5.2 Temperature measurements

Except for a small number of publications, the importance of the temperature measurements of Li and Walker[60] have essentially been ignored. They measured, using a thermocouple in one of the fixed pins of a Mixograph, the temperature of a dough as it was being mixed. Their seminal measurements are reproduced in Fig. 7.6. As identified by Li and Walker,[60] it shows that, in the sense of changing chemicophysical processes, the temperature has four clear phases as a function of the progressive development and breakdown of the dough being mixed:

- **Phase I.** Initially, the temperature increases linearly and rapidly with a 3°C rise occurring during the first 15–20 revolutions of the mixer. As will be explained below, this linear

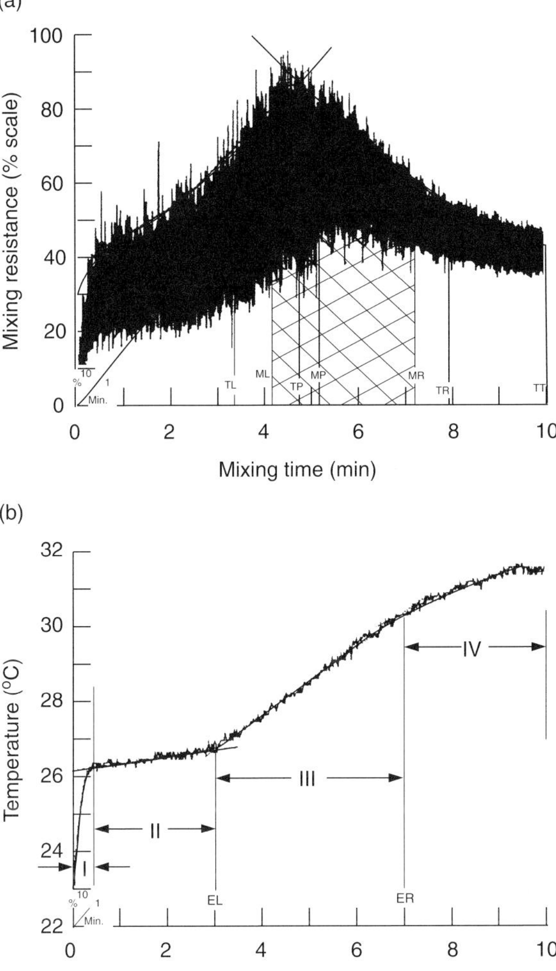



**Fig. 7.6**   The qualitative rheological status of the mixogram of a dough (in (a)) compared with the temperature of that dough (in (b)) recorded in one of the pins of the Mixograph™, with time as the independent variable. From Li & Walker (1992)[60] with permission of the American Association of Cereal Chemists.

increase is indicative of a rapid exothermic reaction occurring between the water and the gluten polymers.

- **Phase II.** The transition from phase I to phase II, though smooth, is quite abrupt, which is clearly indicative of a change in the way that the thermal energy in the dough is being generated. Though the temperature increases during phase II, which lasts from 150 to 200 revolutions, it is very gradual and, consequently, is likely to involve only the dissipation of the thermal energy associated with the mechanical action of the mixing. It is therefore natural to conclude that the abrupt change from phase I to II flags the end of the exothermic reaction of the water with the gluten polymers.

- **Phase III.** The transition of phase II into phase III is quite gradual. During this phase, the temperature has a definite sigmoidal structure, which seems to indicate that, after a more rapid increase than that occurring in phase II, the temperature returns to a behaviour similar to that in phase II. The greater slope that characterizes phase III, compared with that in phase II, is again indicative of a chemicophysical change in the generation of thermal energy within the dough being mixed. As noted by Li and Walker,[60] the slope of phase III decreases as the protein content decreases. This implies that the chemicophysical change associated with phase III has a strong connection to the formation of the gluten network within the dough being mixed.

- **Phase IV.** Consistent with the mentioned sigmoidal structure of phase III, the transition from phase III to IV is also quite gradual, with phase IV essentially corresponding to a counterpart of phase II.

An analysis of the underlying heat flow (diffusion) problems (R.S. Anderssen & C.E. Walker, pers. comm.) indicates that, in order for the thermocouple to record the type of rapid linear temperature increase shown in Fig. 7.6, the temperature in the dough must have increased more or less instantaneously.

Among other things, such results establish that the water must combine exothermically with key components of the wheat flour. Because of the walk-in-refrigerator results about the building of a basic protein network, the circumstantial evidence is that the exothermic reaction is principally between the water and some (unknown) components of the gluten proteins.

Thus, together, the walk-in-refrigerator experiments and the temperature measurements imply that:

- Hydration plays a crucial role in the formation of a dough. In fact, indirect validation can be obtained from a mixogram in terms of how quickly the hydration develops the strength of the dough. In terms of the discussion in Section 7.3, this is the reason for the steep slope of the initial rapid hydration.
- Though the water will also attach to the starch, the circumstantial evidence is that it has its biggest effect on the gluten components as the strength with which the dough resists the mixing on a recording mixer increases rapidly during hydration. Only the opening-out and the cross-linking of the gluten components can achieve this increase in strength.

The importance of such measurements in understanding the molecular dynamics of dough formation might turn out to be far greater than is currently appreciated.

### 7.5.3 Mixograms

For the stress–strain dynamics of dough formation that Mixographs and Farinographs measure and record as, respectively, mixograms and farinograms, there are two important scales – the global and the local. As illustrated in Figs 7.3 and 7.4, the most familiar is the global. As already mentioned and discussed in some detail in Section 7.3, the quantitative structure of the global has been the rheological summary from which the bulk of the current understanding about wheat-flour dough formation, rheology, quality and genetics has been developed, since the invention of recording mixers. Though, in terms of such quantitative measures, the global contains quite important rheological information, such results, because of the smoothing involved to obtain the global from the local dynamics, represent only a qualitative encapsulation of the underlying local molecular dynamics.

On the other hand, as illustrated in Plate 4 and Fig. 7.7, the local dynamics can be recorded as a high-resolution mixogram,[33] which tracks the response of a dough to its elongation by the relative motion of the moving pins away from the fixed pins. In essence, it is such local dynamics that explicitly monitor the molecular interactions of dough formation, which the global summarizes as a low-resolution encapsulation. Furthermore, because, locally, mixograms track the stress–strain dynamics of the elongation-rupture-relaxation events that are an essential part of wheat-flour dough formation, they contain a substantial amount of information about the molecular rheology of wheat-flour dough formation. As a direct consequence, there are various independent ways in which relevant molecular information can be extracted from mixograms.

### 7.5.3.1 Qualitative and quantitative summaries of the global stress–strain dynamics in a mixogram

As already explained in Section 7.3, the popular seven-phase qualitative rheological summary of dough formation (MR, PR, BWPR, RBD, BWBD, RMBW and MBW)[15] has played a key role in developing the current understanding of the relationship between wheat varieties, wheat-flour dough rheology and end-product quality. The reason for this, as discussed

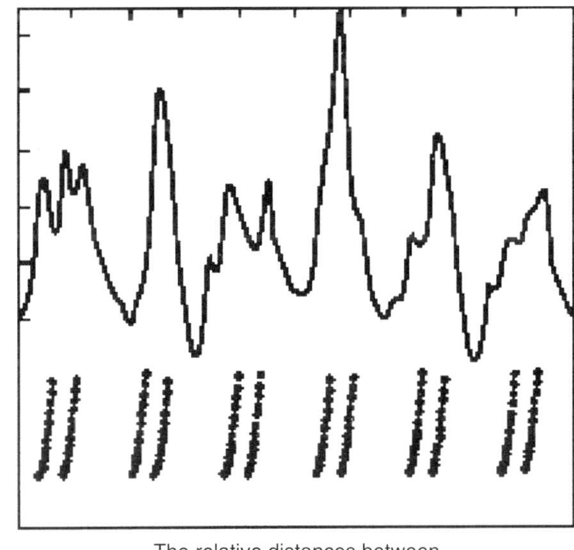


**Fig. 7.7**  A plot of a short section of a high-resolution mixogram for one full revolution of the Mixograph's planetary gear and moving pin system, which involves 12 close encounters between the four moving pins and the three fixed pins. It also illustrates the alignment in the stress with which the dough resists elongation by the moving pins relative to the fixed, and the distances between the moving and fixed pins that are performing the corresponding elongation. From Gras *et al.* (2000)[33] with permission of Elsevier.

in Section 7.1, and in particular in item (vii) of Section 7.1.2, is the fundamental nature of rheology as the link between wheat varieties and end-product quality.

But, as stated, this represents only a qualitative assessment of the rheology and therefore is quite limited in the information it contains other than as a linker. The recovery of key information about the molecular dynamics of dough formation requires a quantitative analysis of the explicit structure in a mixogram. There are various ways in which this can be achieved. A more quantitative five-phase rheological assessment, in terms of rapid hydration, dough development, dough stability, breakdown and asymptotic equilibrium, has been proposed in Section 7.3. On the basis of the discussion above about walk-in-refrigerator experiments and temperature measurements, this five-phase assessment can be given the following molecular dynamic interpretation:

(1)  The strongly elastic-like behaviour during rapid hydration is indicative of the gluten network forming quite rapidly locally within the dough, and is connected with a rapid exothermic reaction, which is implied by the corresponding rapid temperature increase. This is also consistent with the mixing being mainly associated with the distribution of the water in the flour rather than the development of the dough. From a molecular rheological perspective, one is principally seeing the formation of the gluten network with little interaction with the other components within the flour/dough, as this explains the initial narrow width $W(N_r)$.

(2)  The more gradual dough development phase indicates that mixing is required to establish the global network and that this network involves all the components within the dough. The increasing value of $W(N_r)$ during dough development implies that the dough has an increasing local elongational viscosity indicating that the viscosity in the viscoelastic response of the dough is becoming more dominant as more and more components of the original flour are connected into the overall network.

(3)  The existence of the dough stability phase confirms that, once the global network has formed, it is a quite stable molecular structure involving all the components of the original flour.

(4)  The number of revolutions over which the dough stability phase holds and the character of the breakdown contains information about the nature and internal strength of the cross-linking of the gluten network within the dough.

(5)  The asymptotic equilibrium contains additional information about the cross-linking of the degraded gluten network within the dough, which indirectly relates back to the original composition of the flour, the amount of added water and other components and how it is mixed.

### 7.5.3.2 The hysteretic nature of the local structure in a mixogram

An alternative approach to the recovery of quantitative rheological information from a mixogram can be based on the work of Gras *et al.*,[15] where it has been established that the local structure in a mixogram is, in essence, a series of *in situ* planar elongational extension events. Consequently, any rheological analysis of such events will give independent insight about changing molecular dynamics during dough formation. In their assessment,[33] Gras *et al.* explained in some detail why the rheological flow of the mixing action on a Mixograph is predominantly elongational. Even though the elongation events occur in a stochastic manner, the simple and consistent nature of this flow, as imposed by the action of a Mixograph on a dough, has an essentially ergodic structure that can be averaged. The reasons given[15] can be summarized as follows.

Flow imposed on the dough by the motion of moving pins relative to the fixed Plate 4 and Fig. 7.7 illustrate the type of response recorded in a high-resolution mixogram. In addition, the motion of the moving pins away from the fixed pins has been matched with the changing force with which the dough resists its elongation. When a moving pin is close to and moving away from a fixed pin, the response of a dough, as recorded in a high-resolution mixogram, exhibits increasing resistance (Fig. 7.8) similar to that recorded in an extension test.

Visual inspection

A careful visual inspection of a dough, during mixing inside a 35-gram Mixograph, allows one to see the deformation process being performed on the dough, and how, as a result, the dough responds.

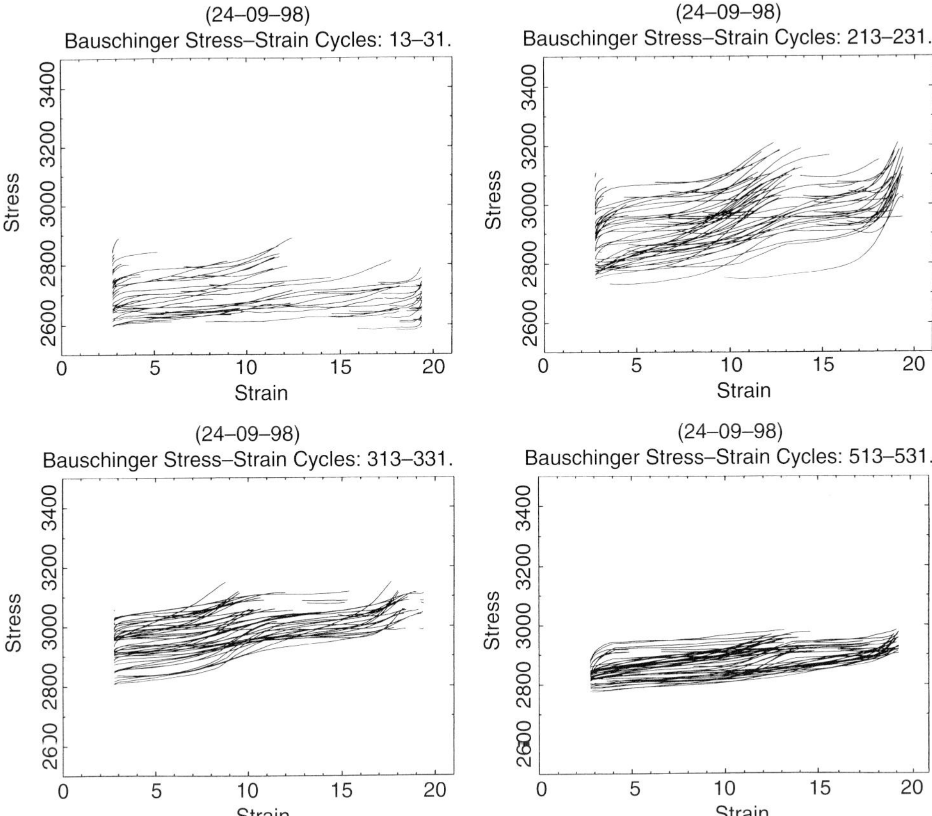

**Fig. 7.8**   The upward branches of the mixing hystereses, for different stages in the development and breakdown of a dough, as determined from the corresponding high-resolution mixogram, through the matching of the stress with the strain as illustrated in Fig. 7.7. From Anderssen & Gras (2000)[61] with permission.

- Because the elasticity in a viscoelastic material follows the streamlines of its rheological flow, the dough forms, after the hydration phase, a thick rubber-like viscoelastic sheet about the fixed pins with little contact with the bottom of the mixing bowl. A good example is shown in Plate 5.
- As the epitrochoidal path of the moving pins of Fig. 7.9 shows, much of the motion of a moving pin is outside the thick sheet of dough stretched (tightly) around the fixed pins.
- A moving pin only passes into a dough when it slows down to come into a close encounter with a fixed pin. This is clear from Fig. 7.9. At this stage, there will be shearing of the dough as a moving pin approaches a fixed pin, as well as minor compression between these two pins.
- As a moving pin passes through its epitrochoidal loop adjacent to a fixed pin, it attaches to dough, which it then elongates as it moves away from the fixed pin. Further dough is collected and elongated as it passes back through the thick rubber-like sheet around the fixed pins.
- During the dough development phase, long, continuous sinew-like structures can be observed in the dough.

Minimal shearing

Though some authors describe the rheological flow in a Mixograph in terms of a shearing action, this, as the above deliberations illustrate, is incorrect. Clearly, a small amount of shearing occurs as a moving pin, travelling towards and then away from a close encounter with a fixed pin, passes through the dough. However, this is a minor consideration, as a

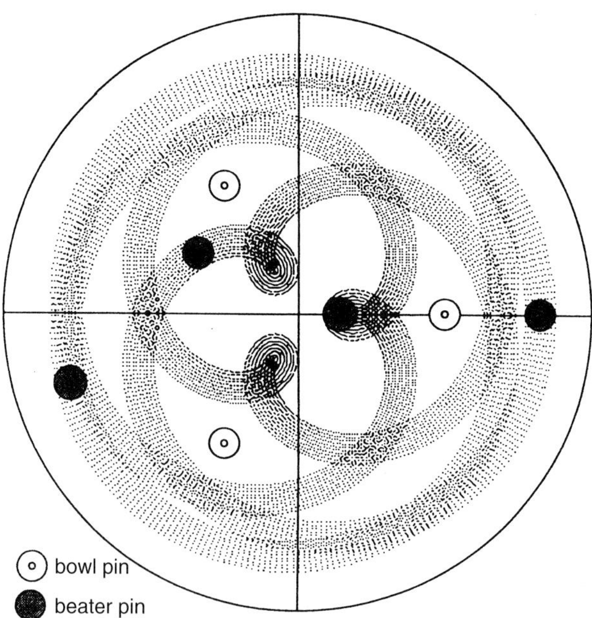


**Fig. 7.9**   A graph of the epitrochoidal path that the moving pins follow, relative to the fixed pins. The moving pins follow the same path but at different phases with respect to each other. From Buchholtz (1990)[47] with permission of the Applied Probability Trust.

moving pin is either slowing down or speeding up as it passes through a close encounter. Furthermore, the above considerations tend to imply that, though shearing does no harm, shearing alone is unable to produce the same structure in a dough as that produced by a succession of elongations. Clearly, in some industrial mixers, the elongations are replaced by compressive events, which are the opposite of elongations and do not correspond to shearing actions.

Independent evidence from other sources
As well as the papers cited above in this section, there is a considerable independent rheological literature that supports the above conclusions.[23,30,36] The mentioned individual stochastic extension events are the upward branches of the mixing hystereses. As explained above, they can be recovered from high-resolution mixograms (Plate 4 and Fig. 7.7).[33,61] Alternatively, they can be assessed indirectly through appropriate extension tests, as discussed below. This is where alternative key molecular information is contained and the reason why the rheological analysis and molecular interpretation of uniaxial and biaxial extensograms are a key to understanding the molecular dynamics of dough formation. The structure of the upward branches of the mixing hystereses are the link between the information recorded on a Mixograph (or recording mixer) and the explicit molecular dynamics that they are recording.

### 7.5.3.3 A hysteretic summary of the global structure in a mixogram

An analysis of the quantitative rheology of a mixogram along the lines outlined in 1–5 of Subsection 7.5.3.1 can in part be summarized by the upper and lower curves of a mixogram. The difference between them thus measures the changing width $W(N_r)$. Though software has been developed to recover such curves, little has been done to date to exploit the molecular rheology information that they contain in the manner just outlined in Section 7.5.3.2. This matter is examined further in Section 7.6, below.

## 7.5.4 Uniaxial and biaxial extensions

It follows from the discussion in Section 7.5.3 that, for the recovery of explicit molecular information about the molecular dynamics occurring during dough formation, it is first necessary to understand how to recover such information from extension tests. Consequently, extension tests, as possible models for the upward branches of the mixing hystereses, become the default link between the hysteretic information recorded on a Mixograph and the molecular dynamics that the hystereses record.

Now, however, a key issue becomes the choice of the extension test. Not only is the information in a biaxial test different from that in a uniaxial test, it is also different between different types of uniaxial tests. In many laboratories, the choice of extension test is made on the basis of the available equipment rather than on the relevance of the matters under examination. As illustrated in Fig. 7.10, the standard extension test,[62] with thick samples, produces a force–displacement plot that is clearly different from that obtained using a micro-extension test,[63] with thin samples (Fig. 7.5), as well as other extension testers, such as the Instron and the TA-XT2i Texture Analyser (Stable Micro Systems, URL

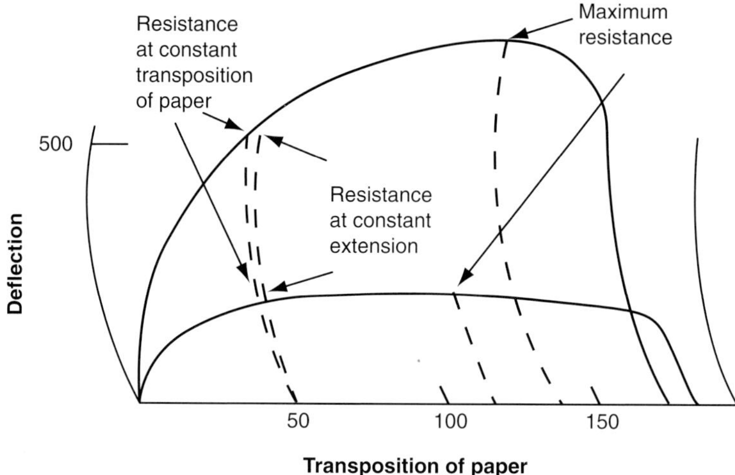


**Fig. 7.10**  The different responses of the same dough when subjected to extension testing on different instruments, depending on the size of the dough sample tested (compare the structure of the force–extension curve in this figure with that in Fig. 7.5). From Rasper & Preston (1991)[62] with permission of the American Association of Cereal Chemists.

www.stablemicrosystems.com). It appears that the form of the force–displacement plot depends on the diameter of the sample being tested, as well as the manner in which it is being extended, with the thicker sample giving much less differentiation that the thinner sample. In addition, the structure of the force–displacement plot from a biaxial extension test (Alveograph) appears to have less differentiation than the plot from a micro-extension test. However, the rheological significance of such observations does not appear to have been pursued in any detail in the literature.

From the plots (Fig. 7.8) of the upward branches of the mixing hystereses,[61] it is clear that they are similar in structure to the force–displacement plots of Fig. 7.5 obtained from a micro-extension tester.[16] Consequently, because of the appropriateness of uniaxial extension testing, as outlined above, attention will focus on a qualitative molecular interpretation of the structure of a micro-extension test.

Figure 7.5 corresponds to Figure 1 in Anderssen *et al.*[16] It plots the force–displacement responses of eight different flours to micro-extension testing. It shows unambiguously that micro-extension testing reveals clear differences between the force–displacement responses of strong, intermediate (medium strength) and weak wheat flours. The micro-extensograms from the intermediate and weak flours show a clear double-hump structure, the phases of which can be given the interpretations of Fig. 7.11. Though the micro-extensograms for the strong flours show a single hump, it could be interpreted, on continuity grounds, as a hidden double-hump. Although the glutenin subunit composition changes discontinuously from weak to intermediate to strong flours, the internal molecular structure within the corresponding doughs will have a not too dissimilar molecular network structure because the relationships among the gliadins, glutenins and the disulfide and hydrogen bonding change in only minor rather than major ways. An analysis of these differences on the basis of glutenin subunit composition is given in Anderssen *et al.*[16] Among other things, they note that:

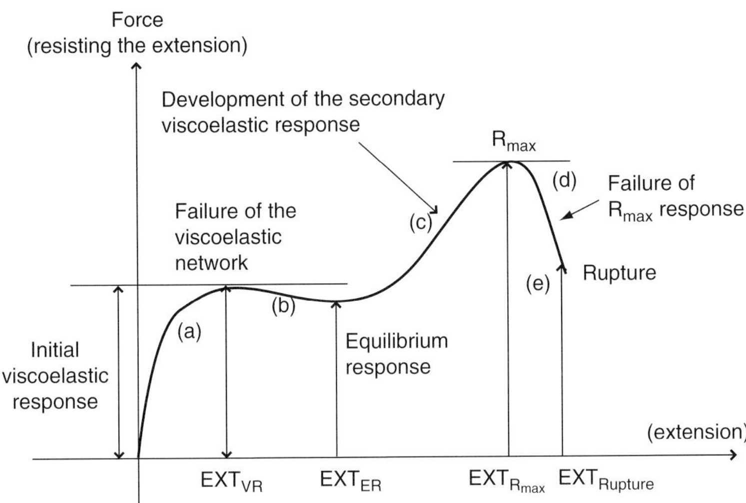


**Fig. 7.11**  Some of the different rheological phases that can be identified in the stress–strain plot obtained from a micro-extension test. From Anderssen *et al.* (2004)[16] with permission of Elsevier.

- It is the (*b,b,d*) alleles that determine the dominant single-hump structure for the strong flours, and this highlights the importance of the *b* allele in Glu-B1 and the *d* allele in Glu-D1 in determining dough strength.
- The mechanisms associated with the *b* allele in Glu-B1 and the *d* allele in Glu-D1 are different; in the former, it relates to the significantly greater expression level of the x-type polypeptide in the allele, while, in the latter, it relates to the extra cysteine residue in the x-type polypeptide.
- Though the $R_{max}$ and $Ext_{Rmax}$ (extension only to $R_{max}$ and not to 'rupture') values vary greatly from strong to intermediate to weak flours, the areas under their force–displacement plots are quite similar indicating that about the same amount of energy is required to extend a dough to $R_{max}$ at least for these eight representative flours.

With respect to the relevance and importance of modelling the double-hump structure in terms of the glutenin subunit composition of the eight flours, Anderssen *et al.*,[16] on p. 199, comment:

All of these differences in graphical features represent clear indicative measures for interpreting the allelic differences of the eight flours in terms of the molecular dynamics occurring during the extension testing. In particular, they are in agreement with the double-network model proposed by Anderssen and Hoffmann (2003) [ref. 64] for the upward branch of the extensional hysteresis (Fig. 3) [Fig. 7.8 in this Chapter] in the elongation of a dough during mixing. A key feature of that model is the recognition that $R_{max}$ occurs when the earlier phases of the extension have maximized the alignment of the macromolecules in the dough. Consequently, the alignment occurs more or less instantaneously in the stronger doughs and more rapidly in the medium-strength doughs than the weak. In terms of this double-network model, it implies that the *b* of Glu-B1 and the *d* of Glu-D1 are associated with a smaller level of entropic contraction than that associated with the corresponding *c* of Glu-B1 and *a* of Glu-D1. This leads to the conclusion that the interrelationship between the two responses of a dough to its elongation is a sensitive measure of the changing rheology induced by the changing chemistry, and, thereby, of the glutenin composition of the dough, along with the role of the other components such as starch, pentosans, gliadins, albumins, etc. This double-network model also represents one way of placing on a more formal footing the concept of a macropolymer introduced by Wang *et al.* [ref. 45], as well as the loop-and-train model of Belton [ref. 44].

**Note:** As already mentioned in item (xi) of Section 7.1.2, strain hardening has been used by various authors to assess and compare extension tests, but it is limited because it is unable to track the developmental rheology of dough formation.

Uniaxial extension testing can also be seen as a model system for the biaxial behaviour of sheets around the bubbles in wheat-flour dough, which relates back to the strength of the bubbles to hold the fermentation gases, given that the diffusivity of such gases through the walls of the bubbles is virtually zero on the timescales of dough mixing, proofing and baking. However, the utility of this approach over the direct measurement of the biaxial extension on an Alveograph needs further investigation.

### 7.5.5 *The modalities that indirectly underpin the rheology*

The other modalities, such as SDS-PAGE, RP-HPLC and glutenin subunit composition, represent the glue that allows a molecular dynamical interpretation to be connected to the quantitative rheological measurements discussed above. However, before one can make significant progress, there is a need to formulate appropriate mathematical models.

## 7.6 Modelling the viscoelasticity of wheat-flour dough formation

Dobraszczyck and Morgenstern[41] explain and illustrate the seminal point that, when the 'property' ('state') of a material is independent of how it is deformed, the goal must be to work with the stress–strain counterpart of any force–displacement (time!) measurements of an applied deformation. The results will be independent of the size and shape of the specimen being tested and how it is deformed. Among other things, this will imply that the underlying constitutive relationship will be linear and allows one to consistently and rigorously compare one material with another. Even when the effects of applied deformations on the material must be taken into account, working with the stress–strain counterpart of any measurements will ensure that any dependence on size and shape will have been removed.

The goal behind formulating representative mathematical models is to have, when the opportunity arises, appropriate frameworks and strategies (consistent with the matters (questions) under examination) for placing on a rigorous and quantitative footing the pragmatic rationale of matching rheological measurements to the corresponding molecular biorheology occurring in the wheat-flour dough being deformed. The utility of such models is that they will complement and supplement the science and technology of the measurements performed in both molecular and processing biorheology applications.

From a molecular perspective, the stress–strain dynamics in a mixogram must be examined at a much finer scale than the current and traditional scales as already discussed in considerable detail in Sections 7.3 and 7.5. Though the phases discussed and explained there have played a fundamental role in placing cereal science and technology on the rigorous footing that it enjoys today (as outlined in Section 7.1) their scale (in terms of the information that they see) is macroscopic. Such phases only monitor an averaging (macroscopic encapsulation) of the molecular dynamics occurring during dough formation. As identified in Anderssen *et al.*[46] (and discussed in some detail in Gras *et al.*[33]) and Anderssen and Gras,[61] the fine scale corresponds to the stress–strain dynamics of the elongation-rupture-relaxation events occurring during the mixing on a Mixograph. An illustration of such fine-scale structure is given in Plate 4 and Fig. 7.7. These clearly illustrate the stochastic nature in which the dough resists, on a Mixograph, its elongation by the moving pins relative to the fixed pins.

In the formation of a dough during mixing, the stress–strain structure evolves with each elongation-rupture-relaxation event. The underlying dynamics is clearly hysteretic, because the stress–strain dynamics of an elongation is not reversible in the subsequent relaxation after the rupture. As illustrated in Fig. 7.8, the upward branches of the stress–strain pattern of an elongation, where the stress with which the dough resists its elongation is increasing, are quite different from what would be expected in the subsequent relaxation of the dough in the downward branch, where the stress with which the dough could resist elongation is decreasing. In addition, the individual stress–strain hystereses, consisting of an upward and its corresponding downward branch, will not form closed loops, as occurs in magnetism.[65] If the loops were closed, this would imply that the stress in the dough, with which it can resist elongation, does not change. The hysteretic structure is open-loop because of the accumulation and subsequent breakdown of a dough's elastic potential energy, which, with respect to a fixed strain (such as the strain at the commencement of each elongation), must correspond, respectively, to an increasing and subsequent decreasing stress.

From a mathematical modelling perspective, the disadvantage of an open-loop structure, compared with the closed-loop,[65] is the lack of a generic mathematical framework in which to perform the modelling. In the closed-loop context, one has the Preisach model, which holds for any closed-loop structure independent of the context in which it has arisen. For open-loop situations, the modelling must be performed on the basis of the context, as only then can the nature of the failure of the loops to close be identified chemicophysically and, accordingly, modelled mathematically.

In summary, from a wheat-flour dough rheology perspective, the importance of the open-loop structure is that, with respect to a fixed strain, it models how the stress, with which the dough resists its elongation, changes. The way in which the open-loop structure evolves is therefore a characterization of the nature of the accumulation and breakdown of the elastic potential energy, which relates directly to the molecular structure of the flour that made the dough and how the various molecular structures interacted with water and other ingredients during the mixing.

Stress–strain patterns are modelled mathematically using constitutive relationships. The underlying science of the chemistry, physics and measurements, which are utilized in the formulation, analysis and interpretation of such models, is called 'rheology'. Because of the huge differences between materials, there is a very wide range of models from which to choose. Consequently, the choice must be made on the basis of the material being examined. For wheat-flour doughs, the appropriate framework is viscoelasticity.

Because of the open-loop hysteretic structure of the stress–strain dynamics, any model for the constitutive relationship must have a different form for the upward and downward branches. In addition, it must, in some appropriate manner, allow for the changing nature of the cross-linking in the polymers as the mixing progresses. Consequently, the associated modelling and mathematics is quite complex and beyond a full discussion in this chapter.

What follows is a brief heuristic summary of the basic concepts, logic and equations that are involved in the formulation of such a set of constitutive relationships. As a first approximation, it will be assumed that the individual upward and downward branches of the hystereses can be modelled using the Boltzmann model of linear viscoelasticity:

$$\sigma(t) = K(t)\gamma(t) + \int_{t_0}^{t} G(t-\tau)\dot{\gamma}(\tau)d\tau, \qquad t_0 \leq t \leq t^*, \tag{7.1}$$

which has the current 'state' of a viscoelastic material encapsulated in the **elasticity** $K(t)$ and the **relaxation modulus** $G(t)$, which are independent of the stress $\sigma(t)$ and the strain $\gamma(t)$ characterizing the nature of the deformation. The limits of integration, $t_0$ and $t^*$, define the time interval over which an individual upward or downward branch is active. The hysteretic nature of the mixing is accommodated by having $K(t)$, $G(t)$ and $\gamma(t)$ change their structure from one branch to the next. Even though the overall model is a linear combination of linear equations for the upward and downward branches of the open-loop hystereses, the resulting constitutive relationship is nonlinear, because the resulting model of the measured stress must satisfy additional conditions that define how the successive branches must be coupled.

As it stands, the measurements of the stress on the successive open-interval of time $(t_0, t^*)$ only determine how the modelling of the relaxation moduli $G(t)$ should change from one branch to the next. Because they do not determine how the unknown $K(t)$ should be modelled, the formulation, as it stands, is under-determined. Consequently, there is a need to define additional conditions to make the recovery of information about $K(t)$ and $G(t)$ fully determined. For a mixogram, information about the changing nature of $K(t)$ is contained in the changing values $(\sigma^*)$ of $\sigma(t)$ that occur at the end of each downward branch, which is the start of the next upward branch. This is where the information in the open-loop structure of the mixing hystereses plays a key role. At the start $t_0$ of an upward branch, $\gamma(t_0)$ is zero, as no elongation is being performed on the dough. A small time $(\Delta t)$ later, the stress (measured by a Mixograph) relative to the residual elastic stress in the dough of $\sigma(t_0)$, can be modelled using Equation 7.1 as:

$$\sigma(t_0 + \Delta t) - \sigma(t_0) = K(t_0)\gamma(t_0 + \Delta t) + \int_{t_0}^{t_0 + \Delta t} G(t-\tau)\dot{\gamma}(\tau)d\tau, \tag{7.2}$$

In situations, like on a Mixograph, where the value of $\gamma$ is initially quite small and the value of the integral in the above formula is essentially zero, the last equation yields the following approximation for $K(t_0)$:

$$K^{**} = (\sigma(t^{**} + \Delta t) - \sigma(t^{**}))/(\gamma(t^{**} + \Delta t)). \qquad (7.3)$$

In terms of the discussion in Sections 7.3 and 7.5.3, the points on a Mixograph's stress–strain plot corresponding to $t_0$ lie on the lower curve of such a plot. Among other things, this establishes the importance of this lower curve and indicates that it is an indirect measure of the changing elastic potential energy during the formation of a dough.

Consequently, for a given extension $x_0$ and cross-sectional area $A(t)$, the elastic potential energy stored within the dough, at time $t$, will be:

$$\frac{1}{2}K(t)A(t)x^2(t). \qquad (7.4)$$

At this point in time, there is no clear understanding about how this energy is stored. However, it is not unreasonable to assume that the 'glutenin' molecules have formed a network of parallel springs. If it is assumed that, at time $t$, this is the result of having $N(t)$ identical springs in parallel with known Young's modulus $g\kappa_0$, then one can estimate $N(t)$ as:

$$N(t) \sim \frac{K(t)A(t)}{\kappa_0}. \qquad (7.5)$$

For the utilization of this formula, not only is it necessary to determine $K(t)$, as outlined above, and independently derive an estimate of $A(t)$, but also to invoke an appropriate assumption about the value of $k_0$.

In a nonlinear form of the Boltzmann model, the $K(t)$ and/or $G(t)$ will have a dependence on $\sigma(t)$ and/or $\gamma(t)$ indicating that, as a result of the deformation, there has been a change in the state of the material. This will take two forms:

- A recoverable (elastic-like) state, when the material returns to its original configuration, after the deformation has returned to its original state.
- A nonrecoverable state, when the material changes, as a result of the deformation, to a new state that becomes permanent independent of any subsequent deformations.


## 7.7 Some future challenges

The bakers of antiquity had an intuitive concept of the viscoelasticity of the various doughs that they kneaded by hand. Experience and regular familiarity allowed them to assess whether a dough was appropriate for the subsequent baking to be performed. There was a comprehensive, subconscious rather than conscious, understanding that the feel and colour of the grain, the texture of the flour milled from the grain and the amount of water added to the resulting flour related to the relationship between the viscoelastic feel and end-product quality of the kneaded dough.

As a natural consequence, 'extensibility' and 'dough strength' became the first indirect assessment of the viscoelasticity of a kneaded dough and its potential end-product quality. As milling technology of wheat became reasonably consistent and predictable, grain hardness, especially in terms of the texture of the flour that various wheats produced, took on the role of a higher-level indirect assessment of the 'extensibility' and 'dough strength'.

Even though, as the discussion in this chapter has aimed to show, we now have a good understanding of the pathway from plant breeding to end-product quality and the fundamental role that rheology plays in connecting plant breeding with end-product quality, there are still major challenges to be resolved before the intuitive understanding of the farmers, millers and bakers of antiquity can be placed on a fully rigorous molecular and process biorheology footing. The challenges include:

- The relative importance of the protein–water and starch–water exothermic reactions in the formation of a dough. It is clear that, because a gluten network is formed in undeveloped doughs (in walk-in-refrigerator experiments), a protein–water interaction is involved. However, the relative importance of the associated starch–water interaction has yet to be assessed.
- The modelling of the exothermic protein–water interaction in the formation of the gluten network within an undeveloped dough.
- The recovery of molecular information from upward/downward hysteresis branches and uniaxial extension tests.
- Dimension analysis of the key measurements in wheat-flour dough rheology.
- Relationship between various extension measurements, based on the differences in their rheological flows. In designing a wheat-flour dough rheometer, the rheological flow should be as simple as possible, as this reduces the complexity of the associated mathematical modelling.
- Recovery of information from the upper and lower points of the upward/downward branches of the mixing hystereses, and their relationship to the rheological phases introduced in Section 7.3.
- What proteins, in particular, are opened out by the water in the initial hydration.
- Discussion of how elastic potential energy might be accumulated molecularly.

## Acknowledgements

In order to give an independent modelling perspective about the rheology of wheat-flour dough, this chapter has been written independently of any direct input from colleagues. Nevertheless, it could not have been written without the indirect input and support received over the last decade from various colleagues and friends. My involvement with wheat-flour dough only occurred because of an initiative and financial support from within CSIRO, and, in particular, the Division of Mathematics and Statistics (now CSIRO Mathematical and Information Sciences), to explore the role of mathematical modelling in food and related sciences, and to identify where it might make a nontrivial contribution to the improvement of current technologies. For CSIRO, the motivation was clear – increase Australia's export income by value-adding to wheat and cereal grains. The initial and subsequent professional support and trust received from Drs Rudi Appels, Murray Cameron, Bob Frater, Matthew Morell, Jim Peacock and Ron Sandland is acknowledged with considerable thanks. My subsequent understanding benefited, first and foremost, from the unselfish advice and support of Dr. Peter Gras, as well as from Drs Fin Macritchie and Chuck Walker. Other colleagues who have assisted, in one way or another, include Drs Frank Bekes, Barbara Butow, Reka Haraszi and Sadiq Rahman. Independently, my collaboration with colleagues at BRI Australia Limited, and, in particular, Drs Brian Osborne and Ian Wesley, has contributed to my current understanding of the subject.

# Appendix 1: A brief literature summary

As mentioned in Section 7.2, in order to avoid unnecessary repetition and duplication, this chapter does not review earlier material where there are excellent recent publications available. In particular, the following papers represent a brief encapsulation of the current status of cereal science as it relates to wheat-flour dough rheology.

(1)  Shewry *et al.*[13] discuss the basic science of HMW subunits and related matters in considerable detail. In terms of the discussion in Section 7.1.2, this article relates quite specifically to issues connected with items (i), (ii), (iii), (v) and (vii).

(2)  Dobraszczyck and Morgenstern,[41] where the relationship between rheology and the bread-making process is examined in detail. The emphasis is strongly 'process biorheology' in terms of matching the rheological measurements with the industrial processing being applied to a wheat-flour dough in the bread-making process. Among other things, it presents a good historical survey as well as a comprehensive view of recent wheat-flour dough technology from a process biorheology perspective.

(3)  The 'Wheat Gluten Book'[66] contains the papers presented at the Gluten Workshop in 2000. They cover a broad spectrum of issues associated with wheat-flour dough rheology and give an excellent comprehensive background to the cereal science of wheat as it relates to wheat-flour dough. The breadth of the matters treated represents validation for the crucial importance of wheat-flour dough rheology to the cereal science and technology of wheat – the point of view discussed and analysed in various ways in Section 7.1 of this chapter. Specific papers of direct relevance to the deliberations of this chapter include the papers listed under the heading 'Viscoelasticity, Rheology and Milling'.

(4)  The subject has a long and distinguished history which dates back to various key events, such as the discovery of gluten more than 250 years ago, and key personalities, such as William Farrer. Web of Science, Google and library searches will quickly find the relevant sources.

(5)  Cereal science is fundamentally chemistry. The importance of genetics and rheology is crucial, but their application and interpretation only have meaning through the chemistry. From the chemistry perspective, key concepts include disulfide bonding,[67,68] redox reactions[69,70] and the chemical properties of the gliadins and glutenins.[71]

(6)  Because of the molecular biology revolution, genetics has become a sophisticated computational science in terms of QTL analysis and related investigations. In many ways, the most appropriate source, from an applications perspective, is the special issue of the *Australian Journal of Agriculture Science* on mapping, which has appeared as a book[72] edited by Appels *et al.*

(7)  The concluding sections of this chapter are reasonably mathematical when they turn to a consideration of hysteresis and viscoelasticity. Background about such concepts can be found in Larson,[73] Mayergoyz,[65] Anderssen *et al.*,[74] Brokate and Sprekels,[75] and Anderssen and Loy.[20]

**Caveat:** Because of the history and the importance of the subject, as partially outlined in Section 7.1, it is not possible to specifically cite all relevant sources, even the most recent ones. Some attempt has been made to cite sources that have good reference lists so that the reader can turn to them as a window on the subject. The availability of search engines, such as Web of Science and Google, can assist greatly with this task when there is a specific matter to be investigated.

## Appendix 2: Symbols and abbreviations

| | |
|---|---|
| DNA | Deoxyribonucleic acid |
| $Ext_{rupture}$ | Extension of a dough to rupture in a uniaxial extension test |
| $G'$ | Storage modulus |
| $G''$ | Loss modulus |
| Gli-B1 | The allele for the gliadin protein on the B genome of wheat |
| Gli-D1 | The allele for the gliadin protein on the D genome of wheat |
| Glu-B1 | The allele for the glutenin protein on the B genome of wheat |
| Glu-D1 | The allele for the glutenin protein on the D genome of wheat |
| HPLC | High-performance liquid chromatography |
| MWD | Molecular weight distribution |
| NIR | Near-infrared spectroscopy |
| PAGE | Polyacrylamide gel electrophoresis |
| PDD | Peak dough development |
| QTL | Quantitative trait loci |
| $R_{max}$ | Maximum resistance of a dough to extension in a uniaxial extension test |
| RP-HPLC | Reverse phase HPLC |
| SDS-PAGE | Sodium dodecyl sulfate PAGE |
| SE-HPLC | Size exclusion HPLC |
| SEM | Scanning electron microscopy |
| SKCS | Single kernel characterization system |
| TA-XT2i | A texture analyser device manufactured by Stable Micro Systems (www.stablemicrosystems.com) |
| TEM | Transmission electron microscopy |
| TPP | Total polymeric protein |
| UPP | Unextractable polymeric protein |

# 7.8 References


1 Naeem, H.A., Darvey, N.L., Gras, P.W. & MacRitchie, F. (2002) Mixing properties, baking potential, and functional changes in storage proteins during dough development of triticale-wheat flour blends. *Cereal Chem.* **79**, 332–339.

2 Duke, T.A.J. (1989) Tube model of field-inversion electrophoresis. *Phys. Rev. Letters* **62**, 2877–2880.

3 Deutsch, J.M. (1988) Theoretical studies of DNA during gel electrophoresis. *Science* **240**, 922–924.

4 Brenton, G., Danyluk, J., Charron, J.B.F. & Sarhan, F. (2003) Expression profiling and bioinformatic analyses of a novel stress-regulated multispanning transmembrane family from cereals and Arabidopsis. *Plant Physiol.* **132**, 64–74.

5 Ogihara, Y., Mochinda, K., Nemoto, Y. *et al.* (2003) Correlated clustering and virtual display of gene expression patterns in the wheat life cycle by large-scale statistical analyses of expressed sequence tags. *Plant J.* **33**, 1001–1011.

6 Schuster, P. (1997) Genotypes with phenotypes: Adventures in an RNA toy world. *Biophys. Chem.* **66**, 75–110.

7 Ancel, L.W. & Fontana, W. (2000) Plasticity, evolvability and modularity in RNA. *J. Exp. Zool.* **288**, 242–283.

8 Fontana, W. (2002) Modelling "evo-devo" with RNA. *Bioessays* **24**, 1164–1177.

9 Parry, M.A.J. & Shewry, P.R. (2003) Genotype-phenotype: narrowing the gap. *Ann. Appl. Biol.* **142**, I–II.

10 Gupta, R.B., Bekes, F. & Wrigley, C.W. (1991) Prediction of physical dough properties from glutenin subunit composition in bread wheats – Correlation studies. *Cereal Chem.* **68**, 328–333.

11 Larroque, O., Gianibelli, M.C. & Macritchie, F. (1999) Protein composition for pairs of wheat lines with contrasting dough extensibility. *J. Cereal Sci.* **29**, 27–31.

12 Rhazi, L., Cazalis, R. & Aussenac, T. (2003) Sulfhydryl-disulfide changes in storage proteins of developing wheat grain: influence on the SDS-unextractable glutenin polymer formation. *J. Cereal Sci.* **38**, 3–13.

13 Shewry, P.R., Halford, N.G., Tatham, A.S. *et al.* (2003) The high molecular weight subunits of wheat glutenin and their role in determining wheat processing properties. *Adv. Food Nutr. Res.* **45**, 219–302.

14 Bekes, F., Gras, P.W., Anderssen, R.S. & Appels, R. (2001) Quality traits of wheat determined by small-scale dough testing methods. *Aust. J. Agric. Res.* **52**, 1325–1338.

15 Gras, P.W., Anderssen, R.S., Keentok, M. *et al.* (2001) Gluten protein functionality in wheat flour processing: a review. *Aust. J. Agric. Res.* **52**, 1311–1323.

16 Anderssen, R.S., Bekes, F., Gras, P.W. *et al.* (2004) Wheat-flour dough extensibility as a discriminator for wheat varieties. *J. Cereal Sci.* **39**, 195–203.

17 Anderssen, R.S., Mead, D.W. & Driscoll IV, J.J. (1997) On the recovery of molecular weight functionals from the double reptation model. *J. Non-Newtonian Fluid Mech.* **68**, 291–301.

18 Anderssen, R.S. & Mead, D.W. (1998) Theoretical derivation of molecular weight scaling for rheological parameters. *J. Non-Newtonian Fluid Mech.* **76**, 299–306.

19 Thimm, W., Friedrich, C,. Marth, M, & Honerkamp, J. (2000) On the Rouse spectrum and the determination of the molecular weight distribution. *J. Rheol.* **44**, 429–438.

20 Anderssen, R.S. & Loy, R.J. (2002) Rheological implications of completely monotone fading memory. *J. Rheol.* **46**, 1459–1472.

21 Rhazi, L., Cazalis, R., Lemelin, T. & Aussenac, T. (2003) Changes in the glutathoine thiol-disulfide status during storage wheat grain development. *Plant Physiol. Biochem.* **41**, 895–902.

22 Gupta, R.B., Khan, K. & Macritchie, F. (1993) Biochemical basis for flour properties in bread wheats. I. Effects in variation in the quality and size distribution of polymeric proteins. *J. Cereal Sci.* **18**, 23–41.

23 Simmonds, D.H. (1989) *Wheat and Wheat Quality in Australia.* CSIRO Australia, Willian Brooks, Queensland.

24 Amend, T. & Belitz, H-D. (1991) Microstructural studies of gluten and a hypothesis on dough formation. *Food Structure* **10**, 277–288.

25 Pagani, M.A., Resmini, P. & Dalbon, G. (1989) Influence of the extrusion process on characteristics and structure of pasta. *Food Structure* **8**, 173–182.

26 Moss, R. (1985) The application of light and scanning electron microscopy during flour milling and wheat processing. *Food Structure* **4**, 135–141.

27 McNabb, A. & Anderssen, R.S. (2006) Pasta drying. In: Hui Y.H. (ed.) *Food Drying*

28 Zghal, M.C., Scanlon, M.G. & Sapirstein, H.D. (1999) Prediction of bread crumb density by digital image analysis. *Cereal Chem.* **76**, 734–742.

29 Phan-Thien, N., Newberry, M. & Tanner, R.I. (2000) Non-linear oscillatory flow of a soft-like viscoelastic material. *J. Non-Newtonian Fluid Mech.* **92**, 67–80.

30 Meissner, J. & Hostettler, J. (1994) A new elongational rheometer for polymer melts and other highly viscoelastic liquids. *Rheol. Acta* **33**, 1–21.

31 Charalambides, M.N., Wanigasooriya, L., Williams, J.G. & Chakrabarti, S. (2002) Biaxial deformation of dough using the bubble inflation technique. I. Experimental. *Rheol. Acta* **41**, 532–540.

32 Newberry, M.P., Phan-Thien, N., Larroque, O.R. *et al.* (2002) Dynamic and elongational rheology of yeasted bread doughs. *Cereal Chem.* **79**, 874–879.

33 Gras, P.W., Carpenter, H.C. & Anderssen, R.S. (2000) Modelling the developmental rheology of wheat-flour dough using extension tests. *J. Cereal Sci.* **31**, 1–13.

34 Voisey, P.W. & Kloek, M. (1980) Note on methods of recording dough development curves from electronic recording mixers. *Cereal Chem.* **57**, 442–444.

35 Anderssen, B. & Monypenny, R. (1993) Link concepts and model partitioning in model formulation. *Math. Comput. Modelling* **17**, 105–113.

36 Butow, B.J., Gras, P.W., Haraszi, R. & Bekes, F. (2002) Effects of different salts on mixing and extension parameters on a diverse group of wheat cultivars using 2-g mixograph and extensigraph methods. *Cereal Chem.* **79**, 826–833.

37 Sliwinski, E.L., Kolster, P. & van Vliet, T. (2004) On the relationship between large-deformation properties of wheat flour dough and baking quality. *J. Cereal Sci.* **39**, 231–245.

38 Sliwinski, E.L., Kolster, P., Prins, A. & van Vliet, T. (2004) On the relationship between gluten protein composition of wheat flour and large-deformation properties of their dough. *J. Cereal Sci.* **39**, 247–264.

39 Osborne, B.G., Turnbull, K.M., Anderssen, R.S. *et al.* (2001) The hardness locus in Australian wheat lines. *Aust. J. Agric. Res.* **52**, 1275–1286.

40 Osborne, B.G. & Anderssen, R.S. (2003) Single-kernel characterization principles and applications. *Cereal Chem.* **80**, 613–622.

41 Dobraszczyk, B.J. & Morgenstern, M.P. (2003) Rheology and the breadmaking process. *J. Cereal Sci.* **38**, 229–245.

42 Shadow, W. (2003) *The Falling Number Method and its Uses.* Perten Instruments Application Report FN1. Perten Instruments, Huddinge, Sweden.

43 Chang, S-Y., Delwiche, S.R. & Wang, N.S. (2002) Hydrolysis of wheat starch and its effect on the falling number procedure: Mathematical model. *Biotech. Bioeng.* **79**, 768–775.

44 Belton, P.S. (1999) On the elasticity of wheat gluten. *J. Cereal Sci.* **29**, 103–107.

45 Wang, M., Hamer, R.J., van Vliet, T. *et al.* (2003) Effect of water unextractable solids on gluten formation and properties: Mechanistic considerations. *J. Cereal Sci.* **37**, 55–64.

46 Anderssen, R.S., Gras, P.W. & Macritchie, F. (1998) The rate-independence of the mixing of a wheat-flour dough to peak dough development. *J. Cereal Sci.* **27**, 167–177.

47 Buchholz, R.H. (1990) An epitrochoidal mixer. *The Mathematical Scientist* **15**, 7–14.

48 Gras, P.W., Hibberd, G.E. & Walker, C.E. (1990) Electronic sensing and interpretation of dough properties using a 35-g Mixograph. *Cereal Foods World* **35**, 568–571.

49 Anderssen, B. (1992) Linking mathematics with applications: The comparative assessment process. *Math. Comp. in Simul.* **33**, 469–476.

50 Anderssen, R.S. & Campbell, M.C.W. (1984) Computational aspects associated with the direct use of indirect measurements: refractive index of biological lenses. In: Noye, J. & Fletcher, C. (eds) *Computational Techniques and Applications: CTAC-83*, pp. 893–902. Elsevier/North Holland. Amsterdam.

51 Anderssen, R.S. (1999) The pragmatics of solving industrial (real-world) inverse problems with exemplification based on the molecular weight distribution problem. *Inverse Probl.* **15**, R1–R40.

52 Basman, A., Koksel, H. & Ng, P.K.W. (2002) Effect of increasing levels of transglutaminase on the rheological properties and bread quality characteristics of two wheat flours. *Euro. Food Res. Technol.* **215**, 419–424.

53 Eagles, H.A., Hollamby, G.J., Gororo, N.N. & Eastwood, R.F. (2002) Estimation and utilization of glutenin gene effects from the analysis of unbalanced data from wheat breeding programs. *Aust. J. Agr. Res.* **53**, 367–377.

54 Eagles, H.A., Hollamby, G.J. & Eastwood, R.F. (2002) Genetic and environmental variation for grain quality traits routinely evaluated in southern Australian wheat breeding programs. *Aust. J. Agr. Res.* **53**, 1047–1057.

55 Wooding, A.R., Kavale, S., Wilson, A.J. & Stoddard, E.L. (2000) Effects of nitrogen and sulfur fertilization on commercial-scale wheat quality and milling requirements. *Cereal Chem.* **77**, 791–797.

56 Dobraszczyk, B.J. & Roberts, C.A. (1994) Strain hardening and dough gas cell-wall failure in biaxial extension. *J. Cereal Sci.* **20**, 265–274.

57 Martin, C.R., Rousser, R. & Barbec, D.L. (1993) Development of a single-kernel wheat characterization system. *Trans. ASAE* **36**, 1399–1404.

58 Daniels, N.W.R. (1974) Some effects of water in wheat flour doughs. In: Duckworth, R.B. (ed.) *Water Relations in Foods. Proceedings of an International Symposium held in Glasgow*, pp. 573–586. Academic Press, London.

59 Campos, D.T., Steffe, J.F. & Ng, P.K.W. (1997) Rheological behaviour of undeveloped and developed wheat dough. *Cereal Chem.* **74**, 489–494.

60 Li, A. & Walker, C.E. (1992) Dough temperature changes during mixing in a Mixograph. *Cereal Chem.* **69**, 681–683.

61 Anderssen, R.S. & Gras, P.W. (2000) The hysteretic behaviour of wheat-flour dough during mixing. In: Shewry, P.R. & Tatham, A.S. (eds) *Wheat Gluten,* pp. 391–395. Royal Society of Chemistry, Cambridge.

62 Rasper, V.F. & Preston, K.R. (1991) *The Extensograph Handbook.* American Association of Cereal Chemists Inc., St. Paul, MN.

63 Rath, C.R., Gras, P.W., Zhonglin, Z. *et al.* (1995) A prototype extension tester for two-gram dough samples. In: Panozzo, J.F. & Downie, P.G. (eds) *Proceedings of the 44th RACI Cereal Chemistry Conference*, pp 122–126. RACI, North Melbourne.

64 Anderssen, R.S. & Hoffmann, K-H. Modelling the stress-strain phenomenology of the extensional hysteresis in the mixing of a wheat-flour dough (in preparation).

65 Mayergoyz, I.D. (1991) *Mathematical Models of Hysteresis.* Springer-Verlag, New York.

66 Shewry, P.R. & Tatham, A.S. (2000) *Wheat Gluten.* Royal Society of Chemistry/MPG Books, Cornwall, UK.

67 Shewry, P.R. & Tatham, A.S. (1997) Disulphide bonds in wheat gluten proteins. *J. Cereal Sci.* **25**, 207–227.

68 Shewry, P.R. & Halford, N.G. (2003) Genetics of wheat gluten proteins. *Adv. Genet.* **49**, 111–184.

69 Grosch, W. & Wieser, H. (1999) Redox reactions in wheat dough as affected by ascorbic acid. *J. Cereal Sci.* **29**, 1–16.

70 Labat, E., Morel, M-H. & Rouau, X. (2000) Effects of laccase and ferulic acid on wheat flour doughs. *Cereal Chem.* **77**, 823–828.

71 Anderson, O.D., Hsia, C.C. & Torres, V. (2001) The wheat gamma-gliadin genes: characterization of ten new sequences and further understanding of gamma-gliadin gene family structure. *Theor. Appl. Genet.* **103**, 323–330.

72 Appels, R., Gustafson, J.P. & O'Brien, L. (2001) *Wheat Breeding in the New Century: Applying Molecular Genetic Analysis to Key Quality and Agronomic Traits.* CSIRO Publishing, Collingwood, Victoria.

73 Larson, R.G. (1988) *Constitutive Equations for Polymer Melts and Solutions.* Butterworths, Boston, MA.

74 Anderssen, R.S., Götz, I.G. & Hoffmann, K-H. (1998) The global behaviour of elasto-plastic materials with hysteresis-type state equation. *SIAM J. Appl. Math.* **58**, 703–723.

75 Brokate, M. & Sprekels, J. (1996) *Hysteresis and Phase Transitions.* Springer, New York.

# Index


acrylamide 21
agar 155–6, 166
ageing 117–18, 119
   staling 171–2
alginates 160–62
amphiphilic materials 5–6
   polysaccharides 6
   proteins 6–7
   small-molecular 7–8
α-amylolysis 56, 57, 58
amylopectin 20, 170
   gels 41, 42, 44
   glass transition 111
   glasses 47
   percentage weight in starch 24
   structure
      analysis and modelling 28–9
      cluster-type models 26–9, 30
      liquid crystal 23–4
amylose 20, 179
   chain flexibility 26
   complexation
      alcohols and aromas 53–5
      lipids 50–53
   gelation 33, 44, 170–72
      mechanism 42–4
      structural changes 40–41
   glass transition 111
   helical structures in starch 24
   iodine binding 26
   lipid complexation 50–53
   molecular weight distribution 25–6
   percentage weight in starch 21, 24
   structures 24–5, 153–4

atomic force micrograph 155
V amylose 50–55
aromas and flavours
   complexation with amylose 53–5
   partitioning 17

biorheology 200–201, 209
biscuits 210, 212
brick pack 138

carbohydrates
   glass formation 109
   solvent properties 126–9
   *see also* polysaccharide gels; starch; sugars
carrageenans 155–6, 157, 172
casein 183
cassava starch 69
cellulose 155
coalescence 11, 13–14
   mechanism 14–15
colloidal glasses 113–14, 119
creaming 11
   and coalescence 13–14
   and jammed structures 119

dextram 111
diffusion, oil–water interface 4–5
dilitancy 138–9
dissipation 136–7
doughs 20
   *see also* wheat-flour dough
durum wheat 200

electrostatic forces 8–10

emulsions 1, 152
  aroma partitioning and flavour 17
  chemical reactivity 17
  coalescence 11, 13–14
    mechanism 14–15
  and colloidal glasses 114
  creaming 11
  droplet concentration 3
  droplet size and size distribution 2–3
  elastic properties at high concentration 3
  flocculation 11, 12–13
  functionality 16–17
  interdroplet potentials 8–10
    attractive/repulsive
      force-distance functions 8, 9
    electrostatic forces 8–10
    hydrophobic forces 10
    steric interactions 10
    van der Waals forces 10
  liquid foam 3
  oil-water interface
    illustrated 4
    protein networks 186–7
  Ostwald ripening 5
  oxidation of lipids 17
  rheology 16–17
  surface properties 4–5
    adsorption energetics 5–6
    amphophilic interfacial materials 6–8
    interfacial tension 4
    sorption isotherms 5–6
  surfactants
    protein 6–7
    small molecules 7–8
  vitrified 125
encapsulation matrices 125–6, 127
extruders 34
extrusion 40

fatty acids
  complexes with amylose 51–3
  in starch 32
flavour
  encapsulation 125–6, 127
  perception 17
flocculation 11, 12
  and coalescence 13–14
  fractal dimension 12, 13, 16

rheology of emulsions 16–17
foams 152, 186, 212
foods, starchy classified 55
furcellaran 164, 165

galactomannans 156–7, 160
gelatin
  gels 95–6, 181–2, 185
  glass transition 112
  water diffusivity 101
gellan 155, 163, 165, 168
gels
  general characteristics summarized 151–3
  jammed structures 119
  *see also* polysaccharide gels; protein gels
genetics, wheat 201–3, 207–8
glasses and glass transitions 108
  ageing phenomena 117–19, 119
  colloidal glasses 11–14
    jammed structures 119
  emulsions, vitrified 125
  flavour encapsulation 125–6
  glassy state dynamics 114
    glucitol 114–15, 116, 126, 127
    glucose 115–16
    nonexponetial primary relaxation 116
  kinetics in single-phase systems 120
    Kramer's theory 123–4
    Smoluchowski theory 120–21, 122
    Stokes–Einstein relationship 122–4
  low molecular weight liquids 109
    glucose 109–110
    sucrose 95–6, 110–111
  maltotriose 126–8
  oxidation of lipids 125
  plasticization and biopolymers 111
    flexible proteins 112
    glucan polymers 111
    water/glycerol plasticizers 112–13
  preservation and biostabilization 119–20,
    124–5
  solvent properties of carbohydrates
    126–8
  starch
    colloidal glasses 114
    melting 37–8, 43
    and plasticization 44–6, 112–13
  structural relaxation 117

time-dependent changes and physical
    ageing 117–19
    Tool–Narayanaswamy method 118–19
    Williams–Landel–Ferry relationship 117
  viscosity
    colloidal glasses 113
    glucose 109–110
glucan polymers 111
glucitol 114–15, 116, 126, 127
glucomannans 156–7, 176–8
glucose
  glassy dynamics 109–110, 115–16
  Maillard reaction 121
  and polysaccharide structures 154
  viscosity and glass transition 110
gluten 98–9
  wheat dough 203, 204, 220–22, 234
glycaemic index 55
granular materials *see* powders
gums 156–7, 167

Hamaker function 10
hoppers 137, 145, 146, 147
hydrocolloids 152
  and starch 48, 172
hydrophobic forces 10

interfacial tension 4

jamming of powders 145–7

konjac mannan 160
Kramer's theory 123–4
Krieger–Dougherty equation 11, 16

β-lactoglobulin 179–80, 186
Laplace Pressure 4
lipids
  complexation with amylose 50–52
  emulsified and oxidation 17, 125
  in starch 31–2
liquid foam 3
lysine 121
lysozyme 89–90

Maillard reaction 121
maltotriose 126–8
mastication 69–70

mayonnaise 3
methyl cellulose 160
microgels 168
milling 32, 208, 211

Ostwald ripening 5

packing of powders 139–42
pasta 210, 212
pastes 20, 40, 44
pectins 156–7, 160–62, 173–4
physical ageing 117–18, 119
phytoglycogen 30
Pickering stabilization 8
plant breeding, wheat 200–201, 208–9
plasticization
  and biopolymer glasses 111, 119
  of starch and glass transition 44–6, 119
  water/glycerol plasticizers 112–33
polysaccharide gels
  algal polysaccharide–galactomannan gels
      177–8
  algal polysaccharide–glucomannan gels
      177–8
  background and summary 151–3
  block structures 160
    alginates and pectins 160–62
    calcium binding 160–62
    egg box model 161
    galactomannans 160
    konjac mannan 160
    long-range structures 162
    methyl cellulose derivatives 160
  carrageenans, semi-refined 172
  fluid gels 166–7
    carrageenan 168
    gellan 168
    microgels 168
    xanthan 167
  higher-order helical aggregates 163
    agar 166
    carregeenan 163–5
    cation binding 164, 165–6
    furcellaran 164, 165
    gellan 163, 165
    X-ray diffraction data 163–4
  networks and structures 158–66
    coupled 173–8

interpenetrating 172
phase-separated 169–72
swollen 172
pectin-alginate gels 173–4
point-cross links 159
polysaccharide mixtures 168–9
polysaccharide–protein gels 189–91
starch gels
    formation and model 170–72
    as phase-separated network 169–70
structure of polysaccharides 153, 154
    agar 155–6
    alginate 156
    amylose 24–5, 153–4
    carrageenans 155–6, 157
    celluslose 155
    galactomannans 156–7
    gellan 155
    glucomannans 156–7
    pectin 156–7
    xanthan 155
xanthan–galactomannan gels 176–7
xanthan–glucomannan gels 174–6
powders and granular materials
    background and discussion 136–7, 148
    jamming 145
        cohesive arches 145–6
        force chains 147
        interlocking arches 146
        jamming phase diagram 147
        particle bridges 146–7
    packing
        consolidation 140
        dynamics of shaking and tapping 140–41
        interparticle cohesive forces 142
        powder density and fractional volume
            139–40
        rods and ellipsoids 142
    segregation 142–3
        convective mechanism 143–4
        Monte Carlo simulations 143–4
        operational experiences 143, 144
        rotating drums 144–5
        scale-up of blenders 145
    terms described and defined
        cohesive forces 142
        dilatancy 138–9
        dissipation 136–7

funnel flows and hopper angles 137
repose, angle of 137
segregation 142–3
static friction 137
wall friction 137, 138
preservation and vitrification 119–20, 124–5
prolamins 112
protein gels 178–9
    background and summary 151–3
    gelation mechanisms
        β-lactoglobulin 179–80
        casein 183
        fibrous proteins 181–2
        gelatin 181–2
        globular proteins 179–81
    interfacial networks 184–5
        β-lactoglobulin 186
        food foams and emulsions 186–9
        gelatin 185
        globular protein 185–6
        orogenic displacement model 188
        protein–surfactant mixtures 187–8
    polysaccharide–protein gels 189–91
    protein mixtures 184
proteins 178–9
    amphophilic properties in emulsions 6–7
    conformational changes under pressure 102
    glass transition 112
    globular 7, 100, 112
        colloidal glasses 114
        gelation mechanisms 179–81
        nonfreezing water 100
    hydration 220
        dynamics of 90–94, 100, 102
    interaction with starch 49
    in starch 31
    surfactants in emulsions 6–7
    wheat-flour dough 202, 204, 212, 234
pullulan 111

repose, angle of 137
resistant starches 5–7, 56–7
rheology
    emulsions 16–17
    *see also* wheat-flour dough
rice flour 44

salt 17, 46

segregation of powders 142–5
silo design 138
Smoluchowski theory 12, 120–21, 122
sorbitol *see* glucitol
starch 20
    acrylimide formation 21
    ageing of plasticized material 119
    α-amylolysis 56, 57, 58
    amylopectin 22
        analysis and modelling 28–9
        cluster-type models 28–9, 30
        gels 170
        internal granule structure 23
        percentage weight 24
        structure and chain organization 26–30
    amylose 20
        alcohol, aroma and flavour complexation
            53–5
        chain flexibility 26
        gelation 33, 40–45, 170–72
        iodine binding 26
        lipid complexation 50–52
        molecular weight distribution 25–6
        percentage weight 21, 24
        structures 24–5, 153–4
    biopolymers
        phase separation in mixtures 47
        variability and thermal stability 20–22
    cassava 69
    and colloidal glasses 114
    composition 21
        free fatty acids and lipids 31–2
        intermediate macromolecules 30
        proteins 31
    extrusion 34, 40
    gelatinization 32–3
        IR studies 35–6
        and nutrition 55
        side-chain liquid crystal model 38–9
        solubilization 34
        structural changes and gelation 33–4
        and sugars 49–50
        synchroton radiation studies 34–5, 53
        wheat and cassava starches 35
    gelation 33, 169–72
        amylopectin 41
        mechanisms 42–4
        and melting 43–4
        modelling 41, 42
        opaque gels 42
        structural changes 40–41
        texture and nutritional effects 44–5
        two-phase mixed gels 42
    glass transition 45
        melting 37–8, 43
        and plasticization 45–6, 112–13
    gluten 48
    glycaemic index 55
    granular structure 21
        A-type lenticular 22
        B-type spherical 22
        and botanical source 22
        crystallinity 22–4, 25
        growth rings 23
    hydrocolloid interactions 48, 172
    melting 35
        and gelation 43
        glass transition 37–8, 43
        side-chain liquid crystal model 38–9
        viscoelastic melt 34
        water content effects 36–7
    milk and derivative interactions 48
    minerals 32
    mungo starch vemicelli 44
    native 59
    pastes 40, 44
    pea starch/egg albumen matrix 70–74
    physical ageing 46, 47
    protein interactions 49
    puffing 40
    resistant 5–7, 56–7
    rice flour noodles 44
    solid foam 40
    staling 171–2
    sugar interactions 49–50
    thickening 40
    variability 21
    X-ray diffraction 24, 54
    *see also* water transport and dynamics
Stokes law 11
Stokes Einstein relationship 122–4
sugars
    and starch gelatinization 49–50
    sucrose glass formation 95–6, 110–111
surface properties of emulsion droplets 4–8
surfactants

and emulsions 5–6
fine particles 8
polysaccharides 6
protein 6–7
small-molecular 7–8

Tool–Narayanaswamy method 118–19
tryptophan 93

van der Waals forces 10
viscosity
    colloidal glasses 113
emulsions 16–17
low molecular weight liquids/glasses 109–110
vitrification and preservation 119–20, 124–5
    *see also* glasses and glass transitions
Vogel–Tammann–Fulcher law 109

water transport and dynamics
    background and challenges 68–70
    conclusion and summary 103–4
    high pressure water dynamics 101–2
        molecular dynamics simulations 102–3
    low water-content systems 95
        gelatin gels 95–6
        gluten and glutenin 98–9
        multistate theory 96–8
        NMR proton transverse relaxation 95–6
        sugar glass 95–6, 97
    microscopic water distribution 74
        gas phase consideration 77
        multi-dimensional correlation NMR
            75–7
        relaxation spectrum for starch granules
            74–5
        statistical mechanics 70–74
    multistage theory 97–8
    nanopores, state of water in 82
        Biogel-P investigations 85–6
        infrared spectroscopy (ATR) of cellulose
            membranes 84–5, 86
        magnetic relaxation dispersion techni-
            ques 83–4
        microosmosis 85
        modelling synthetic membranes 82–3
    nonequilibrium microheterogeneous sys-
        tems 80–82
    nonfreezing water 99–100

    diffusivity measurements 100–101
    pea starch/egg albumen (exemplar) 70, 74
    protein hydration dynamics 90–94
    statistical mechanics 70–74
    surface water diffusion studies 100–101
    water self-diffusion propagator 77–8
        experimental measurement by NMR
            78–9
    water–biopolymer interactions 86
        hydration/dehydration mechanism 88
        infrared hydration spectra 87–9
        molecular dynamic simulations 94
        NMR methods (NOESY and ROESY)
            90–91
        oxygen-17 magnetic relaxation disper-
            sion 91–3
        tryptophan time-resolved fluorescence
            spectrum 93–4
        X-ray diffraction of lysozyme 89–90
wheat
    hard 200, 207–8, 212
    soft 200
wheat-flour dough rheology 199–200
    analytical techniques 203
    biorheology
        molecular 200–201, 209
        process 201, 209
    biscuits 210, 212
    bubble rheology 209–210
    cereal science key issues 201–3
    dough formation and properties 212
        developed/underdeveloped dough 220
        mixogram phases 213–14
        visual inspection 225
    durum wheat 200
    exothermic reactions 220–22
    extension testing 227–30
        allelic differences 229–30
        choice of test/instrument 227–8
        force resisting measurement 206
        micro-extension testing 228–9
        uniaxial/biaxial 227
    Farinograph 206, 222
    fermentation 209–210
    genetics
        end-product quality 207–8
        and expression of proteins 203
        and molecular biorheology 200–201

and pathway to end-product schematic
202
gliadins 203
gluten 220, 221, 222, 234
analysis of structure 203–4
glutenins 203
hard wheats 200, 207–8
dough formation 212
hardness index 208
hardness phenotype 208
hydration 220, 222
loss modulus 206
microscopy 203–5
mixograms 212, 214–16, 222–3
asymtotic equilibrium phase 214
breakdown phase 214
dough development phase 213
dough stability phase 213–14
global stress-strain dynamics 223–4
hysteric nature of local structure 224–7
rapid hydration phase 213
shearing 226–7
stress-strain modelling 215–16
temperature measurements and phases
220–22
visible inspection 225–6
Mixograph 206
modelling 211, 215–19
molecular structure
characterization 203–4
microscopy studies 204
role of MWD, TPP and UPP 25, 204

pasta 210, 212
peak dough development (PDD) 205, 206
plant breeding 208–9
and molecular biorheology 200–201
proteins
dough formation 212, 234
molecular weight distribution (MWD)
202, 204
total polymeric protein (TPP) 202, 204
unextractable polymeric protein (UPP)
202, 204
and wheat genome 203
raw materials 210–211
receival station testing 210–211
recording mixers 206
rheological measurements 205–7
linking genetics to end-product 207–8
and modelling 211
recording mixers 206
soft wheats 200
stereological assessments 204–5
storage modulus 206
stress-strain dynamics and hysteresis
223–7
temperature measurements 220–22
viscoelasticity 205
dough formation 212
modelling 230–33
walk-in-refrigerator experiments 219–20
Williams–Landel–Ferry relationship 110, 117

xanthan 155, 167, 174–7